移动互联网开发技术丛书

Android 零基础入门到实战

App项目开发 · 鸿蒙开发入门 · 微课视频版

赵圣鲁 胡颖辉 余燕萍 汪宗伟 吴微微 著

清華大學出版社
北京

内 容 简 介

本书围绕 Android 初学者从零基础到实战达人的过程进行设计，采用项目教学法，以作者开发的“欢乐购商城”App 为例，以一个完整的项目开发为主线，将项目开发分解为 9 个教学模块，分别为 App 应用体验、Android 基础界面控件、Android 高级界面控件、列表控件、页面跳转与切换、Android 中的数据存取、Android 客户端与服务器端交互、综合项目“欢乐购商城”实现等。读者在学习基础知识过程中将熟悉 App 综合项目开发流程，逐步培养独立开发综合项目的能力，并最终实现综合项目。同时本书引入鸿蒙开发入门知识，供有兴趣的读者参考学习。

本书可作为应用型本科计算机专业、软件专业、高职软件技术专业及相关专业的教材，也可作为 Android 和鸿蒙开发爱好者以及初、中级 Android 应用开发人员的参考工具书。

图书在版编目（CIP）数据

Android 零基础入门到实战：App 项目开发・鸿蒙开发入门・微课视频版/赵圣鲁等著. —北京：清华大学出版社，2022.2(2023.1重印)
（移动互联网开发技术丛书）
ISBN 978-7-302-60067-1

Ⅰ. ①A… Ⅱ. ①赵… Ⅲ. ①移动终端－应用程序－程序设计 Ⅳ. ①TN929.53

中国版本图书馆 CIP 数据核字(2022)第 016650 号

责任编辑：王冰飞　薛　阳
封面设计：刘　键
责任校对：焦丽丽
责任印制：朱雨萌

出版发行：清华大学出版社
网　　址：http://www.tup.com.cn，http://www.wqbook.com
地　　址：北京清华大学学研大厦 A 座　　**邮　　编**：100084
社 总 机：010-83470000　　**邮　　购**：010-62786544
投稿与读者服务：010-62776969，c-service@tup.tsinghua.edu.cn
质量反馈：010-62772015，zhiliang@tup.tsinghua.edu.cn
课件下载：http://www.tup.com.cn，010-83470236
印 装 者：三河市君旺印务有限公司
经　　销：全国新华书店
开　　本：185mm×260mm　　**印　　张**：18.5　　**字　　数**：448 千字
版　　次：2022 年 3 月第 1 版　　**印　　次**：2023 年 1 月第 3次印刷
印　　数：2701～4200
定　　价：59.80 元

产品编号：094283-01

前言

FOREWORD

Android是Google公司推出的专为移动设备开发的平台，从2008年9月推出以来，在短短的几年时间里就超越了称霸10年的诺基亚Symbian系统和苹果iOS系统，成为用户量最大的手机操作系统。近年来，移动互联网的影响力越来越大，各种新的App层出不穷。很多高校的计算机专业和IT培训学校等都将Android应用开发作为教学内容之一。Android学习对于初学者难度相对较大，大多数在学习Android过程中放弃或者最终不能独自完成综合项目，这就需要一本好的教材围绕初学者从零基础到实战达人的过程进行设计。

本书围绕教学团队开发的综合项目"欢乐购商城"进行设计，围绕"欢乐购商城"项目进行讲解。本书先让读者体验最终要实现的项目效果，然后把整个项目拆分成若干个Android案例，在每章中分别进行实现。其中每个案例在每章中拆分成若干个知识点，由浅入深逐步实现项目案例。在客户端和服务器端交互章采用本书提供的一键启动JTM集成包，解决数据库和服务器开发难题，为学习者提供了最大便利。在第9章中，将前面讲解的知识点和案例综合运用实现"欢乐购商城"综合案例。

在2021年6月2日的华为新品发布会中，鸿蒙手机操作系统正式登场，引起社会媒体的广泛关注，同时也掀起了国内华为手机更新鸿蒙系统和学习鸿蒙开发的浪潮。由于HarmonyOS开发和Android开发较为类似，为推动国产手机系统发展，本书在第10章为App开发爱好者提供了鸿蒙开发入门知识。

全书共分为10章。第1章导读，介绍App开发流程；第2章App应用体验，运行JVM服务器端和客户端，体验"欢乐购商城"项目；第3章为Android基础界面编程；第4章为Android高级界面控件；第5章为列表控件；第6章为页面跳转与切换；第7章为Android中的数据存取；第8章为Android客户端与服务器端交互；第9章为综合项目——"欢乐购商城"实现；第10章为鸿蒙系统开发入门。

为了方便读者学习，本书配套了微课教学视频、源码、课件、试题、课程大纲等教学资源。所有配套资源均可在清华大学出版社官方网站下载。其中安卓项目案例源码，基于Java JDK 1.8+Android Studio Arctic Fox+Android SDK 11.0+Gradle：7.0.0-beta04开发；鸿蒙项目案例源码，基于Open JDK 1.8+DevEco Studio 2.1 release+SDK(API version 5)+Gradle 6.3开发。

本书由江西信息应用职业技术学院Android教学团队共同编写。副校长胡颖辉负责全书的方案设计、内容策划、案例分解、细节把握、质量控制和统编定稿工作。各章分工如下：赵圣鲁负责第1～4章和第10章，同时指导和参与了其余各章的编写、修改、初稿编排工作；

余燕萍负责第 5、7、9 章；汪宗伟、赵圣鲁负责第 6、8 章。吴微微进行了大量课件制作、课程大纲制作、试题编写、微课视频录制等工作。管银枝、罗凌云、邹贤芳、夏侯斌、袁定治、周香庆、刘大勇、邓丽萍、薛兰等参与了本书内容设计和策划工作。

在此特别感谢支持本书编写、出版工作的江西信息应用职业技术学院软件工程系及清华大学出版社的帮助。

希望本书的出版有助于 Android 任课教师更好地讲授 Android 课程，也能帮助使用本书的读者更快、更扎实地掌握 Android 应用开发技能。

编　者

2022 年 2 月

目录

CONTENTS

随书资源

第1章

导　读

在本章中，我们将开门见山地迈入移动开发的大门，通过"欢乐购商城"App来了解软件开发的知识模块与流程，并且从功能和界面的角度阐述"欢乐购商城"从设计到开发的全过程。在此基础上，从高校教师教学的形式出发，以项目案例驱动形式辅助教学，摒弃以往的知识点讲解模式，提高学习效率。

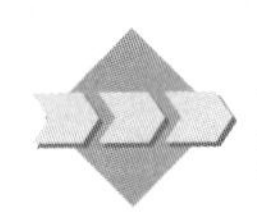

1.1　App开发流程

视频详解

一个完整的软件开发分为网页端与移动端(App)，而移动端又包括Android端、iOS端和鸿蒙端，使用的开发模式是移动应用开发常用的MVC开发模式。本书主要讲解Android端的开发和鸿蒙开发入门。

1. 服务器框架搭建

服务器端与数据库开发流程如图1-1所示。

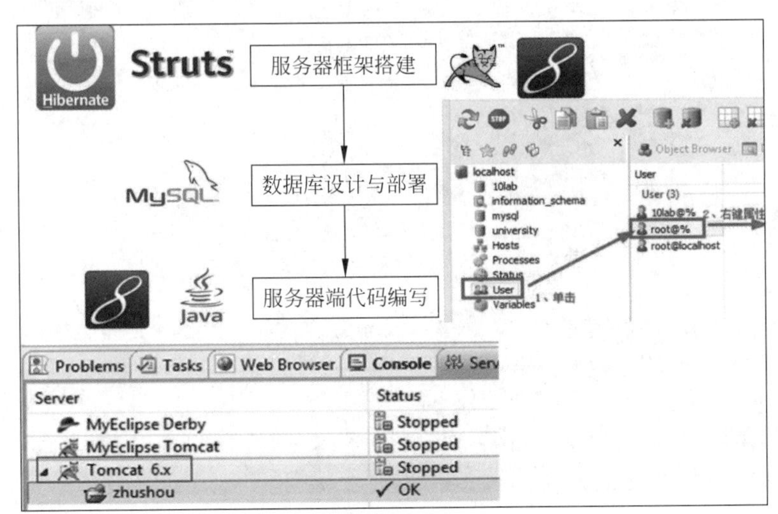

图1-1　服务器端与数据库开发流程

服务器端设计在一个移动项目设计中是很重要的部分，如果说客户端负责为用户展现数据，那么服务器端可以说是为用户准备数据的。只有服务器端将用户所需要的数据准备完毕后，客户端才有可能去展示。

2. 数据库设计与部署

对于移动 App 开发，数据库的选择显得至关重要，因为数据存储结构和读取速度直接影响到用户体验，所以要尽量使用轻量级数据库，这里使用的是 MySQL。

数据库开发分为逻辑结构设计和物理结构设计。逻辑结构设计是对数据库实体属性的确定、E-R 图的确定以及数据库存储结构进行设计，需要确定项目中所用到的字段并确定数据类型。物理结构设计就是在 MySQL 中创建相应的数据库表，对数据进行录入存储。最后需要将数据库与后台服务器相连接。

3. 服务器端实现

服务器端是在 MyEclipse 10 中实现的，首先是要进行开发环境的配置。配置完成后，在 MyEclipse 10 中创建“欢乐购商城”工程，并将工程集成在 JTM 中，为读者学习客户端提供服务数据。

4. 客户端框架搭建

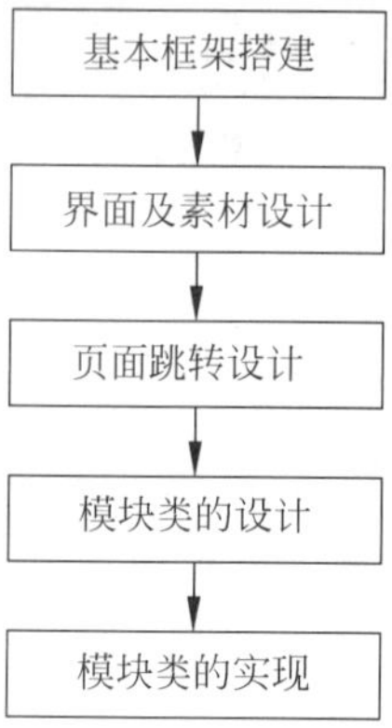

客户端开发流程如图 1-2 所示。

基本框架搭建 → 界面及素材设计 → 页面跳转设计 → 模块类的设计 → 模块类的实现

图 1-2　客户端开发流程

如图 1-3～图 1-6 所示为开发时的主要工作界面。

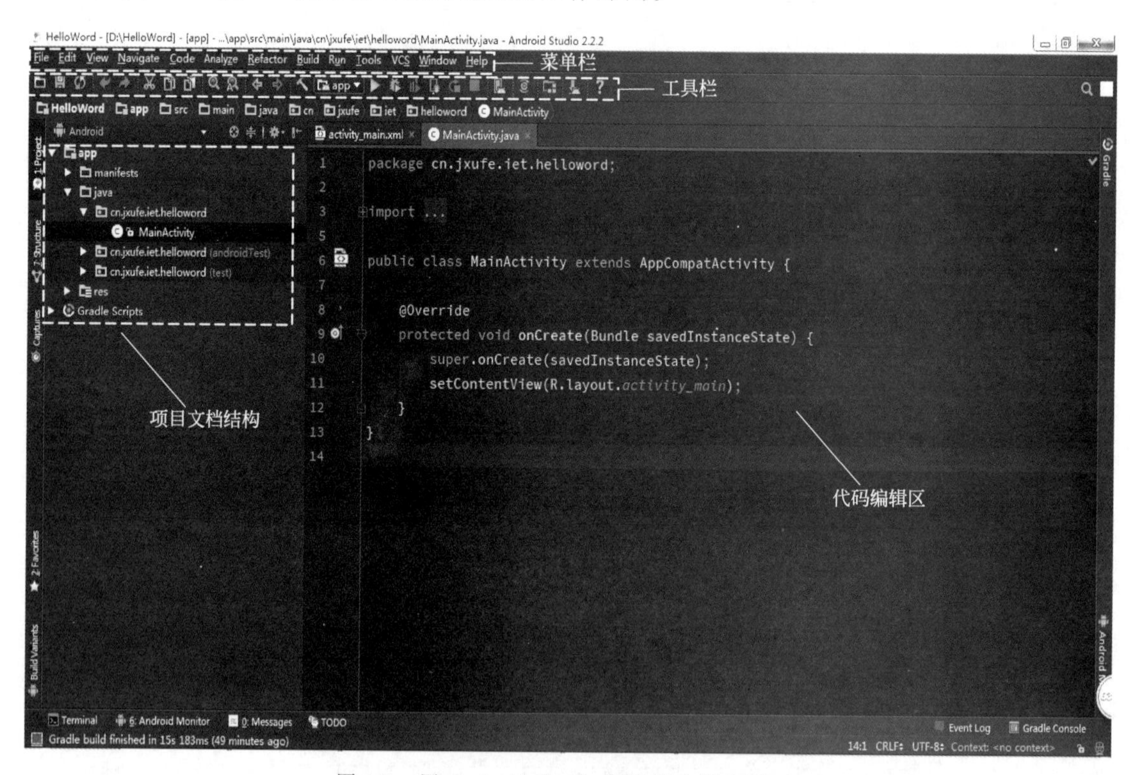

图 1-3　用 Android Studio 开发客户端主界面

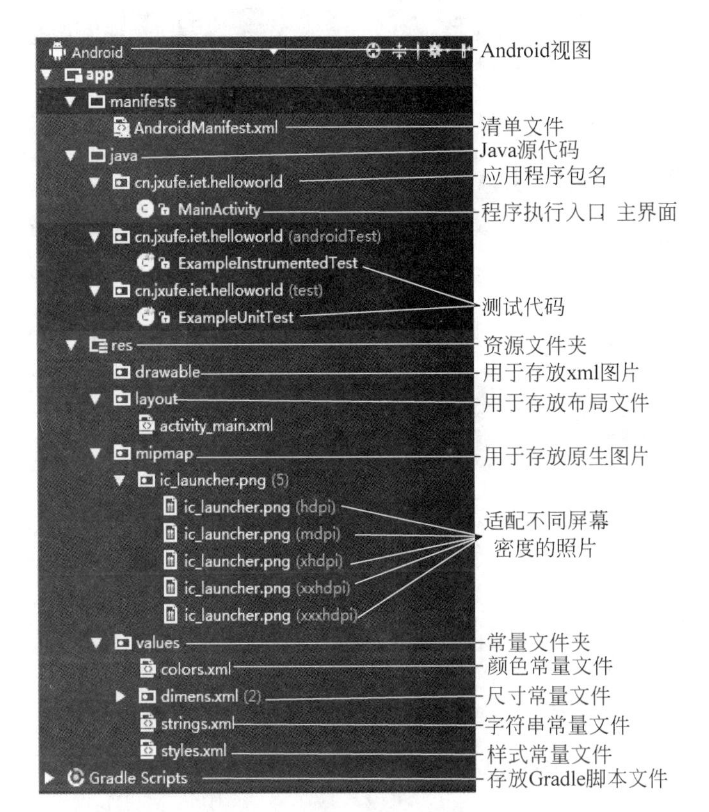

图 1-4　Android 视图中项目文档结构及其含义

图 1-5　Project 视图中项目文档结构及其含义

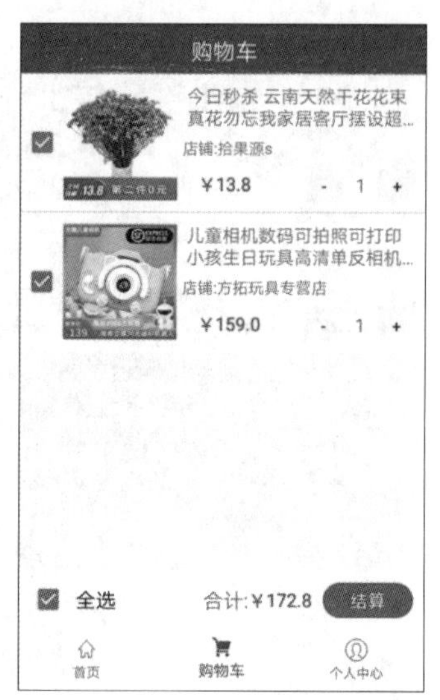

图 1-6 “欢乐购商城”客户端页面

5. 界面及素材设计和界面跳转设计

对于界面设计与模块类实现，可以分工同时进行。

项目素材由美工组制作完成，制作的标准需要按照 Android 开发图标要求制作，在开发前，项目负责人需要指定一个详细的设计文档，指出需要的素材以及规格。而界面与界面跳转设计，也需要通过讨论制定出想要的界面设计文档。其中需要注意的是，界面跳转不宜过多，除了必须跳转的界面之外，可以在当前界面中实现的功能尽量在当前页面实现，这也是影响用户体验的重要指标之一。

6. 模块类的实现

根据项目的需求分析文档中列出的功能模块，在项目中分别创建模块文件夹，然后分别对功能模块进行设计与实现。

7. 网络数据读取接口

网络数据读取接口由服务器端提供，包括网址和需要的参数，客户端发送相应的参数请求，从服务器端获取编译结果，最后由客户端呈现给用户。网络数据接口形式如下。

http://localhost:8080/HappyShop/shopSer? keyword=-1

8. 编译打包及测试

移动应用编译打包、测试及发布上线的简易流程如图 1-7 所示。

对于 Android 端应用，需要在项目中打包生成 APK 文件，编译打包完成后，可以在其他成员的手机上安装测试，看有没有问题出现。如果有，立即对程序进行调试；如果没有，就可以在商店中发布上线了。

9. 发布上线

Android 应用的发布有多种途径，可以选择在多个应用商城发布。在发布前需要对应用进行编译、签名、加密混淆、应用安全加固等操作。在选择的第三方应用商城中上传 apk 文件包、App 应用的截图、应用简介等信息，等待审核。审核通过后，用户可通过该应用商城安装该应用软件。

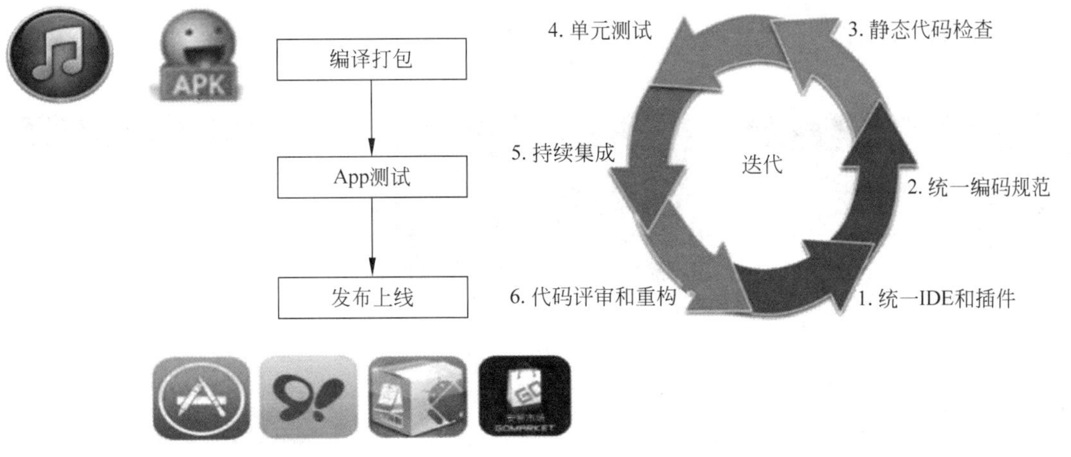

图 1-7　移动应用编译打包、测试及发布上线流程

1.2　案例介绍

本书从 Android 初学者学习角度设计，围绕“欢乐购商城”项目进行开发。先让读者体验最终要实现的项目效果，然后把整个项目拆分成若干个 Android 案例，在每章中分别进行实现。其中每个案例在每章中拆分成若干个知识点，由浅入深逐步实现项目案例。在第 9 章中，将前面讲解的知识点和案例综合运用，实现“欢乐购商城”综合案例。

本书以最小的案例尽可能容纳开发中常用到的知识点，“欢乐购商城”项目中包含首页、购物车、个人中心三个模块。

首页效果和产品列表如图 1-8 和图 1-9 所示。

图 1-8　首页效果

图 1-9　产品列表

购物车效果如图 1-10 所示。

个人中心效果如图 1-11～图 1-14 所示。

图 1-10　购物车效果

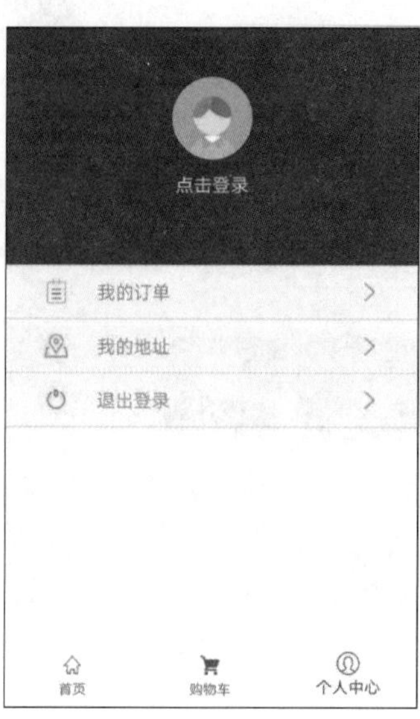

图 1-11　个人中心页面

图 1-12　登录页面

图 1-13　注册页面

图 1-14　订单中心页面

案例中访问的服务器为 JTM 封装包，本书提供的 JTM 包中包含 JDK 运行环境、MySQL 数据库、Tomcat 软件，并且在 JTM 中导入了服务器数据、服务端运行程序。学习者一键启动即可访问服务器，为学习者极大地节约了学习成本，学习者可以把学习重心放在 Android 应用开发上，提高学习效率。图 1-15 为客户端服务器网络连接示意图。

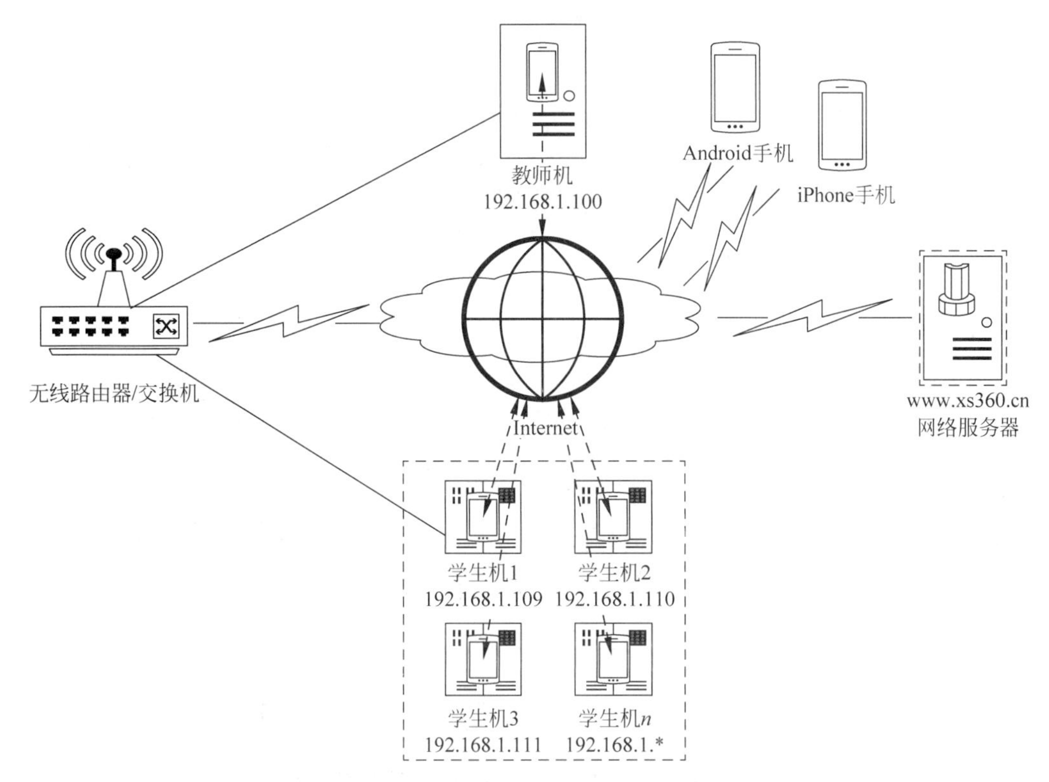

图 1-15　客户端服务器网络连接示意图

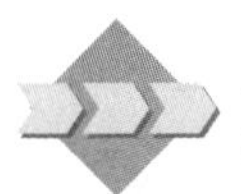

1.3　配置局域网运行和开发环境

视频详解

客户端访问网络数据，可以通过搭建临时局域网服务器的 IP 地址来访问本地服务器获取数据。本节将介绍如何搭建支持 App 客户端可访问的本地服务器和如何修改客户端的请求 URL 链接。

由于"欢乐购商城"服务器端代码是由 JSP 编写的，如果需要将此服务器代码在本地机上运行，则需用到的软件包括 Tomcat、JDK 和 MySQL。为了便于教学和测试，本书采用绿色运行环境 JTM-v1.1(一键搭建运行 JavaEE 应用的工具)为例，将"欢乐购商城"Web 端部署在本地计算机上并运行。

将实验室的教师端计算机与学生端计算机互联在一个局域网内，由教师或每组组长在计算机上将"欢乐购商城"本地服务器运行起来(启动方法参见第 2 章)，其他学生机进行调用，如图 1-16 所示。不同的小组既可以连接本组搭建的服务器，如小组 1 使用服务器 1、小组 2 使用服务器 2 等，也可以连接教师端的服务器，只需将请求的 URL 地址修改为教师端或任何一台学生机服务器的 IP 地址即可，如 192.168.1.100(教师机)，192.168.1.101(小组 1 服务器)。完成上述 URL 地址修改后，将"欢乐购商城"软件安装在学生机中 Android 模拟器上就可以访问本地搭建的服务器数据，如果安装到 Android 手机中测试，只需将 Android 手机连接到与服务器同一局域网内就可以成功获取请求(具体方法详见第 2 章)。

如果实验室的机房网络拓扑图如图 1-17 所示，由于各小组之间都被不同的无线路由器分

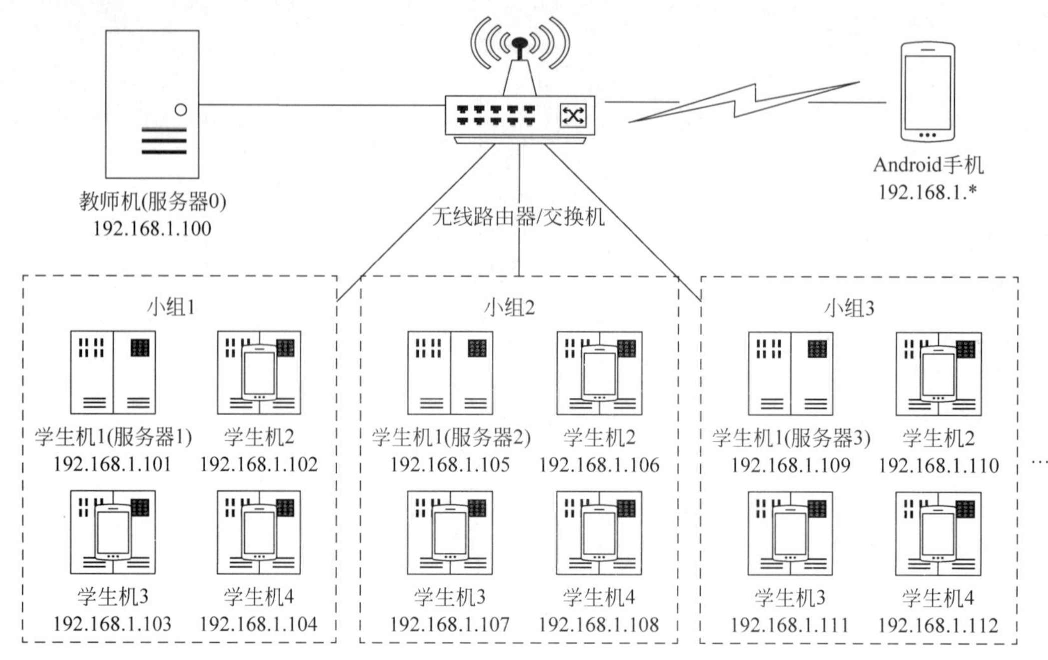

图 1-16 访问本地服务器网络连接拓扑图(一)

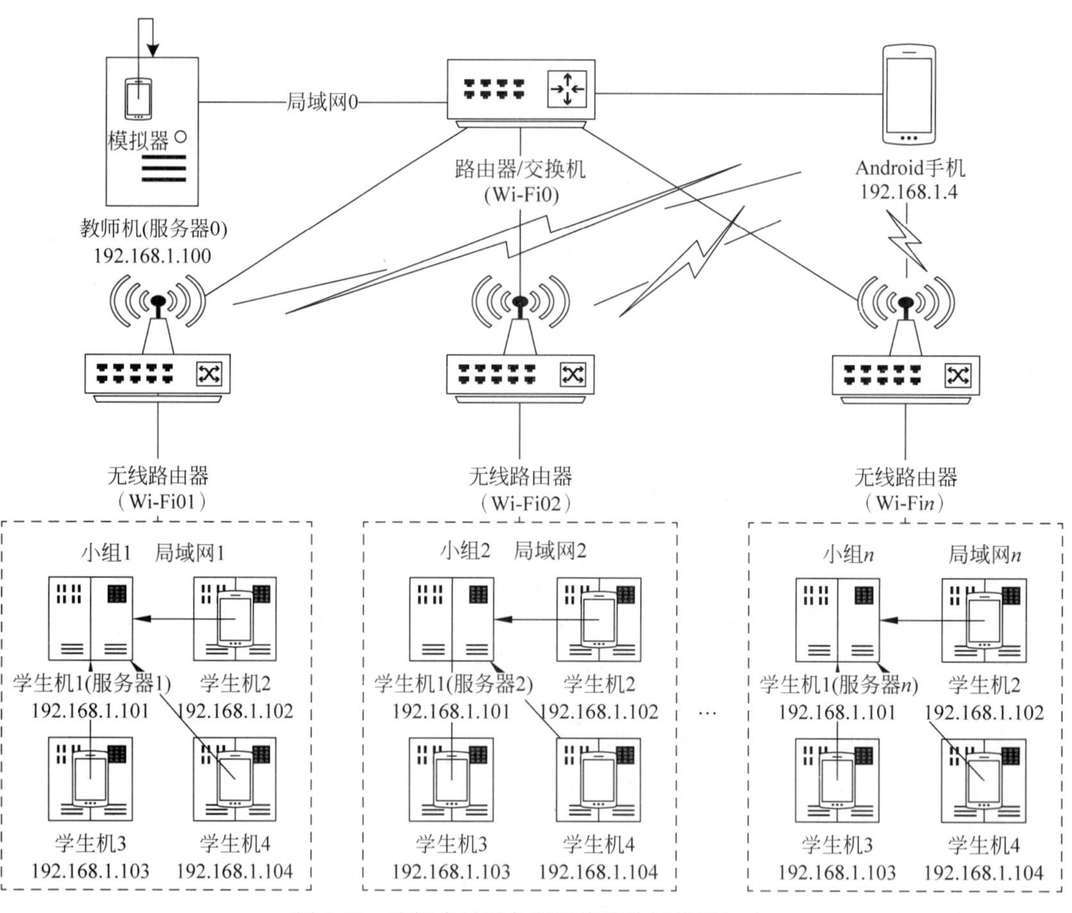

图 1-17 访问本地服务器网络连接拓扑图(二)

割成不同的局域网，则整个访问规则服务器的只能是小组1使用服务器1，小组2使用服务器2，即本组只能访问本组局域网的服务器，而教师机的模拟器也只能访问本机的IP地址。

本地服务器的运行对交换机或路由器的设置没有特殊要求，也不需要联网，只要按照默认配置使用即可。

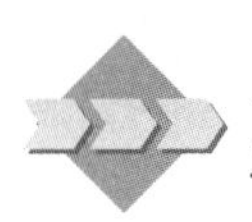

1.4 本书内容简介

视频详解

本书第1章为App开发流程和案例介绍；第2章为App实战应用体验，运行JVM服务端和客户端，体验“欢乐购商城”项目；第3章为Android基础界面编程；第4章为Android高级界面编程；第5章为列表控件；第6章为页面跳转与切换；第7章为Android中的数据存取；第8章为Android客户端与服务器端交互；第9章为综合项目——“欢乐购商城”实现；第10章为鸿蒙系统开发入门。

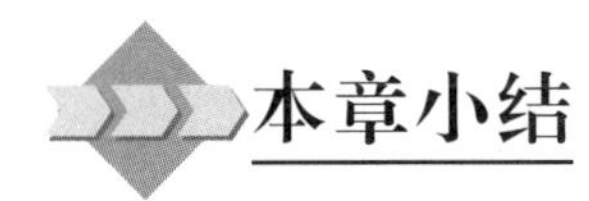

本章小结

本章主要带领读者进入Android的开发世界，介绍了一般App的开发流程，让用户认识Android开发工具和工作页面。然后对本书所用到的“欢乐购商城”功能和界面进行介绍，分析了客户端与服务器端的关系，认识如何配置局域网运行环境和开发环境，认识App运行体系。本书把“欢乐购商城”融入每章中，最后分析了每章的讲解内容。

自测习题

1. 移动App开发一般包括哪儿部分，简要描述每部分的流程。
2. 下载“欢乐购商城”App，思考其界面模块切换的思路，并提出自己的见解。
3. 搭建本地服务器，并在网页中试用Web版应用。

第2章

App应用体验

本章以“欢乐购商城”项目为例，主要介绍“欢乐购商城”搭建开发环境、开发工具下载安装过程和虚拟模拟器配置，以及运行服务器端和客户端运行过程和 Android 应用目录结构文件的作用，达到体验“欢乐购商城”功能效果。

学习目标

本章要点

（1）掌握 Java 环境的安装和配置。

（2）掌握 Android Studio 下载和安装方法。

（3）掌握 Android 虚拟机创建方法。

（4）掌握 Android 服务器端搭建方法。

（5）掌握“欢乐购商城”客户端导入方法。

（6）熟悉“欢乐购商城”的项目结构。

视频详解

2.1 开发环境配置

任务陈述

Android 应用项目除了需要开发软件 Android Studio，还需要一定的开发环境。因为 Android 应用开发是基于 Java 语言开发的，所以在正式安装 Android 开发工具前还要安装 Java 开发环境。

分析：Java 的程序安装软件为 JDK，需要从官网下载安装并配置系统环境变量，Android Studio 才能使用。

相关知识

视频详解

2.1.1 Java 语言

“欢乐购商城”项目由 Android 客户端、服务器端和数据库三大部分组成。Android 客户端和服务器端是基于 Java 语言进行开发，所以项目开发之前需要配置 Java 开发环境。

1996 年，Sun 公司发布了 JDK 1.0 和 HotJava。通过嵌入在网页中的 Applet 运行 Java 程序，这一特性相继被 Microsoft 公司和 Netscape 公司的 Web 浏览器所支持。IBM、Apple、DEC、Adobe、Silicon Graphics、HP、Oracle 和 Microsoft 等公司相继购买了 Java 技术许可证，从此 Java 成为日益流行的计算机语言。

JDK(Java Development Kit，Java 开发工具包)是一个编写 Java 的 Applet 小程序和应用程序的程序开发环境。它是整个 Java 的核心，包括 Java 运行环境(Java Runtime Environment，JRE)、一些 Java 工具和 Java 的核心类库(Java API)。无论什么 Java 应用服务器，实质上都是内置了某个版本的 JDK。主流的 JDK 是 Sun 公司发布的 JDK，除 Sun 公司之外，还有很多公司和组织都开发了自己的 JDK。例如，IBM 公司的 JDK、BEA 公司的 JRocket 及 GNU 组织开发的 JDK。

另外，可以把 Java API 类库中的 Java SE API 子集和 Java 虚拟机这两部分统称为 JRE (Java Runtime Environment)，JRE 是支持 Java 程序运行的标准环境。JRE 是运行环境，JDK 是开发环境。因此，写 Java 程序时需要 JDK，而运行 Java 程序时就需要 JRE。而 JDK 里面已经包含 JRE，因此只要安装了 JDK，就可以编辑 Java 程序，也可以正常运行 Java 程序。但由于 JDK 包含许多与运行无关的内容，占用的空间较大，因此运行普通的 Java 程序无须安装 JDK，而只需要安装 JRE 即可。

视频详解

2.1.2 Java 环境配置

1. 下载 JDK

首先需要下载 Java 开发工具包 JDK，下载地址为 https://www.oracle.com/java/technologies/javase-downloads.html，单击“下载”按钮，如图 2-1 所示。

图 2-1 Oracle 官网 JDK 下载跳转按钮

在下载页面中需要选择接受许可并根据自己的系统选择对应的版本，如果计算机操作系统为 64 位，可以安装 64 位 JDK，也可以安装 32 位 JDK；如果计算机操作系统为 32 位，则只能安装 32 位 JDK（见图 2-2）。注意，本书后面使用的 Eclipse 和 MyEclipse 的位数需要和 JDK 保持一致，本书以 Windows 64 位系统为例。

Java SE Development Kit 8u101

You must accept the Oracle Binary Code License Agreement for Java SE to download this software.

Accept License Agreement　　Decline License Agreement

Product / File Description	File Size	Download
Linux ARM 32 Hard Float ABI	77.77 MB	jdk-8u101-linux-arm32-vfp-hflt.tar.gz
Linux ARM 64 Hard Float ABI	74.72 MB	jdk-8u101-linux-arm64-vfp-hflt.tar.gz
Linux x86	160.28 MB	jdk-8u101-linux-i586.rpm
Linux x86	174.96 MB	jdk-8u101-linux-i586.tar.gz
Linux x64	158.27 MB	jdk-8u101-linux-x64.rpm
Linux x64	172.95 MB	jdk-8u101-linux-x64.tar.gz
Mac OS X	227.36 MB	jdk-8u101-macosx-x64.dmg
Solaris SPARC 64-bit	139.66 MB	jdk-8u101-solaris-sparcv9.tar.Z
Solaris SPARC 64-bit	98.96 MB	jdk-8u101-solaris-sparcv9.tar.gz
Solaris x64	140.33 MB	jdk-8u101-solaris-x64.tar.Z
Solaris x64	96.78 MB	jdk-8u101-solaris-x64.tar.gz
Windows x86	188.32 MB	jdk-8u101-windows-i586.exe
Windows x64	193.68 MB	jdk-8u101-windows-x64.exe

32位　64位

图 2-2　JDK 下载列表

2. 安装文件

单击下载好的 JDK 安装包，会出现如图 2-3 所示的安装界面，单击“下一步”按钮。

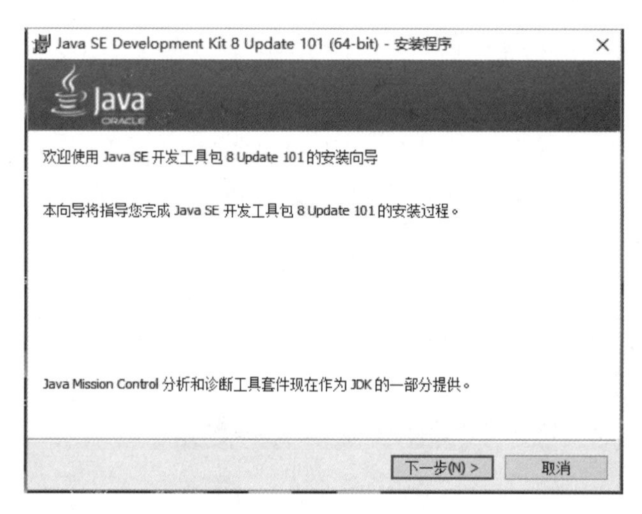

图 2-3　JDK 安装引导页面

在如图 2-4 所示的 JDK 安装地址对话框中单击“更改”按钮，可随意更改 JDK 的安装目录。本文以修改为 D:\Java\jdk1.8.0_101\为例（见图 2-5）。

单击“下一步”按钮（见图 2-6），完成 JDK 安装，如图 2-7 所示。

在 JDK 安装过程中，系统会自动安装 JRE，单击“更改”按钮可进行安装目录更改（见图 2-8）。建议将 JRE 和 JDK 放在同一目录下，本文以 D:\Java\jre1.8.0_101 为例，在 D 盘 Java 文件夹下新建 jre1.8.0_101 文件夹，单击“更改”按钮后，在弹出的对话框中选择 jre1.8.0_101 文件夹完成目录更改（见图 2-9）。

在图 2-9 中单击“下一步”按钮，完成 JRE 安装，如图 2-10 和图 2-11 所示。

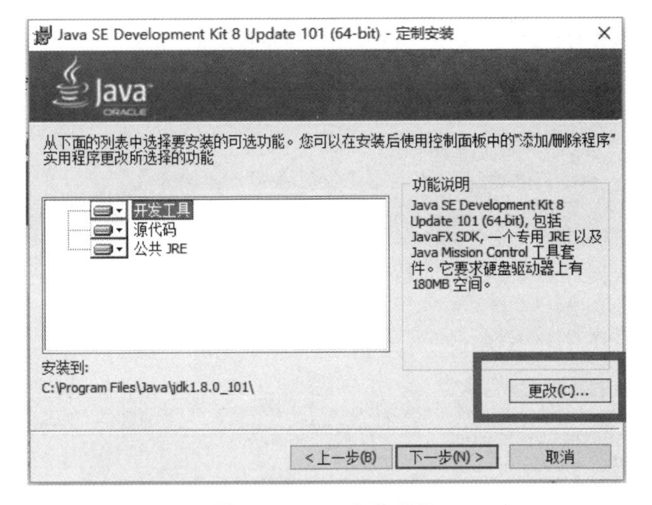

图 2-4　JDK 安装地址

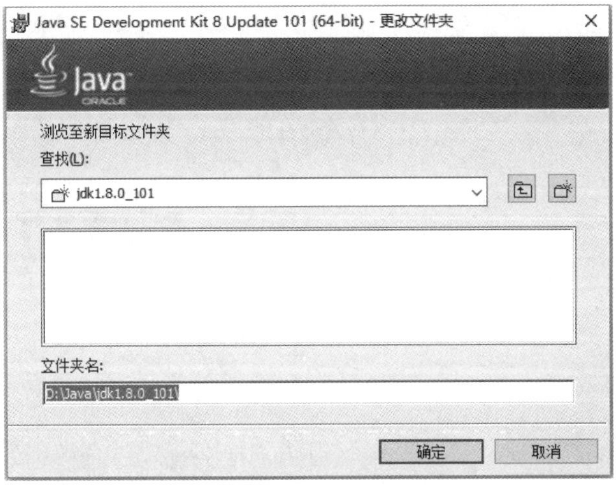

图 2-5　JDK 修改安装目录

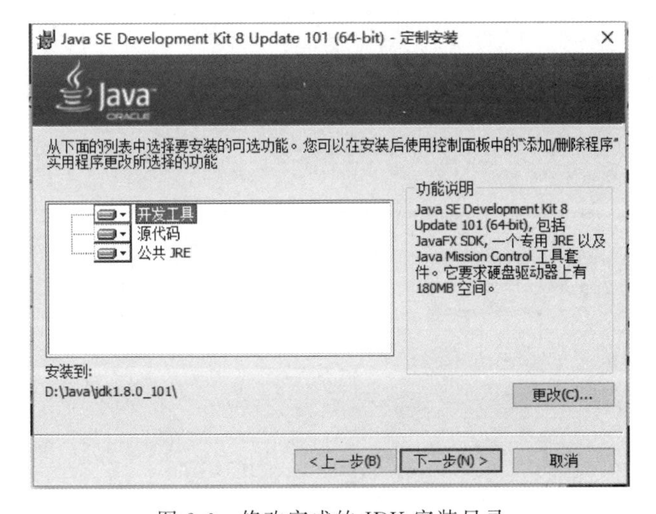

图 2-6　修改完成的 JDK 安装目录

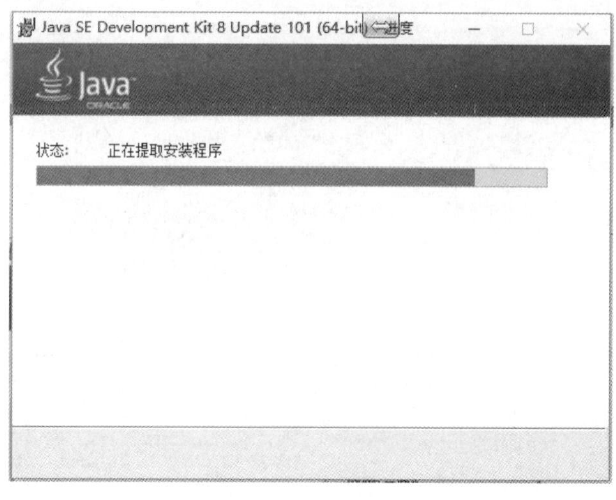

图 2-7 JDK 安装进度

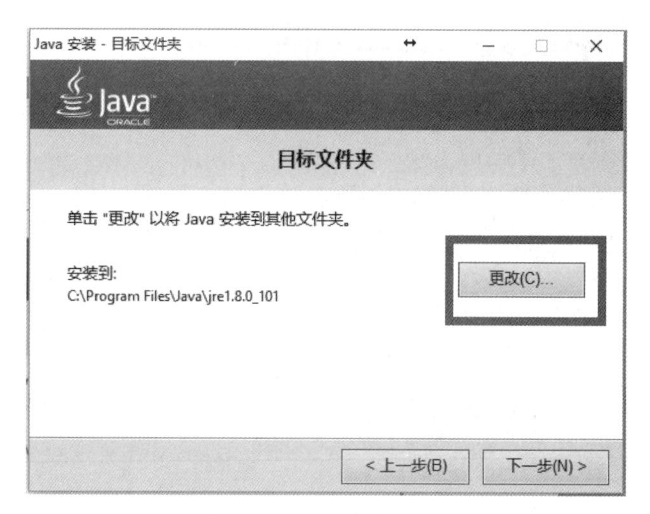

图 2-8 JRE 安装地址

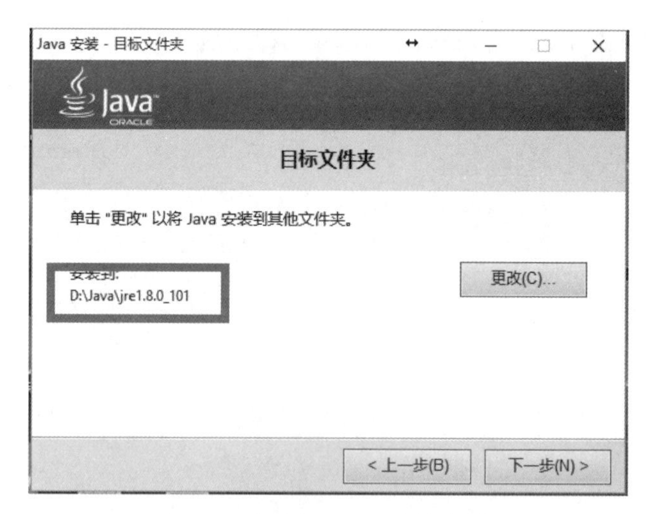

图 2-9 修改好的 JRE 安装目录

图 2-10 JRE 安装进度

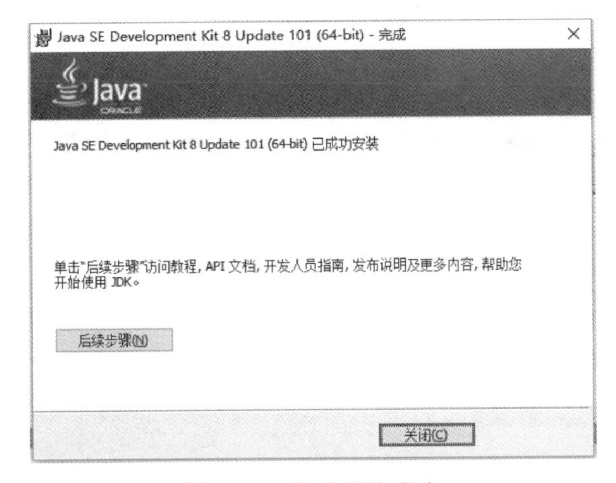

图 2-11 JDK 安装成功

3. 配置 Java 环境

在 Java JDK 1.5 之前，Java JDK 安装完成后并不能立即使用，还需要配置相关环境变量。Java JDK 1.5 之后，系统会有默认的配置，但建议手动进行配置。右击“我的电脑”，选择“属性”→“高级系统设置”选项，如图 2-12 所示。

在如图 2-13 所示的系统属性对话框中，单击“环境变量”按钮。

在系统变量中新建一个 JAVA_HOME 变量，单击“新建”按钮，如图 2-14 所示。

在“变量名”文本框中输入 JAVA_HOME、“变量值”文本框中输入 D:\Java\jdk1.8.0_101(要根据自己 JDK 的实际路径配置)，然后单击“确定”按钮，如图 2-15 所示。

建议 JAVA_HOME 变量名大写，表示常量。但 Windows 系统不区分大小写，即大写、小写、大小写混合表示同一个变量名，虽不会出错，但不符合规范。

注意：变量值后不需要加任何符号。

选择 Path 选项，在变量值后面追加输入%JAVA_HOME%\bin;%JAVA_HOME%\jre\bin;，%JAVA_HOME%代表的路径就是 D:\Java\jdk1.8.0_101\。

注意，输入时需要输入法是英文状态，然后单击“确定”按钮，如图 2-16 和图 2-17 所示。

图 2-12 高级系统设置

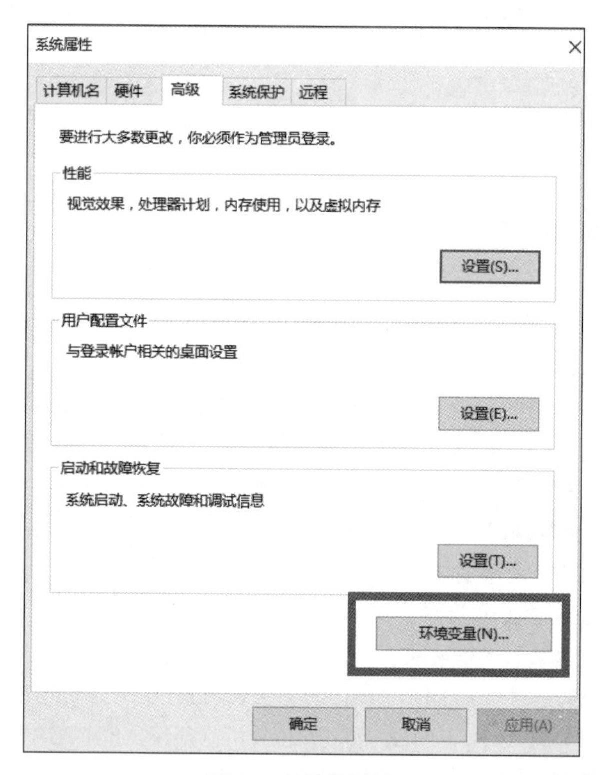

图 2-13 系统属性

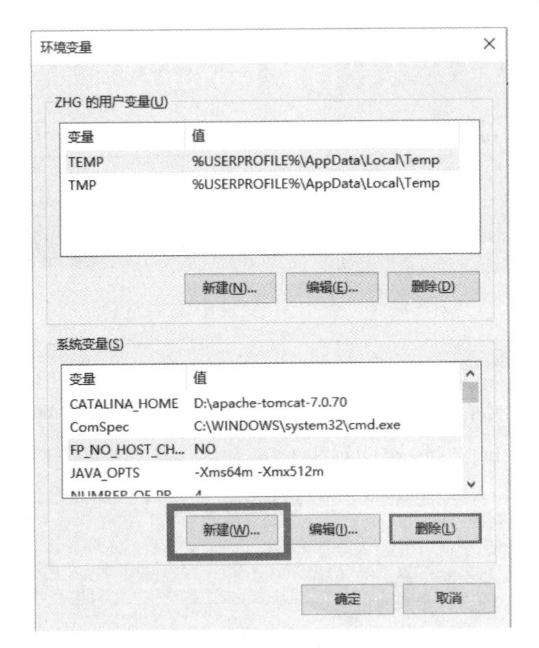

图 2-14 “环境变量”对话框

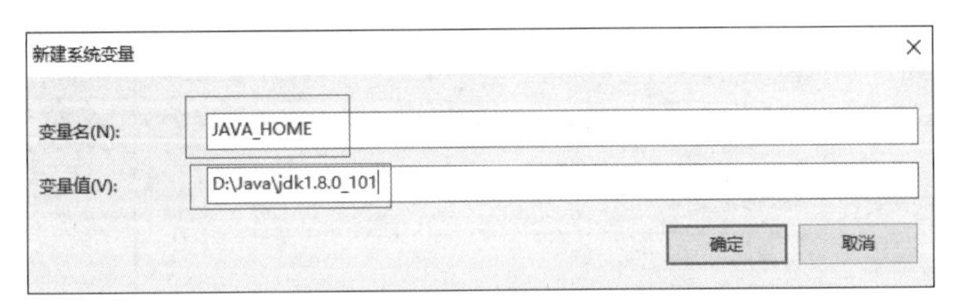

图 2-15 JAVA_HOME 环境变量设置

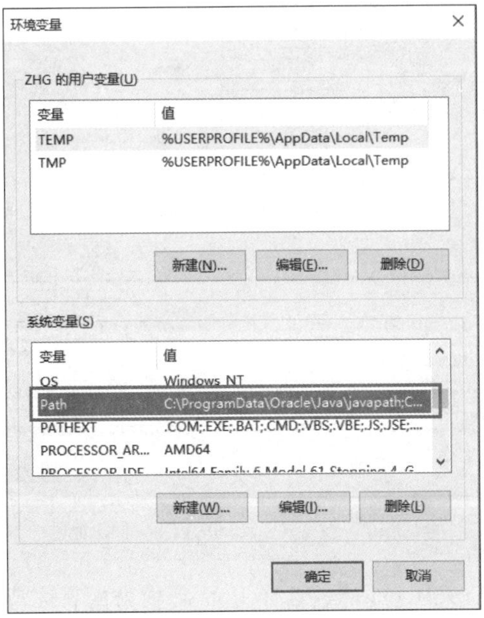

图 2-16 Path 选项

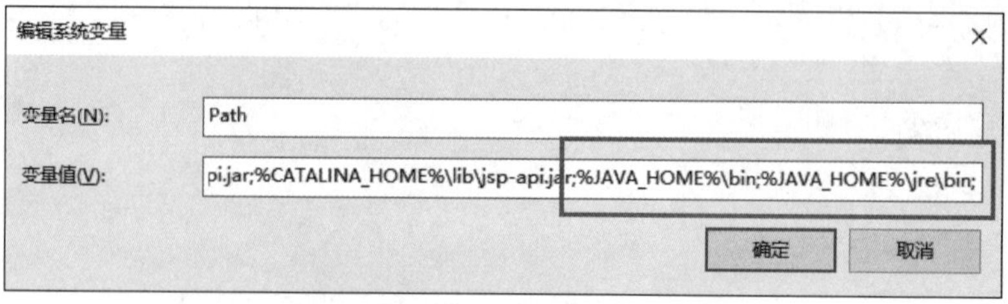

编辑系统变量
变量名(N): Path
变量值(V): pi.jar;%CATALINA_HOME%\lib\jsp-api.jar;%JAVA_HOME%\bin;%JAVA_HOME%\jre\bin;
确定 取消

图 2-17　在 Path 变量中添加路径

单击“新建”按钮，变量名输入 CLASSPATH，变量值输入“.；%JAVA_HOME%\lib\dt.jar；%JAVA_HOME%\lib\tools.jar；”，其中，点(.)表示当前目录，分号表示多个路径之间的分隔符，然后单击“确定”按钮(见图 2-18 和图 2-19)。

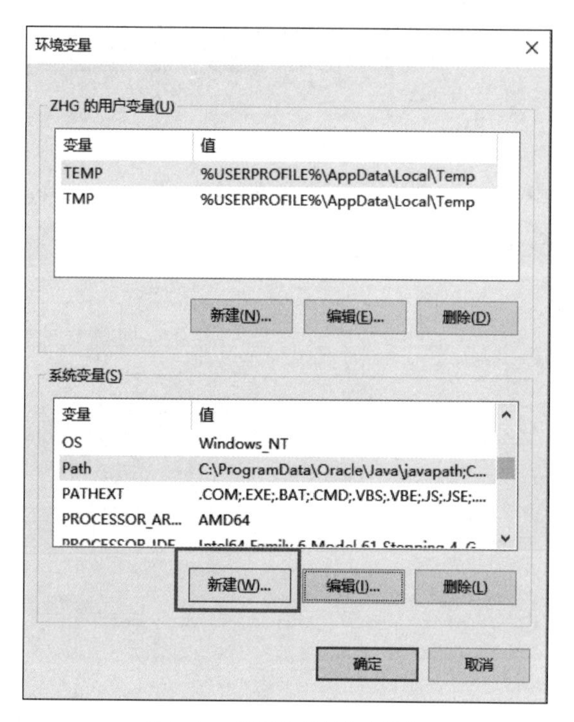

图 2-18　“环境变量”对话框

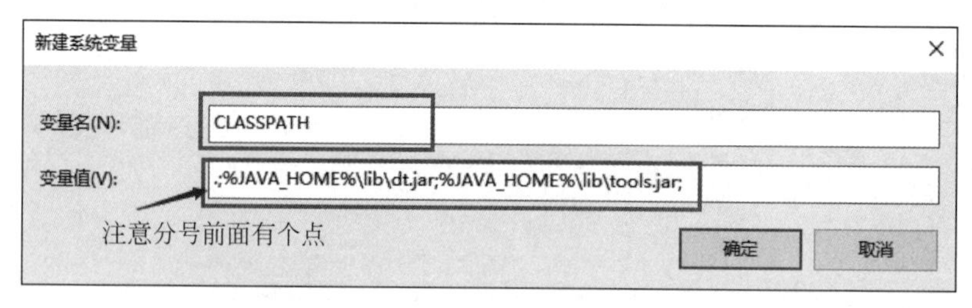

图 2-19　设定 CLASSPATH 环境变量

连续单击“确定”按钮，关闭对话框，这样 Java 环境就配置完成了。

4. 环境变量配置测试

（1）选择“开始”→“运行”，输入 cmd。

（2）输入 java -version、java、javac 等命令，如果出现如图 2-20 所示信息，则说明环境变量配置成功。

```
C:\Users\ZHG>javac
用法: javac <options> <source files>
其中, 可能的选项包括:
  -g                         生成所有调试信息
  -g:none                    不生成任何调试信息
  -g:{lines,vars,source}     只生成某些调试信息
  -nowarn                    不生成任何警告
```

图 2-20 在命令行窗口中输入 javac 命令

任务实施

（1）官网下载 JDK 软件并安装。

（2）在 Windows 系统中配置系统环境变量。

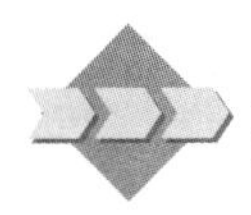

2.2 Android Studio 下载和安装

视频详解

任务陈述

Android 开发运行环境 JDK 安装和配置之后，下一步安装开发工具。Android 常用开发工具包括 Android Studio、Eclipse＋ADT＋SDK 三合一集成包等。

分析：Eclipse＋ADT＋SDK 集成安装包安装比较方便、占用内存小，在早期开发中是最常用的开发工具，近几年 Google 公司停止 Eclipse 集成开发环境更新。发布 Android Studio 工具，虽然 Android Studio 安装较为复杂、占用内存大，但是功能更为强大、界面较为优美，已经被程序员青睐。

相关知识

2.2.1 Android Studio 下载

视频详解

Android Studio 是一个 Android 开发环境，基于 IntelliJ IDEA，类似 Eclipse ADT，提供了集成的 Android 开发工具用于开发和调试。

Google 公司已宣布，为了简化 Android 的开发力度，重点建设 Android Studio 工具，现在已经停止支持 Eclipse 等其他集成开发环境。而随着 Android Studio 正式版的推出和完善，Android 开发者们大多已转向 Android Studio 开发平台。

下载 Android Studio 很简单，利用搜索引擎搜索 Android Studio 就可以轻松找到 Android Studio 最新版本的下载路径。也可以在浏览器中打开 developer. android. com/sdk/installing/studio. html 站点下载。

2.2.2 Android Studio 安装

视频详解

（1）双击 Android Studio 的安装文件，进入如图 2-21 所示的安装界面。

(2) 选择要安装的插件，如图 2-22 所示。其中，第一个是 Android Studio 主程序，必选。第二个是 Android SDK，会安装 Android 的 SDK，需选上。第三个是 Android Virtual Device，用于安装 Android 虚拟设备。完成后单击 Next 按钮。

图 2-21　Android Studio 安装界面

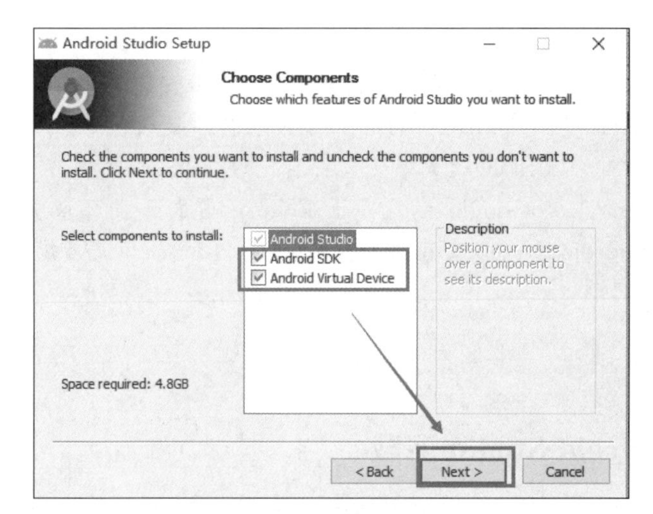

图 2-22　安装插件

(3) 同意条款，单击 I Agree 按钮，如图 2-23 所示。

(4) 选择 Android Studio 和 SDK 的安装目录，如图 2-24 所示。

(5) 设置虚拟机硬件加速器可使用的最大内存，如图 2-25 所示。

如果计算机配置还不错，默认设置 2GB 即可；如果计算机配置比较差，则选择 1GB，否则会影响运行其他软件。

(6) 之后就进入自动安装模式，如图 2-26 所示。

如果没出什么意外，一小段时间后就会看到如图 2-27 所示的界面，也就说明安装成功了，如图 2-28 所示。

(7) 打开 Android Studio 后，进入相关配置界面，如图 2-29 所示。

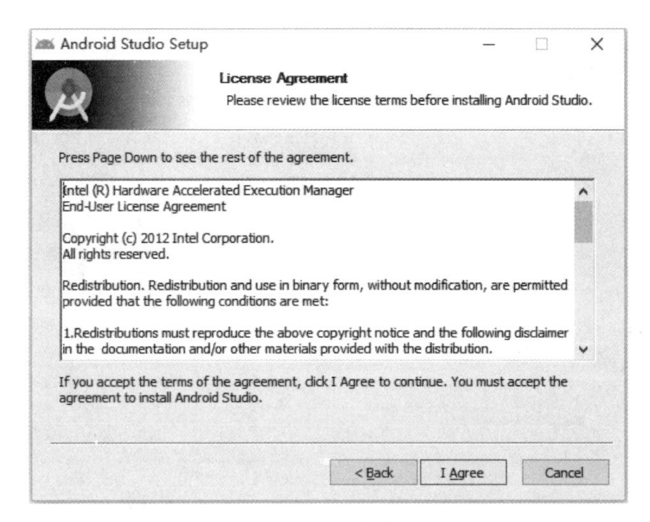

图 2-23　同意条款

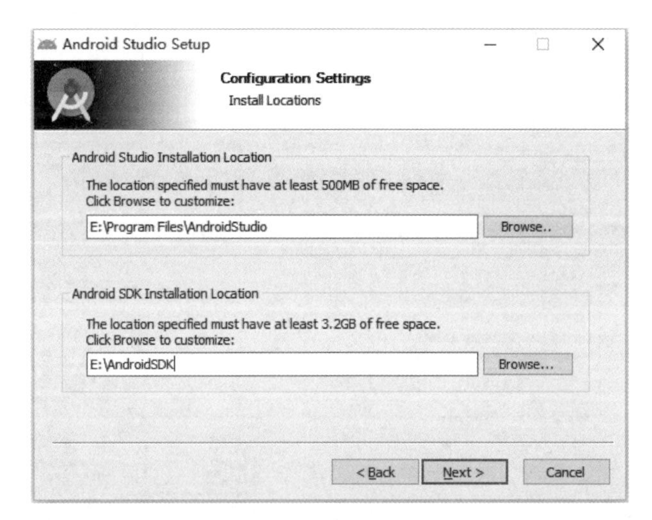

图 2-24　选择安装目录

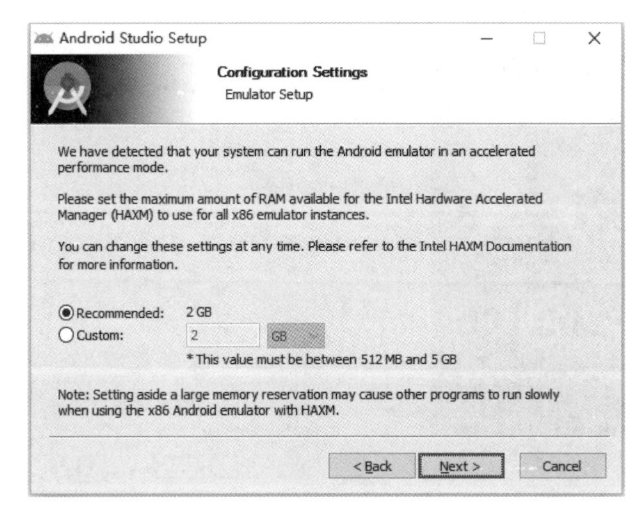

图 2-25　选择虚拟加速器内存

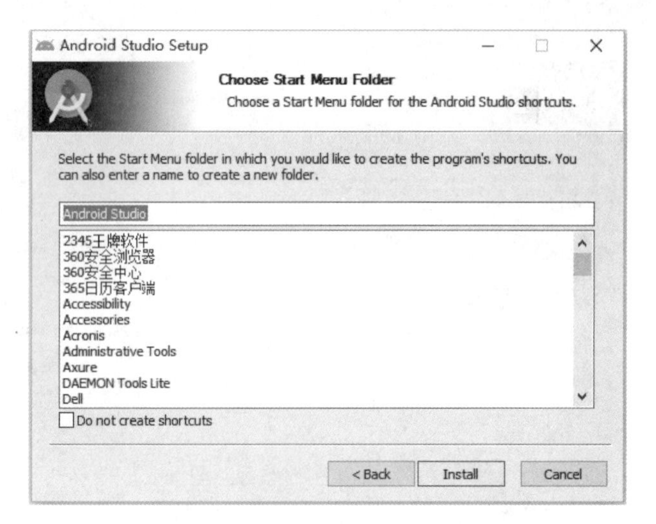

图 2-26　自动安装

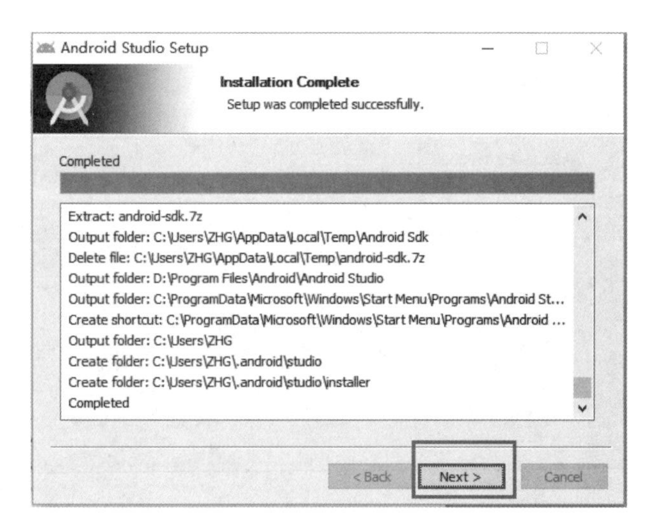

图 2-27　自动安装结束

图 2-28　安装成功页面

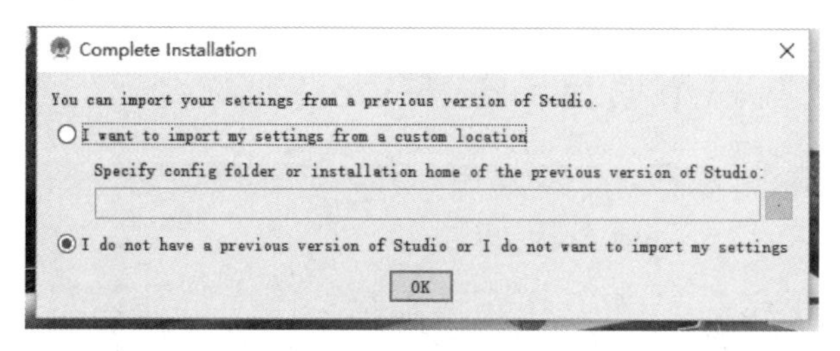

图 2-29　导入配置界面

图 2-29 用于导入 Android Studio 的以前安装版本配置文件，如果是第一次安装，选择最后一项，即不导入配置文件，然后单击 OK 按钮即可。

（8）完成后就会进入如图 2-30 所示页面，这是程序在检查 SDK 的更新情况，然后开始下载组件，如图 2-31 所示。

图 2-30　检查 SDK 更新情况

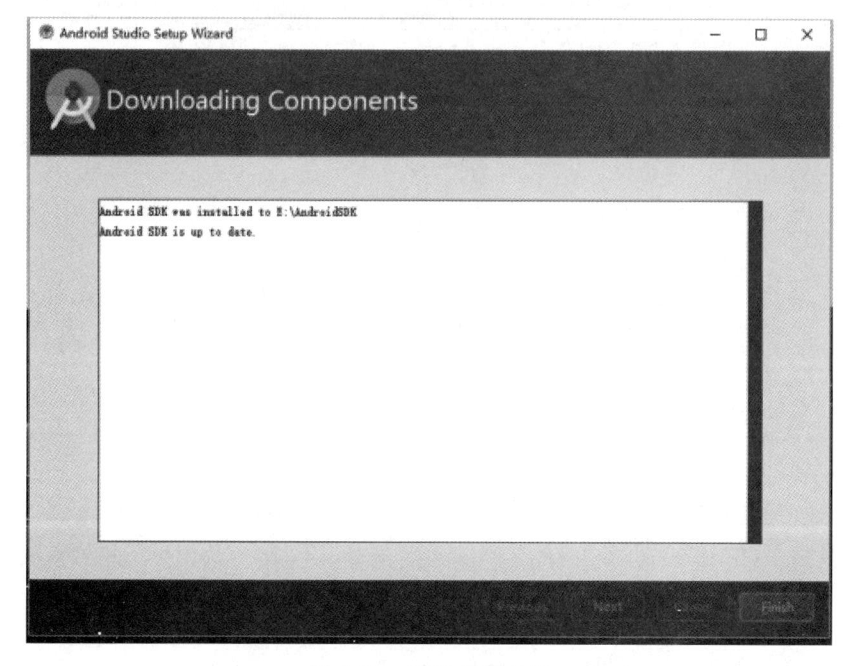

图 2-31　SDK 下载界面

视频详解

2.2.3 创建第一个项目：HelloWorld

(1) 创建 HelloWorld 项目。和其他平台类似，在这之前已经配置好了 JDK 和 Android SDK 环境，并且第一次安装 Android Studio。

当更新完 Android SDK 后，将会看到如图 2-32 所示界面。

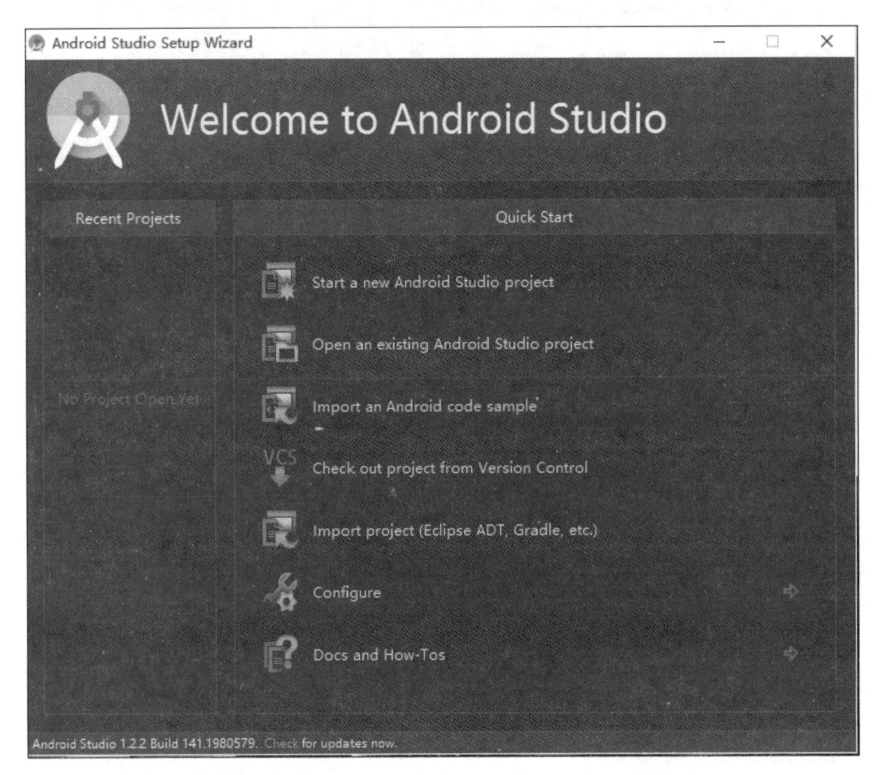

图 2-32 Android studio 初始打开界面

选项 1：创建一个 Android Studio 项目。

选项 2：打开一个 Android Studio 项目。

选项 3：导入官方样例，从网络上下载代码。此功能在以前的测试版本中是没有的，建议查看官方给出的范例。

选项 4：从版本控制系统中导入代码，支持 CVS、SVN、Git、Mercurial 以及 GitHub。

选项 5：导入非 Android Studio 项目，例如原生的 IDEA 开发的项目。

选项 6：设置。

选项 7：帮助文档。

(2) 填写应用名和包名，如图 2-33 所示。

(3) 选择安卓版本，如图 2-34 所示。

(4) 选择合适的界面，如图 2-35 所示。

在这个页面中选择一个 Activity 模板，和 Eclipse 一样，直接选择一个 Empty Activity。

(5) 确认 Activity Name 和 Layout Name 后，完成创建，如图 2-36 所示。

至此，一个简单的 Android Studio 项目就完成了，如图 2-37 所示。

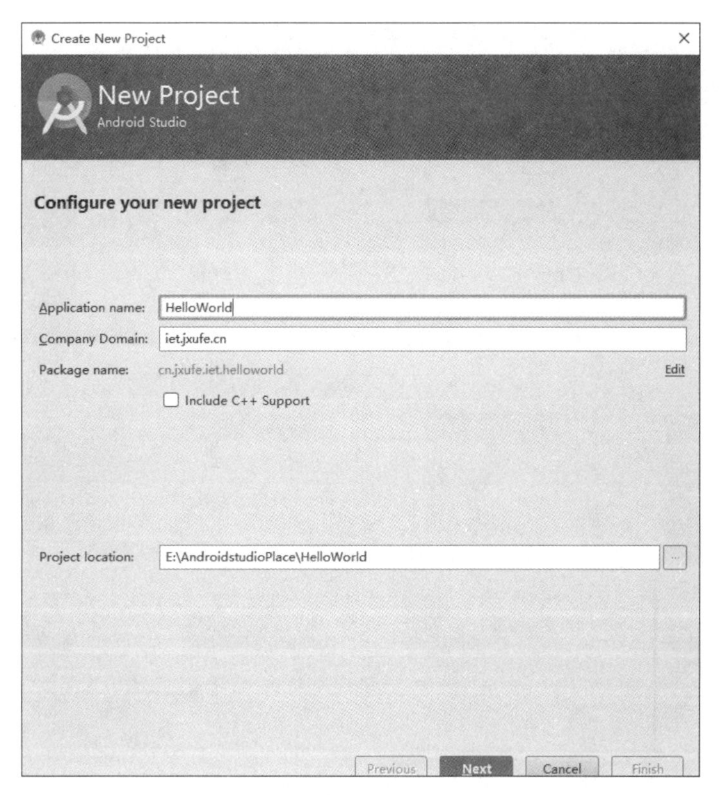

图 2-33　填写应用名和包名

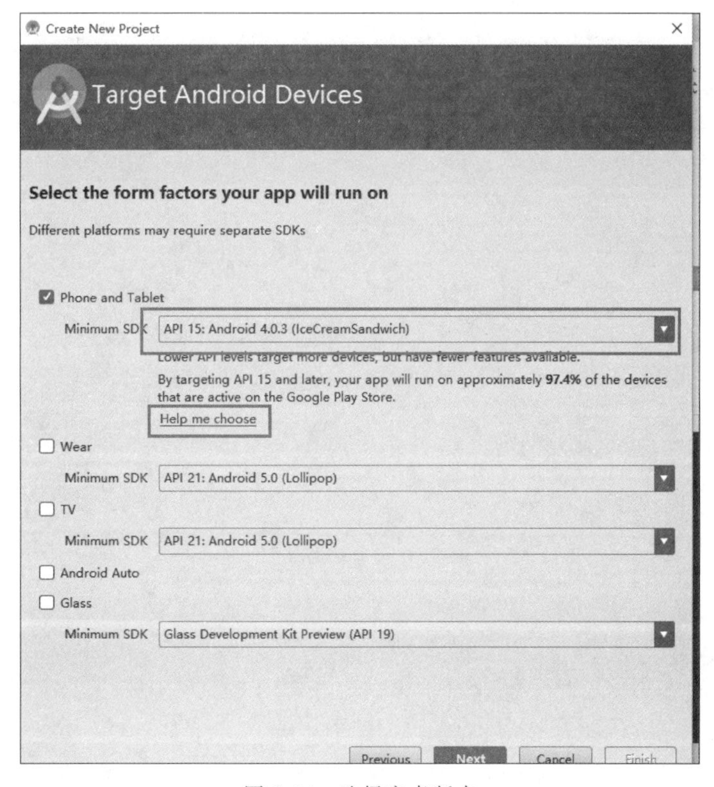

图 2-34　选择安卓版本

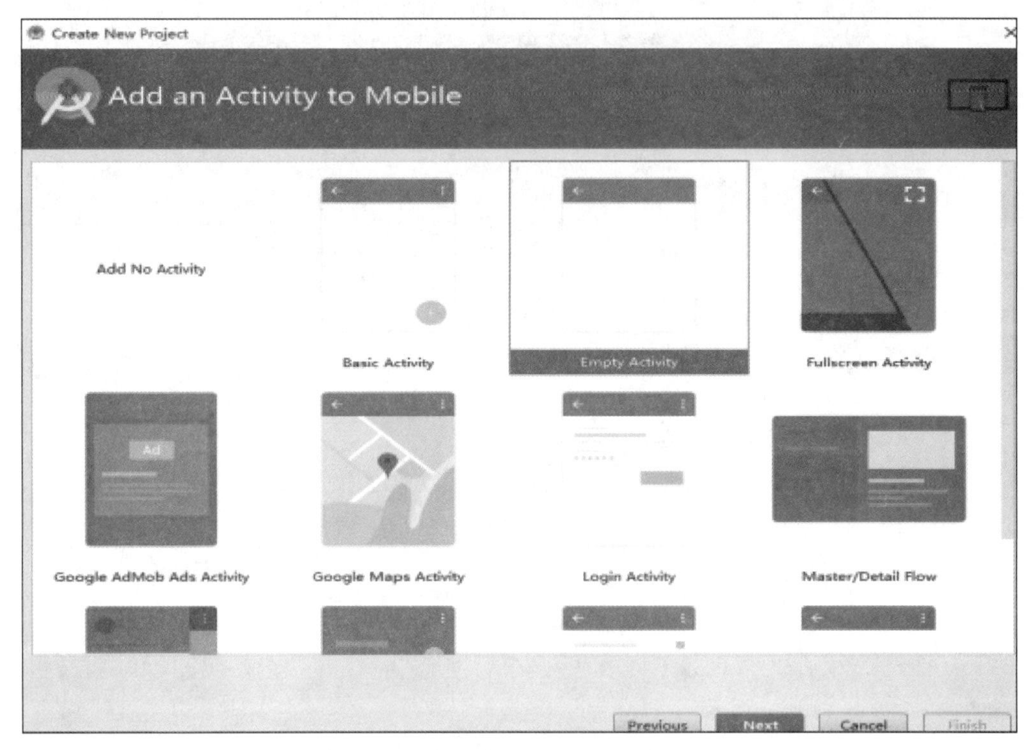

图 2-35　选择合适的界面

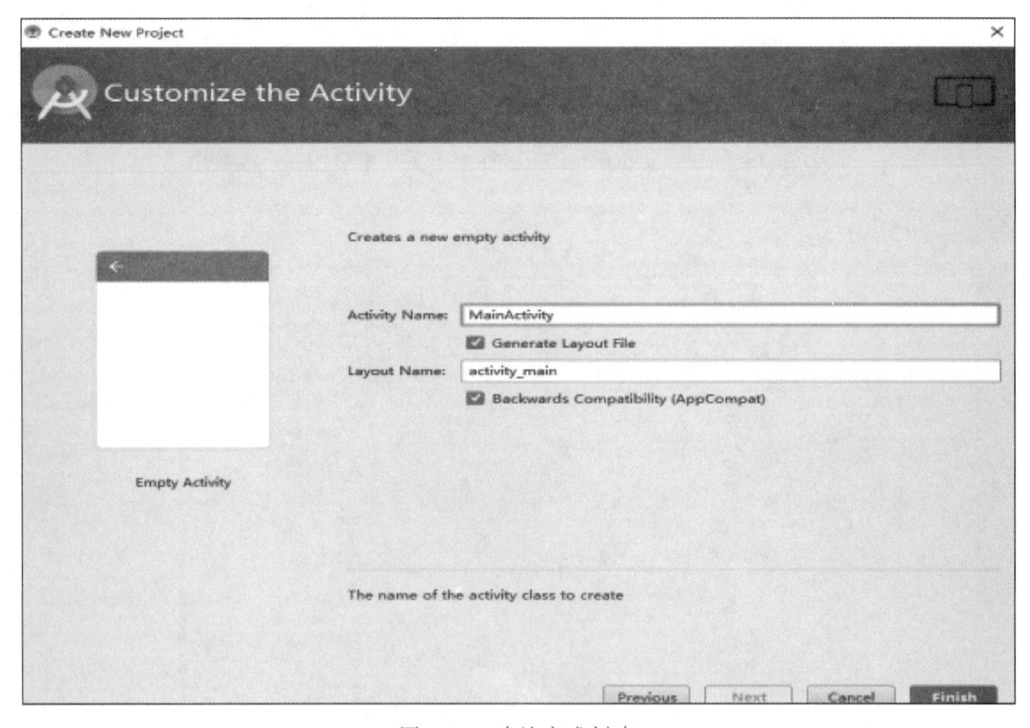

图 2-36　确认完成创建

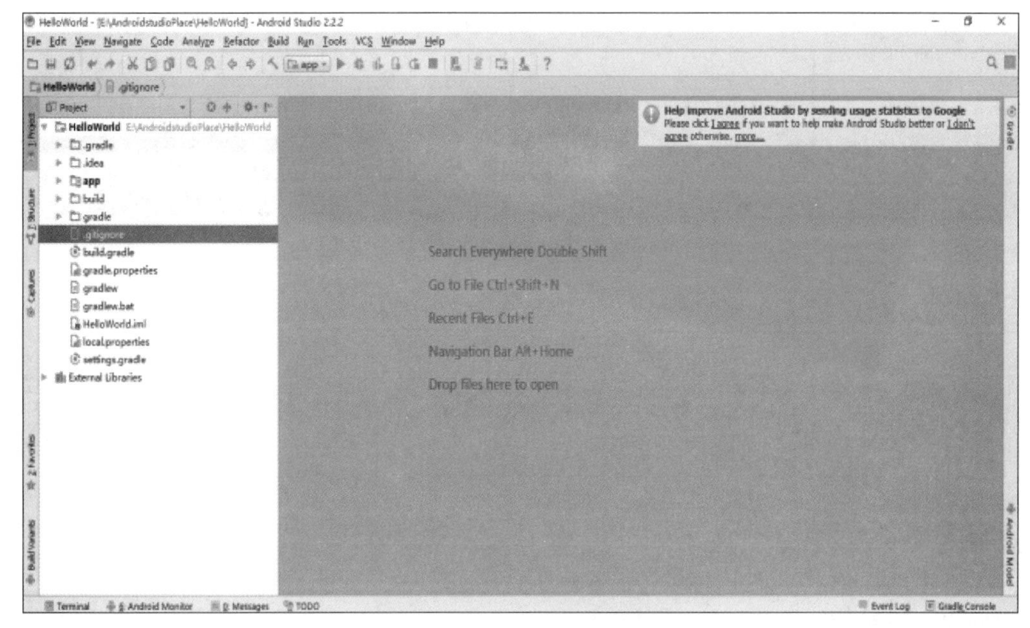

图 2-37 项目创建完成

2.2.4 创建一个虚拟机设备

视频详解

(1) 单击菜单栏 AVD Manager 按钮后，在弹出的对话框中单击 Create Virtual Device 按钮，创建模拟器，如图 2-38 所示。

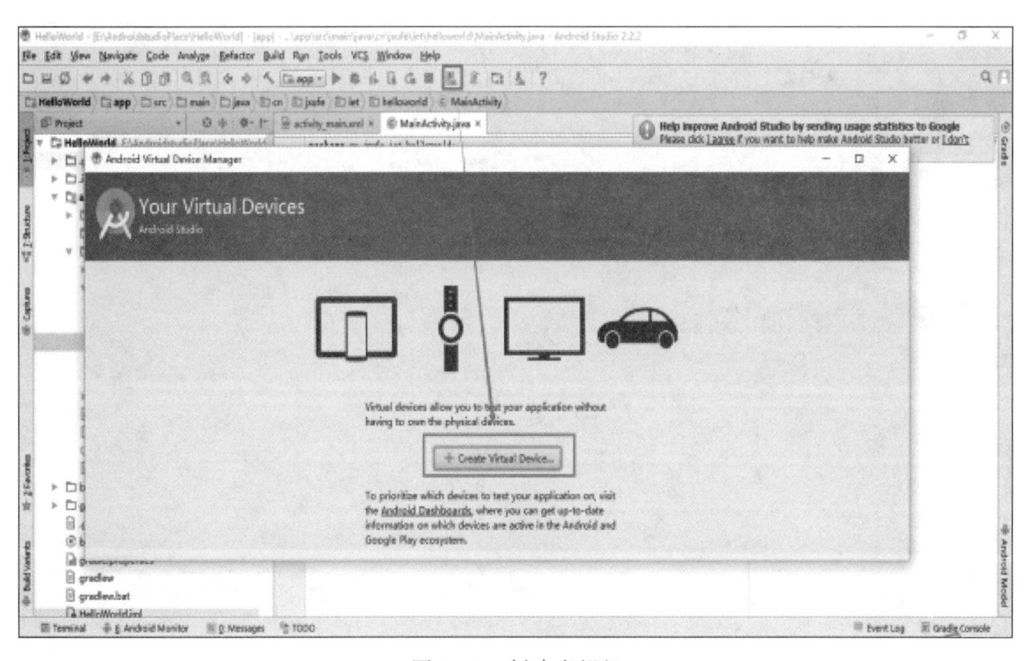

图 2-38 创建虚拟机

（2）在 Category 栏中选择 Phone，然后选择合适的屏幕尺寸和分辨率，单击 Next 按钮，如图 2-39 所示。

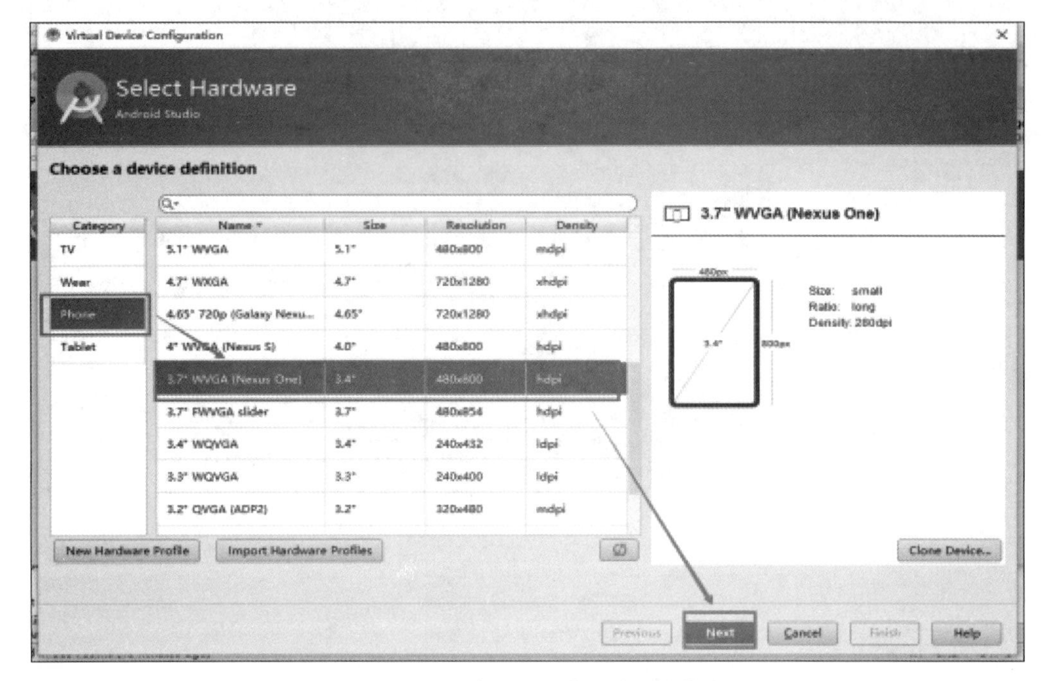

图 2-39　选择合适的屏幕尺寸和分辨率

（3）选择合适的镜像文件，如果没有可单击 Download 按钮下载对应镜像文件(见图 2-40)。

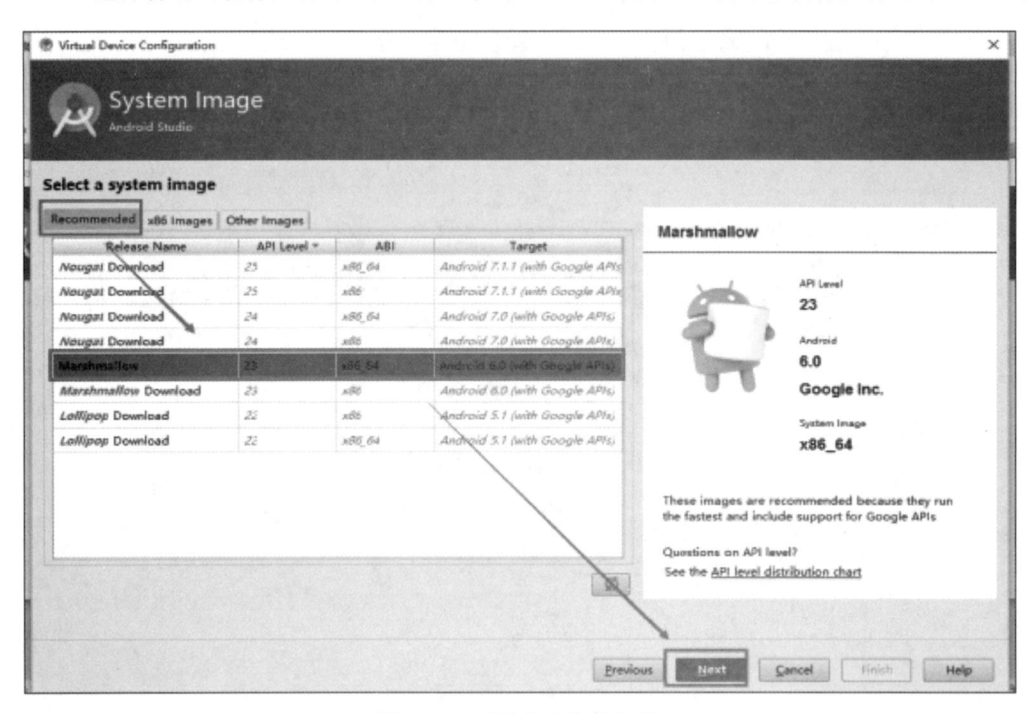

图 2-40　选择合适镜像文件

（4）输入虚拟模拟器名，然后单击 Finish 按钮，这样虚拟模拟器就创建完成了，如图 2-41 所示。

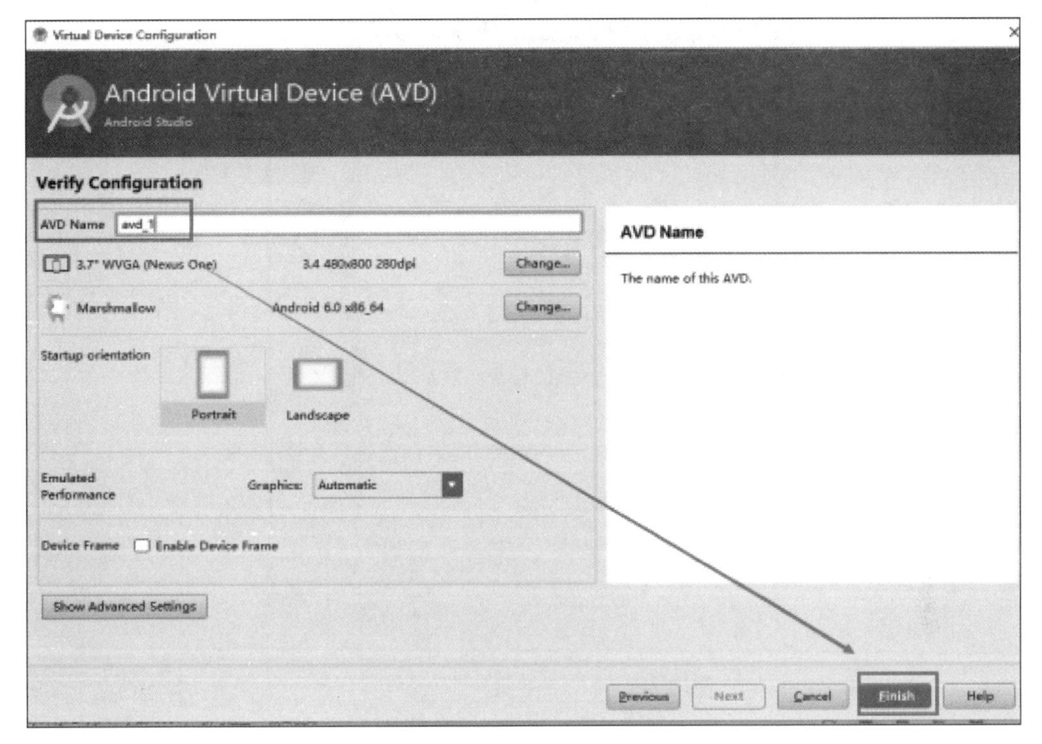

图 2-41　输入模拟器名称

（5）单击“启动”按钮，启动模拟器，如图 2-42 所示。

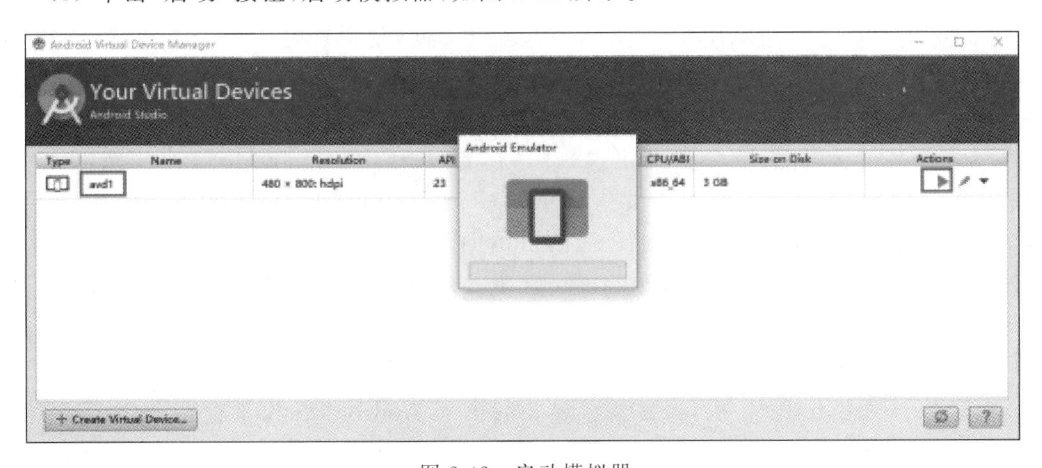

图 2-42　启动模拟器

（6）单击导航栏的“运行”按钮，在模拟器中运行项目，效果如图 2-43 所示。

提示：

（1）在 Android Studio 中启动一个模拟器运行程序时，有可能会出现错误提示“Intel HAXM is required to run this AVD，VT-x is disabled in BIOS”，这里给出解决方案。

图 2-43 项目运行效果

① 在 SDK Manager 中检查是否安装 Intel x86 Emulator Accelerator (HAXM Installer),如果没有安装,则在 SDK Manager 中下载并安装。

② 在 SDK 目录(一般为 C:\Users\Administrator\AppData\Local \Android\sdk)下沿着 extras\intel\Hardware_Accelerated_ Execution_Manager 目录找到 intelhaxm-android.exe 文件,安装并运行。

③ 在安装过程中,可能会遇到错误提示"Intel virtualization technology (vt,vt-x) is not enabled",不用担心,这时只需要进入 BIOS 界面,在 Configuration 选项中找到 Intel Virtualization Technology 并将其设置成 Enable 即可。

④ 重新启动 Android Studio,然后再次启动 AVD 即可。如果成功,可能要花几分钟的时间去显示虚拟器的窗口。

(2) 如何设置 JDK 或者 SDK?

在运行 Android Studio 时,经常会提醒 JDK 或者 Android SDK 不存在,这时需要通过全局的 Project Structure 页面进行重新设置。选择 Configure→Project Defaults→Project Structure,进入全局的 Project Structure 页面。在此页面下设置 JDK 或 Android SDK 目录即可,如图 2-44 所示。

任务实施

(1) 完成 Android Studio 下载任务。

(2) 完成 Android Studio 安装,包括 gradle、SDK 等工具安装。

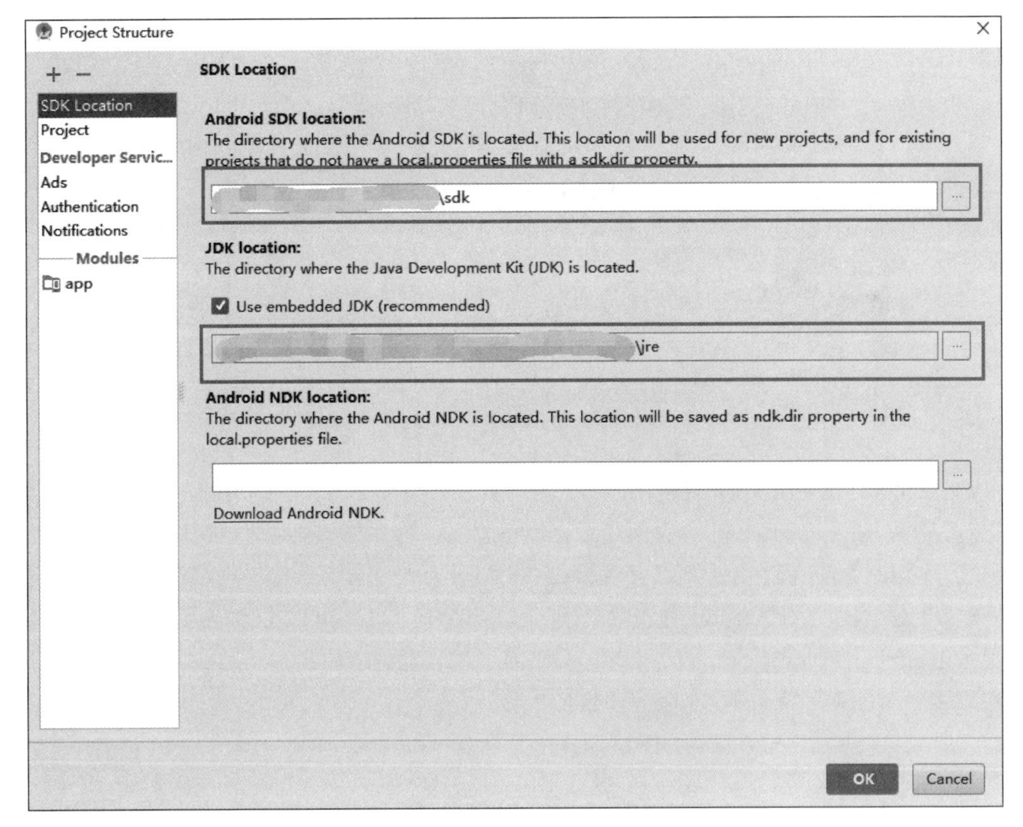

图 2-44　配置 SDK 和 JDK 路径

（3）创建第一个 Android 应用项目 HelloWorld。

（4）创建虚拟机，并把 HelloWorld 项目运行在模拟器上。

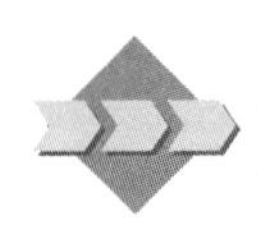

2.3　“欢乐购商城”项目体验

视频详解

任务陈述

前面已经进行开发环境配置和工具安装，下一步将进行项目体验。“欢乐购商城”教学版项目包括服务器端、数据库和移动客户端三部分，体验客户器端之前需要在本地安装服务端和数据库。

分析：本书的任务主要是学习 Android 客户端开发，整套开发又离不开服务器端和数据库。为了方便读者学习 Android，本书将服务器和数据库封装在 JTM 中，一键启动即可。客户端本节在 Android Studio 中导入，调整服务器地址即可运行。

视频详解

2.3.1　服务器端集成包运行

为了让初学者迅速体验 App 的功能效果，本书把服务器端和数据库放在 JTM 集成环境中运行。JTM是绿色免费的JDK＋Tomcat＋MySQL环境集成工具。通过JTM，用户无须对JDK、Tomcat、MySQL进行任何安装和配置即可迅速搭建支持JSP＋MySQL的服务器运行环境。下面介绍如何使用“欢乐购商城”教学版JTM。

（1）从提供的资源网站下载JTM压缩包，解压到不带有空格的路径上（见图2-45和图2-46）。

图2-45　“欢乐购商城”JTM压缩包

图2-46　“欢乐购商城”JTM解压文件

（2）双击JTM.exe文件，启动后的效果如图2-47所示。此时，Tomcat和MySQL前面的图标都是灰色的，表示本机的这两个服务暂未启动，为正常现象。如果这两个图标为绿色的，则表示本机已启动相关服务或者端口被占用，需要停止相应的服务或进程。

（3）单击“启动（调试模式）”按钮，将会弹出两个命令行窗口，分别表示Tomcat的启动和MySQL的启动，此时Tomcat和MySQL前面的图标将会变为绿色，如图2-48所示。同时还会启动一个浏览器显示Tomcat的首页。

图2-47　JTM未启动时正常的状态

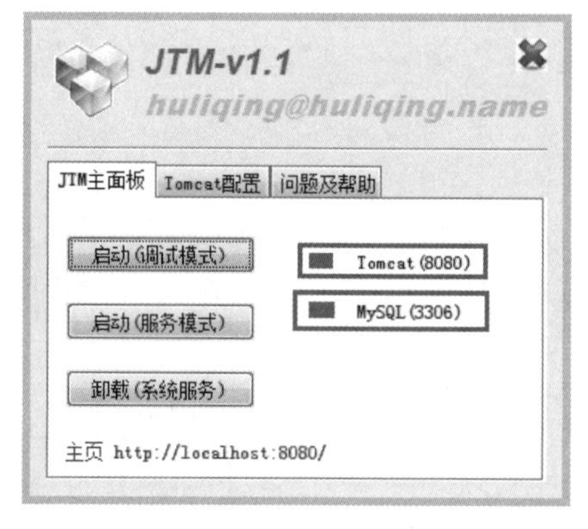

图2-48　JTM启动后正常的状态

（4）在浏览器的地址栏中输入http://localhost：8080/HappyShop，此时，将会出现本地服务器测试成功页面，如图2-49所示。在使用过程中要始终保持这两个窗口处于开启状态，否则将无法访问和获取相关数据。

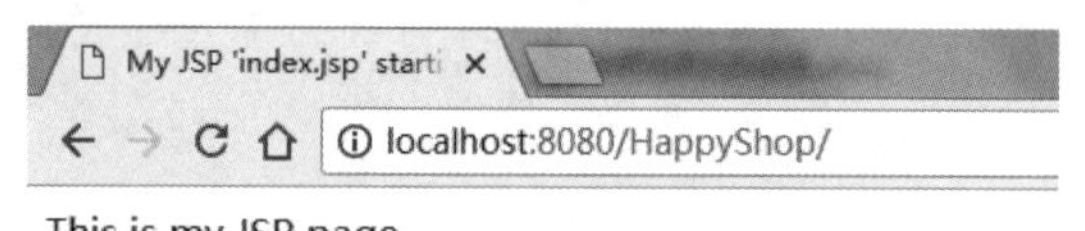

图2-49　本地服务器搭建成功的测试页面

2.3.2 “欢乐购商城”Android项目导入

(1) 打开 Android Studio 和虚拟模拟器,如图 2-50 所示。

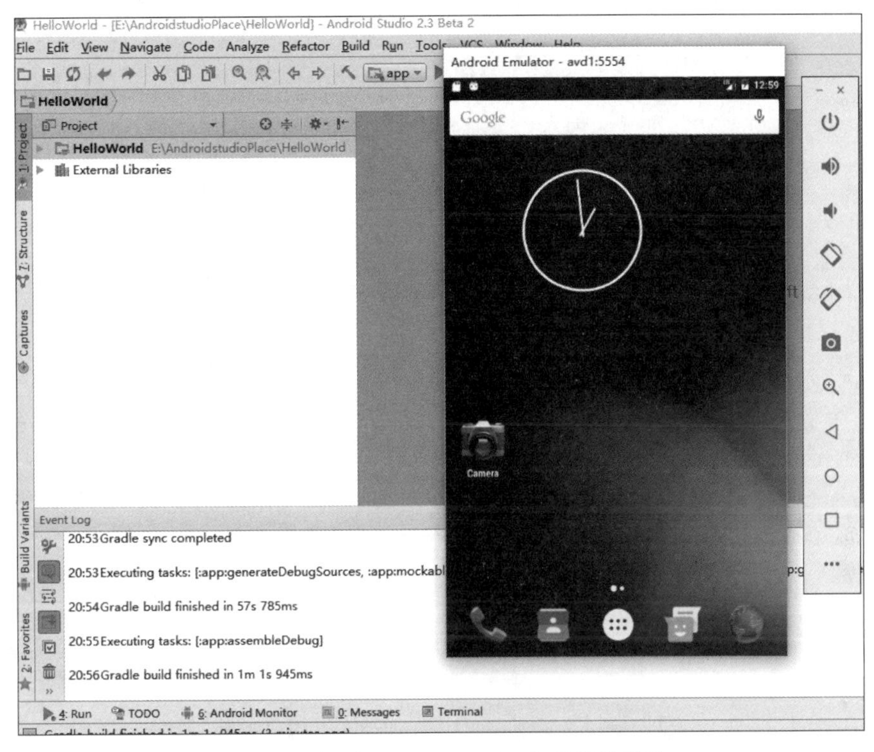

图 2-50 打开 Android Studio 和虚拟模拟器

(2) 从提供的资源网站下载客户端代码包,并进行解压。

(3) 选择 File 菜单下 Open 选项,如图 2-51 所示。

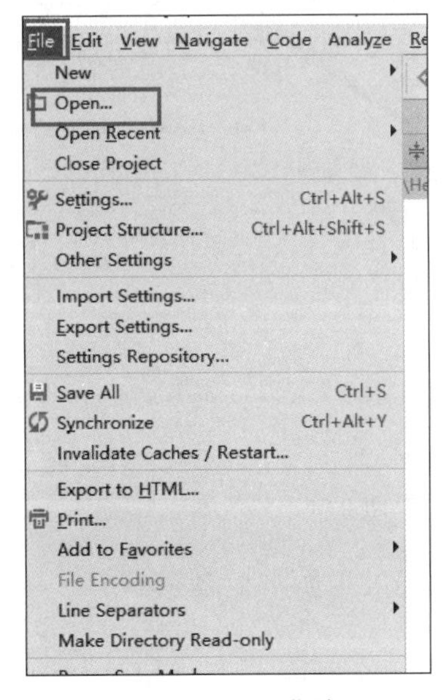

图 2-51 File 菜单

（4）在弹出的列表中选择客户端所在位置，单击 OK 按钮导入项目，如图 2-52 所示。

（5）导入项目后，打开 constant 类，找到 URL 语句，修改 IP 为计算机本地 IP，然后运行项目，如图 2-53 所示。（温馨提示：导入项目时，如果中文字符显示乱码的话，可以将 UTF-8 编码改成 GBK 编码。）

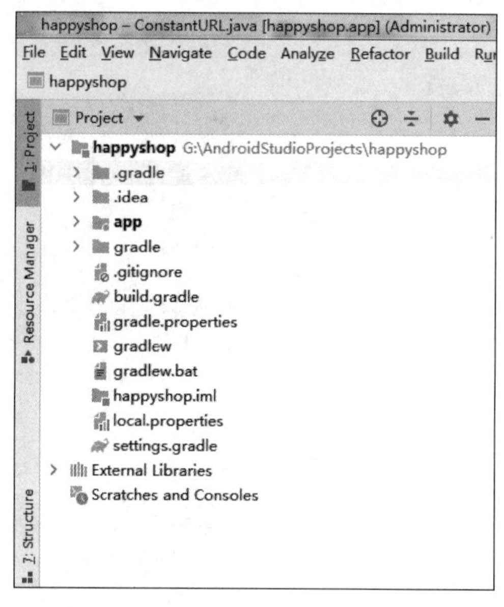

图 2-52 选择客户端所在目录导入项目

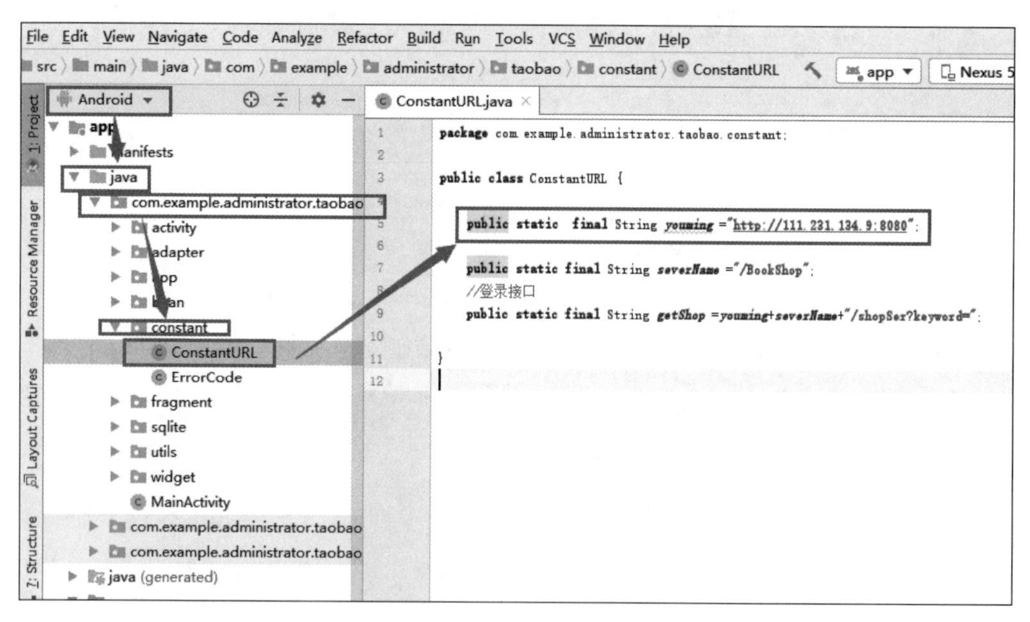

图 2-53 修改本地 IP 地址

导入成功后，客户端成功运行效果如图 2-54 所示。

图 2-54　客户端成功运行效果

2.3.3 “欢乐购商城”项目客户端结构图

视频详解

Android 客户端主要用于发送请求信息、获取结果数据，然后以一种用户容易理解的形式显示，主要涉及界面设计、用户事件处理、网络访问、数据传递与解析等。

Android 客户端部分程序结构如图 2-55 所示。

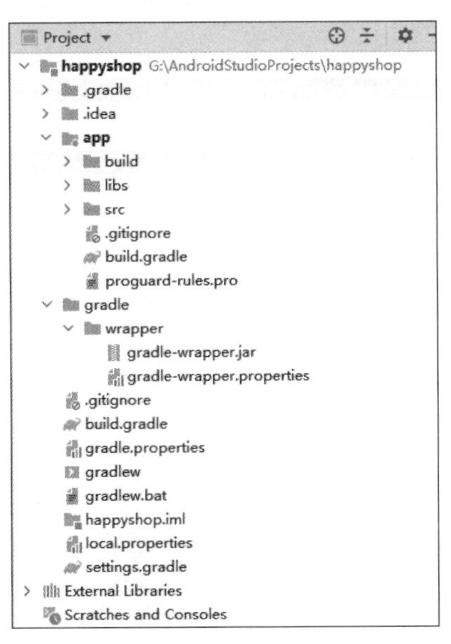

图 2-55　“欢乐购商城”客户端部分程序结构

可以发现有很多文件，表 2-1 介绍了各文件(夹)及其用途。

表 2-1 Android 客户端程序文件(夹)及其用途

文件(夹)名	用　途
. gradle	Gradle 编译系统，版本由 wrapper 指定
. idea	Android Studio IDE 所需要的文件
build	代码编译后生成的文件存放的位置
gradle	wrapper 的 jar 和配置文件所在的位置
. gitignore	git 使用的 ignore 文件
build. gradle	Gradle 编译的相关配置文件(相当于 Makefile)
gradle. properties	Gradle 相关的全局属性设置
gradlew	* nix 下的 Gradle wrapper 可执行文件
graldew. bat	Windows 下的 Gradle wrapper 可执行文件
local. properties	本地属性设置(key 设置，Android SDK 位置等属性)，这个文件是不推荐上传到 VCS 中
settings. gradle	设置相关的 Gradle 脚本

这里介绍了一些外部相关文件，下面来看更重要的 app 模块中的部分文件，如图 2-56 所示。

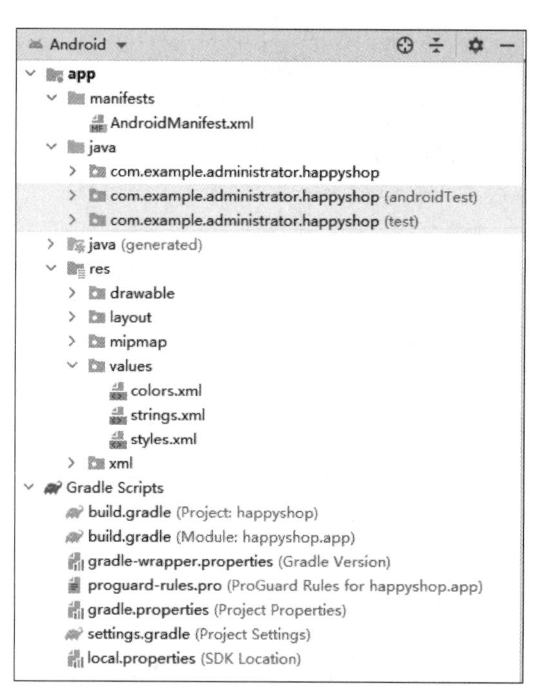

图 2-56 app 模块中的文件目录结构

app 模块中的文件(夹)及其用途如表 2-2 所示。

表 2-2　app 模块中的文件(夹)及其用途

文件(夹)名	用　　途
build	编译后的文件存在的位置(包括最终生成的 apk 也在这里面)
libs	依赖的库所在的位置
src	源代码所在的目录
src\main	主要代码所在位置
src\main\assets	Android 中附带的一些文件
src\main\java	最重要的 Java 代码所在的位置
src\main\res	Android 资源文件所在位置
src\main\AndroidManifest. xml	开发清单文件
build. gradle	和这个项目有关的 Gradle 配置,相当于这个项目的 Makefile,一些项目的依赖就写在这里面
proguard. pro	代码混淆配置文件

(1) 启动 JTM 封装软件运行"欢乐购商城"项目服务器端。

(2) 在 Android Studio 中导入"欢乐购商城"客户端,修改服务器地址,实现客户端和服务器端访问。

(3) 熟悉"欢乐购商城"Android 客户端结构。

本章小结

本章主要让读者体验教学版"欢乐购商城"的整体结构,为后面章节的学习打下基础。如果要体验整个程序需要安装客户端、服务器、数据库三部分,客户端运行前需要安装开发环境和软件,本章为此讲解了 Java 环境搭建、Android Studio 安装和创建项目。而服务器端和数据库运行借助 JTM 软件封装一键启动,为初学者学习 Android 提供了条件。

自测习题

(1) 新建 Android 项目 HelloWorld 项目,并描述程序的执行过程。

(2) 下载并使用"欢乐购商城"真实版 App。

(3) 移动 App 开发流程一般分为哪几部分?简要说明。

(4) 描述 Android 项目下各个文件夹的作用。

第3章 Android基础界面编程

Android 程序开发主要分为三部分：界面设计、代码流程控制和资源建设。代码和资源主要是由开发者进行编写和维护的，对于大部分用户来说是不关心的，展现在用户面前最直观的就是界面设计。作为一个程序设计者，必须首先考虑用户的体验，只有用户满意了开发的产品，应用才能推广，才有价值，因此界面设计尤为重要。

控件是 Android 项目开发的基本组成单位，通过使用组件可以高效地开发 Android 程序，所以熟练掌握控件的使用是进行 Android 程序开发的重要前提。本章对"购物商城"基本组件进行分析，认识常用基础控件和布局管理器的使用方法。通过对基础控件和布局管理器的综合运用实现"欢乐购商城"登录、注册、首页表格等页面。

学习目标

本章要点

(1) 掌握文本显示框的功能和用法。

(2) 熟悉文本编辑框的常用属性。

(3) 掌握按钮的简单用法。

(4) 掌握图片的使用方法。

(5) 掌握线性布局的功能和用法。

(6) 熟悉表格布局的功能和用法。

(7) 掌握相对布局的功能和用法。

(8) 掌握网格布局的功能和用法。

(9) 掌握布局的嵌套使用。

视频详解

3.1 基础 View 组件简介

任务陈述

"欢乐购商城"项目中每个页面中都运用了 Android 基础界面控件，本节运用基础界面控件实现"欢乐购商城"登录页面、注册页面。

分析：登录页面整体采用垂直方向线性布局，从上往下摆放图片（ImageView）、文本编辑框（EditText）、按钮（Button）、文本显示框（TextView）基础控件，如图 3-1 所示。

注册页面整体采用垂直方向线性布局，从上往下摆放图片（ImageView）、文本编辑框（EditText）、按钮（Button），如图 3-2 所示。

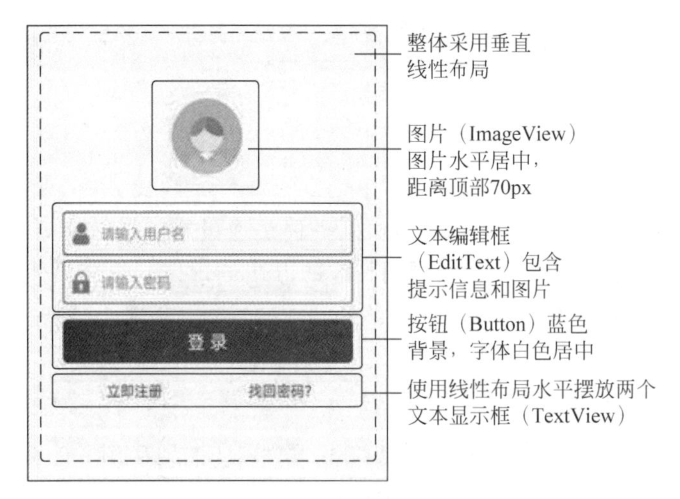

图 3-1　登录页面

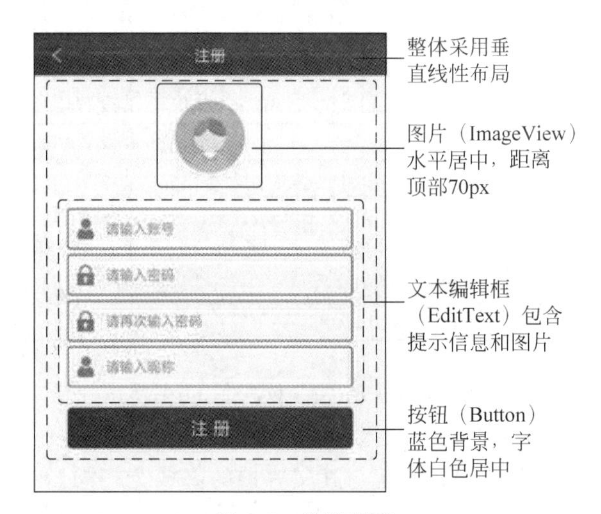

图 3-2　注册页面

相关知识

Android 中所有的组件都继承于 View 类，View 类代表的就是屏幕上的一块空白的矩形区域，该空白区域可用于绘画和事件处理。不同的界面组件，相当于对这个矩形区域做了一些处理，如文本显示框、按钮等。

View 类有一个重要的子类：ViewGroup。ViewGroup 类是所有布局类和容器组件的基类，它是一个不可见的容器，它里面还可以添加 View 组件或 ViewGroup 组件，主要用于定义它所包含的组件的排列方式，例如，网格排列或线性排列等。通过 View 和 ViewGroup

的组合使用，从而使得整个界面呈现一种层次结构。ViewGroup 内包含的组件如图 3-3 所示。

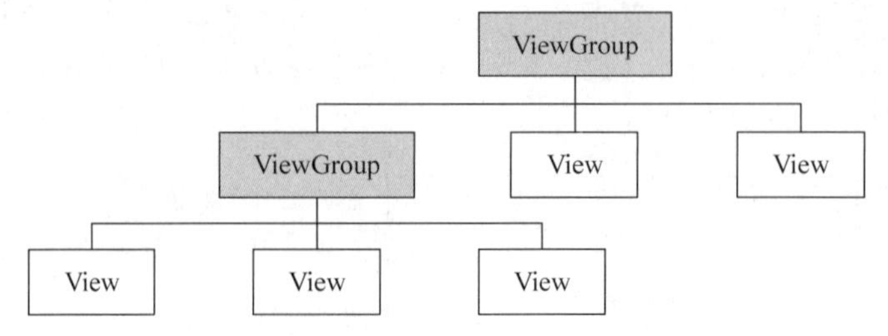

图 3-3 ViewGroup 组件的层次结构

Android 中控制组件的显示有两种方式：一种是通过 XML 布局文件来设置组件的属性进行控制；另一种是通过 Java 代码调用相应的方法进行控制。这两种方式控制 Android 界面显示的效果是完全一样的。实际上，XML 文件的属性与 Java 代码中方法之间存在着一一对应的关系。从 Android API 文档中 View 类的介绍中，可查看所有的属性与方法之间的对应关系，在此只列出一些常用的属性供参考，如表 3-1 所示。

表 3-1 View 类的常见 XML 属性、对应方法及说明

XML 属 性	对 应 方 法	说 明
android：alpha	setAlpha(float)	设置组件的透明度
android：background	setBackgroundResource(int)	设置组件的背景
android：clickable	setClickable(boolean)	设置组件是否可以触发单击事件
android：focusable	setFocusable(boolean)	设置组件是否可以得到焦点
android：id	setId(int)	设置组件的唯一 ID
android：minHeight	setMinimumHeight(int)	设置组件的最小高度
android：minWidth	setMinimumWidth(int)	设置组件的最小宽度
android：padding	setPadding(int,int,int,int)	在组件四边设置边距
android：scaleX	setScaleX(float)	设置组件在 X 轴方向的缩放
android：visibility	setVisibility(int)	设置组件是否可见

几乎每个界面组件都需要设置 android：layout_height、android：layout_width 这两个属性，用于指定该组件的高度和宽度，主要有以下三种取值。

(1) fill_parent：表示组件的高或宽与其父容器的高或宽相同。

(2) wrap_content：表示组件的高或宽恰好能包裹内容，随着内容的变化而变化。

(3) match_parent：该属性值与 fill_parent 完全相同，Android 2.2 之后推荐使用 match_parent 代替 fill_parent。

虽然两种方式都可以控制界面的显示，但是它们又各有优缺点。

(1) 完全使用 Java 代码来控制用户界面不仅烦琐，而且界面和代码相混合，不利于解耦、分工。

(2) 完全使用 XML 布局文件虽然方便、便捷，但灵活性不好，不能动态改变属性值。

因此，我们经常会混合使用这两种方式来控制界面，一般来说，习惯将一些变化小的、比较固定的、初始化的属性放在 XML 文件中管理，而对于那些需要动态变化的属性则交给 Java 代码控制。例如，可以在 XML 布局文件中设置文本显示框的高度和宽度以及初始时的显示文字，在代码中根据实际需要动态地改变显示的文字。

3.1.1　文本显示框 TextView

视频详解

TextView 类直接继承于 View 类，主要用于在界面上显示文本信息，类似于一个文本显示器，从这个方面来理解，有点儿类似于 Java 编程中的 JLable 的用法，但是比 JLable 的功能更加强大，使用更加方便。TextView 可以设置显示文本的字体大小、颜色、风格等属性，TextView 的常见属性如表 3-2 所示。

表 3-2　TextView 类的常见 XML 属性、对应方法及说明

XML 属性	对应方法	说　明
android：gravity	setGravity(int)	设置文本的对齐方式
android：height	setHeight(int)	设置文本框的高度(以 pixel 为单位)
android：text	setText(CharSequence)	设置文本的内容
android：textColor	setTextColor(int)	设置文本的颜色
android：textSize	setTextSize(int,float)	设置文本的大小
android：textStyle	setTypeface(Typeface)	设置文本的风格
android：typeface	setTypeface(Typeface)	设置文本的字体
android：width	setWidth(int)	设置文本框的宽度(以 pixel 为单位)
Android：drawableLeft	setCompoundDrawablesWith IntrinsicBounds(int,int,int,int)	要绘制在文本左侧的可绘制对象

这些是文本显示控件都具有的功能。除此之外，Android 中的 TextView 还具有自动识别文本中的各种链接、能够显示字符串中的 HTML 标签的格式。识别自动链接的属性为 android：autoLink，该属性的值有以下几种。

(1) none：不匹配任何格式，这是默认值。

(2) web：只匹配网页，如果文本中有网页，网页会以超链接的形式显示。

(3) email：只匹配电子邮箱，电子邮箱会以超链接的形式显示。

(4) phone：只匹配电话号码，电话号码会以超链接的形式显示。

(5) map：只匹配地图地址。

(6) all：匹配以上所有。

当匹配时，相应部分会以超链接显示，单击超链接，会自动运行相关程序。例如，单击电话号码超链接会调用拨号程序，单击网页超链接会打开网页等。

而显示 HTML 标签格式，则需要通过 Java 代码来控制。首先为该文本框添加一个 id 属性，然后在 onCreate()方法中，通过 findViewById(R. id. ***)获取该文本框，最后通过 setText()方法来设置显示的内容。

【例 3-1】　使用 TextView 显示“欢迎使用《Android 零基础入门到实战》”文本，其中，“《Android 项目开发实战教程》”为蓝色字体。

注意：本书代码中，右箭头所指内容为编者对读者的提示，实际编译时无须添加。

程序清单 3-1：chart0301\app\src\main\res\layout\activity_main. xml

```
1    <TextView
2        android:id = "@ + id/myText"                →为 TextView 添加 id 属性
```

```
3       android:textSize = "18sp"                    →字体大小设置为 18sp
4       android:layout_width = "wrap_content"        →组件宽度为内容包裹
5       android:layout_height = "wrap_content"       →组件高度为内容包裹
6       />
```

程序清单 3-2：chart0301\app\src\main\java\com\jxcia\chart0301\MainActivity. java

```
1    TextView tv = (TextView) findViewById(R.id.myText); //myText 为 id
2    tv.setText(Html.fromHtml("欢迎使用<font color = blue>《Android 项目开发实战教程》
     </font>"));
```

该代码的显示效果是：《Android 零基础入门到实战》这几个字为蓝色，其他字的颜色为布局文件中设置的颜色，如图 3-4 所示。

欢迎使用《Android零基础入门到实战》

图 3-4　TextView 实现效果

视频详解

3.1.2　文本编辑框 EditText

TextView 的功能仅是用于显示信息而不能编辑，好的应用程序往往需要与用户进行交互，让用户进行输入信息。为此，Android 中提供了 EditText 组件，EditText 是 TextView 类的子类，与 TextView 具有很多相似之处。它们最大的区别在于，EditText 允许用户编辑文本内容。使用 EditText 时，经常使用到的属性有以下几个。

（1）android：hint：设置当文本框内容为空时，文本框内显示的提示信息，一旦输入内容，该提示信息立即消失，当删除所有输入的内容时，提示信息又会出现。

（2）android：password：设置文本框是否为密码框，值为 true 或者 false，设置为 true 时，输入的内容将会以点替代，但已不推荐使用了。

（3）android：inputType：设置文本框接收值的类型，例如，只能是数字、电话号码等。

【例 3-2】 以"欢乐购商城"登录页面输入手机号编辑框为例，介绍文本编辑框使用方法，界面程序代码如下。

程序清单 3-3：char0302\app\src\main\res\layout\activity_main. xml

```
1    <EditText
2        android:layout_width = "match_parent"             →组件宽度为填充父容器
3        android:layout_height = "40dp"                    →组件高度为 40dp
4    android:drawableLeft = "@drawable/user_name_icon"     →编辑框左侧设置图片
5        android:drawablePadding = "10dp"                  →图片内边距为 10dp
6        android:gravity = "center_vertical"               →编辑框内容垂直居中
7        android:hint = "请输入手机号"                      →提示内容
8        android:paddingLeft = "8dp"                       →编辑左侧内编辑为 8dp
9        android:singleLine = "true"                       →单行输入
```

```
10        android:textColor = "#000000"              →字体颜色为白色
11        android:textColorHint = "#a3a3a3"          →提示字体颜色为灰色
12        android:textSize = "14sp" />               →字体大小设置为14sp
```

程序运行效果如图3-5所示。

图3-5 EditText运行效果

3.1.3 按钮Button

视频详解

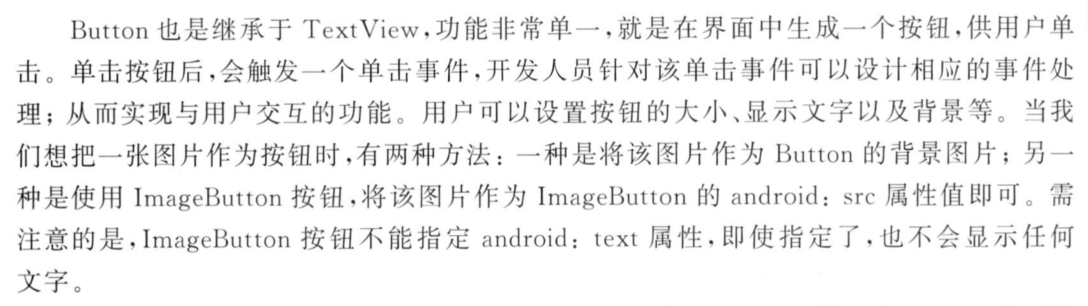

Button也是继承于TextView,功能非常单一,就是在界面中生成一个按钮,供用户单击。单击按钮后,会触发一个单击事件,开发人员针对该单击事件可以设计相应的事件处理;从而实现与用户交互的功能。用户可以设置按钮的大小、显示文字以及背景等。当我们想把一张图片作为按钮时,有两种方法:一种是将该图片作为Button的背景图片;另一种是使用ImageButton按钮,将该图片作为ImageButton的android:src属性值即可。需注意的是,ImageButton按钮不能指定android:text属性,即使指定了,也不会显示任何文字。

【例3-3】 以"欢乐购商城"登录页面"登录"按钮为例,介绍按钮使用方法,界面布局代码如下。

程序清单3-4:char0303\app\src\main\res\layout\activity_main.xml

```
1     <Button
2         android:layout_width = "fill_parent"              →组件宽度为填充父容器
3         android:layout_height = "40dp"                    →组件高度为40dp
4         android:layout_gravity = "center_horizontal"      →组件对外水平居中
5         android:layout_marginLeft = "35dp"                →组件左边外边距35dp
6         android:layout_marginRight = "35dp"               →组件右边外边距35dp
7         android:layout_marginTop = "15dp"                 →组件顶部外边距15dp
8         android:background = "#6200EE"                    →背景为蓝紫色
9         android:text = "登 录"                            →组件文字
10        android:textColor = "#FFFFFF"                     →字体颜色为白色
11        android:textSize = "18sp" />                      →字体大小18sp
```

程序运行效果如图3-6所示。

登 录

图3-6 按钮界面运行效果

3.1.4 图片视图ImageView

视频详解

ImageView(图片视图)的作用与TextView类似,TextView用于显示文字,ImageView则用于显示图片,既然是显示图片,那就要设置图片的来源,ImageView中有一个src属性用于指定图片的来源。显示图片还存在另外一个问题,就是当图片比ImageView的区域大的时候如何显示呢?在ImageView中有一个常用并且重要的属性scaleType,用于设置图片的缩放类型。该属性值主要包含以下几个。

fitCenter：保持纵横比缩放图片，直到该图片能完全显示在 ImageView 中，缩放完成后将该图片放在 ImageView 的中央。

fitXY：对图片横向、纵向独立缩放，使得该图片完全适应于该 ImageView，图片的纵横比可能会改变。

centerCrop：保持纵横比缩放图片，以使得图片能完全覆盖 ImageView。

【例 3-4】 以“欢乐购商城”登录页面登录按钮为例，介绍按钮使用方法，界面布局代码如下。

程序清单 3-5：char0304\res\layout\activity_main.xml

```
1   <ImageView
2       android:layout_width = "70dp"                  →组件宽度为 70dp
3       android:layout_height = "70dp"                 →组件高度为 70dp
4       android:layout_gravity = "center_horizontal"   →组件对外水平居中
5       android:layout_marginTop = "70dp"              →组件顶部外边距 70dp
6       android:scaleType = "fitCenter"                →图片缩放类型保持纵横比
7       android:src = "@drawable/head" />              →引用图片地址
```

程序运行效果如图 3-7 所示。

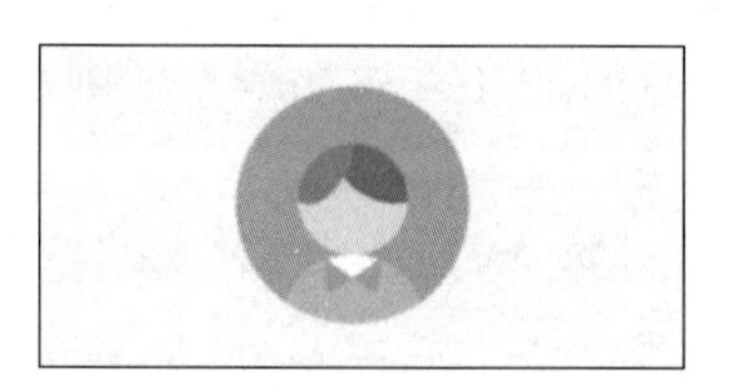

图 3-7　图片控件运行效果

视频详解

3.1.5　实战演练——登录页面

通过前面对基础界面控件 TextView、EditText、Button 和 ImageView 的介绍，可以初步实现“欢乐购商城”的登录页面。如图 3-1 所示，在“欢乐购商城”登录页面中整体采用垂直方向线性布局，从上往下摆放图片(ImageView)、文本编辑框(EditText)、按钮(Button)、文本显示框(TextView)基础控件，以下介绍登录界面具体实现过程。

在布局文件中多次用到@string/ *** 作为 android：text 的属性值，表示引用 R.java 中 string 内部类的成员变量所代表的资源。这些常量值是在 strings.xml 文件中定义的。查看 strings.xml 文件的内容如下。

程序清单 3-6：chart0305\app\src\main\res\values\strings.xml

```
1   <resources>
2       <string name = "app_name"> chart0205 </string>
3       <string name = "inputname">请输入手机号</string>
4       <string name = "inputpassword">请输入密码</string>
5       <string name = "btnLogin">登 录</string>
6       <string name = "registerNow">立即注册</string>
7       <string name = "findPassword">找回密码?</string>
8   </resources>
```

在布局文件中多次用到@color/ *** 作为 android：textColor 的属性值，表示引用 R.java 中 Color 内部类的成员变量所代表的资源。这些常量值是在 colors.xml 文件中定义的。查看 colors.xml 文件的内容如下。

程序清单 3-7：chart0305\app\src\main\res\values\colors.xml

```
1  <resources>
2      <color name="colorPrimary">#6200EE</color>
3      <color name="colorPrimaryDark">#3700B3</color>
4      <color name="colorAccent">#03DAC5</color>
5      <color name="white">#FFFFFF</color>
6  </resources>
```

其实在设置 android：text 和 android：textColor 属性时，可以直接将这些字符串常量赋值给该属性，但是建议不要这么做。因为一些字符串常量可能会在多处被使用，如果都在属性里写，不仅占用更多的内存，而且修改起来也比较麻烦，需要一个个进行修改；另一方面，统一放在 strings.xml 和 colors.xml 文件中，日后如果要修改，只需要修改相应的资源文件就可以了，而不用去更改别的文件，可扩展性比较好。

登录页面主布局实现，代码如下。

程序清单 3-8：char0305\app\src\main\res\layout\activity_main.xml

```
1   <LinearLayout xmlns:android="http://schemas.android.com/apk/res/android"
2       xmlns:app="http://schemas.android.com/apk/res-auto"
3       xmlns:tools="http://schemas.android.com/tools"
4       android:layout_width="match_parent"
5       android:layout_height="match_parent"
6       android:background="@color/white"                  →背景引用颜色文件中白色
7       android:orientation="vertical"                     →线性布局方向为垂直
8       tools:context=".MainActivity">
9       <ImageView
10          android:layout_width="70dp"
11          android:layout_height="70dp"
12          android:layout_gravity="center_horizontal"     →图片控件对外水平居中
13          android:layout_marginTop="70dp"                →图片控件顶部外边距 70dp
14  android:background="@drawable/defaultuser_icon" />
15      <EditText
16          android:layout_width="fill_parent"
17          android:layout_height="40dp"
18          android:layout_gravity="center_horizontal"     →控件对外水平居中
19          android:layout_marginLeft="35dp"
20          android:layout_marginRight="35dp"
21          android:layout_marginTop="35dp"
22  android:drawableLeft="@drawable/user_name_icon"        →引用图片地址
23          android:drawablePadding="10dp"                 →图片内边距 10dp
24          android:gravity="center_vertical"              →控件内容垂直居中
25          android:hint="@string/inputname"            →文字引用 string.xml 中文字地址
26          android:paddingLeft="8dp"
27          android:singleLine="true"                      →单行输入
28          android:textColor="@color/white"
29          android:textColorHint="#a3a3a3"                →提示文字颜色为灰色
30          android:textSize="14sp" />                     →字体大小为 14sp
```

```
31        <EditText
32            android:layout_width = "fill_parent"
33            android:layout_height = "40dp"
34            android:layout_gravity = "center_horizontal"
35            android:layout_marginLeft = "35dp"
36            android:layout_marginRight = "35dp"
37            android:layout_marginTop = "5dp"
38            android:drawableLeft = "@drawable/psw_icon"
39            android:drawablePadding = "10dp"
40            android:hint = "@string/inputpassword"   →文字引用 string.xml 中文字地址
41            android:inputType = "textPassword"
42            android:paddingLeft = "8dp"
43            android:singleLine = "true"
44            android:textColor = "@color/white"
45            android:textColorHint = "#a3a3a3"
46            android:textSize = "14sp" />
47        <Button
48            android:layout_width = "fill_parent"
49            android:layout_height = "40dp"
50            android:layout_gravity = "center_horizontal"   →按钮对外水平居中
51            android:layout_marginLeft = "35dp"
52            android:layout_marginRight = "35dp"
53            android:layout_marginTop = "15dp"
54            android:background = "@color/colorPrimary"   →背景引用颜色文件中颜色
55            android:text = "@string/btnLogin"   →文字引用 string.xml 中文字地址
56            android:textColor = "@color/white"   →字体颜色引用颜色文件中白色
57            android:textSize = "18sp" />
58        <LinearLayout
59            android:layout_width = "fill_parent"
60            android:layout_height = "fill_parent"
61            android:layout_marginLeft = "35dp"
62            android:layout_marginRight = "35dp"
63            android:layout_marginTop = "8dp"
64            android:gravity = "center_horizontal"
65            android:orientation = "horizontal">   →线性布局方向为水平
66            <TextView
67                android:layout_width = "0dp"
68                android:layout_height = "wrap_content"
69                android:layout_weight = "1"   →线性布局中权重设置为 1
70                android:gravity = "center_horizontal"
71                android:padding = "8dp"
72                android:text = "@string/registerNow"→文字引用 string.xml 中文字地址
73                android:textColor = "@color/colorPrimary"   →背景引用颜色文件中颜色
74                android:textSize = "14sp" />
75            <TextView
76                android:layout_width = "0dp"
77                android:layout_height = "wrap_content"
78                android:layout_weight = "1"   →线性布局中权重设置为 1
```

```
79                android:gravity = "center_horizontal"
80                android:padding = "8dp"
81                android:text = "@string/findPassword"→文字引用 string.xml 中文字地址
82                android:textColor = "@color/colorPrimary"   →背景引用颜色文件中颜色
83                android:textSize = "14sp" />
84        </LinearLayout>
85    </LinearLayout>
```

登录页面实现后，编辑框（如图 3-8 所示）与图 3-1 对比会发现用户名和密码编辑框未实现边框，其中边框需要自定义图片作为背景，具体实现将在 4.1.2 节中讲解。

图 3-8　登录页面实现效果

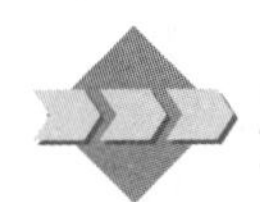

3.2　布局管理器

视频详解

任务陈述

“欢乐购商城”项目中每个页面都使用了布局管理器，例如，在登录和注册页面中整体采用了线性布局。本节重点讲解布局管理器中的线性布局、表格布局、相对布局、层布局、网格布局。本节以实现“欢乐购商城”首页作为案例（不包括列表），先以表格布局实现，然后采用线性布局和网格布局综合使用实现，如图 3-9 所示。同时对相对布局、层布局的布局管理器进行讲解，完成常用布局管理器的学习。

相关知识

3.1 节中学习了几种简单的界面组件，其中在登录页面实现中已经使用到了布局管理器线性布局，如果缺少布局管理器，组件排列杂乱，影响美观。本节将学习 Android 中提供的几种管理界面组件的布局管理器。

图 3-9　首页面表格布局

Android 中的布局管理器本身也是一个界面组件，所有的布局管理器都是 ViewGroup 类的子类，都可以当作容器类来使用。因此，可以在一个布局管理器中嵌套其他布局管理器。Android 中布局管理器可以根据运行平台来调整组件的大小，具有良好的平台无关性。Android 中用得最多的布局主要有：线性布局、表格布局、相对布局、层布局。

视频详解

3.2.1　线性布局

线性布局是最常用也是最基础的布局方式。在前面的示例中，就使用到了线性布局，它用 LinearLayout 类表示。线性布局和 Java 编程中 AWT 编程里的 FlowLayout 有些相似，它们都会将容器里的所有组件一个挨着一个排列。

它提供了水平和垂直两种排列方向，通过 android:orientation 属性进行设置，默认为垂直排列。

(1) 当为水平方向时，不管组件的宽度是多少，整个布局只占一行，当组件宽度超过容器宽度时，超出的部分将不会显示。

(2) 当为垂直方向时，整个布局文件只有一列，每个组件占一行，不管该组件宽度有多小。

在线性布局中，除了设置高度和宽度外，主要设置如下属性。

(1) android:gravity：设置布局管理器内组件的对齐方式，可以同时指定多种对齐方式的组合，多个属性之间用竖线隔开，但竖线前后不能出现空格。例如，bottom | center_horizontal 代表出现在屏幕底部，而且水平居中。

(2) android:orientation：设置布局管理器内组件的排列方向，可以设置为 vertical(垂直排列)或 horizontal(水平排列)。

(3) android:id：用于给当前组件指定一个 ID 属性，在 Java 代码中可以应用该属性单独引用该组件。为组件指定 ID 属性后，在 R.java 文件中，会自动派生一个对应的属性。在 Java 代码中，可以通过 findViewById()方法获取该属性。

(4) android:background：用于为该组件设置背景，可以是背景图片，也可以是背景颜色。为组件指定背景图片时，可以将准备好的图片复制到目录下，然后使用下面的代码进行设置。

```
android:background = "@drawable/background"
```

如果想指定背景颜色时，可以使用颜色值，例如，想要指定背景颜色为白色，可以使用下面的代码。

```
android:background = "#FFFFFF"
```

线性布局与 AWT 编程中 FlowLayout 的最明显的区别：在 FlowLayout 中组件一个个地排列到边界就会自动从下一行重新开始；在线性布局中如果一行的宽度或一列的高度超过了容器的宽度或高度，那么超出的部分将无法显示，如果希望超出的部分能够滚动显示，则需在外边包裹一个滚动组件，ScrollView（垂直滚动）或 HorizontalScrollView（水平滚动）。

在使用 LinearLayout 时，子控件可以设置 layout_weight。layout_weight 的作用是设置子控件在 LinearLayout 的重要度（控件的大小比重）。如果在一个 LinearLayout 里面放置两个 Button：Button1 和 Button2，Button1 的 layout_weight 设置为 1，Button2 的 layout_weight 设置为 2，且两个 Button 的 layout_width 都设置为 fill_parent，则 Button1 占据屏幕宽度的三分之二，而 Button2 占据三分之一。如果两个 Button 的 layout_width 都设置成 wrap_content，则情况刚好相反，Button1 占三分之一，Button2 占三分之二。

【例 3-5】 下面通过一个实例讲解线性布局管理使用方法，界面布局代码如下，程序运行效果如图 3-10 所示。

图 3-10　线性布局管理器实现效果

程序清单 3-9：char0306\app\src\main\res\layout\activity_main.xml

```
1   <ScrollView  →垂直滚动条
    xmlns:android="http://schemas.android.com/apk/res/android"
2       xmlns:app="http://schemas.android.com/apk/res-auto"
3       xmlns:tools="http://schemas.android.com/tools"
4       android:layout_width="match_parent"
5       android:layout_height="wrap_content"
6       tools:context=".MainActivity">
7       <LinearLayout
8           android:layout_width="match_parent"
9           android:layout_height="wrap_content"
10          android:orientation="vertical">           →线性布局方向为垂直
11              <LinearLayout
12                  android:layout_width="match_parent"
13                  android:layout_height="130dp"
14                  android:orientation="horizontal"> →线性布局方向为水平
15                  <TextView
16                      android:layout_width="wrap_content"
17                    android:layout_height="match_parent"
18                      android:gravity="center"         →文本编辑框内部居中
19                      android:layout_weight="1"        →权重比为 1
20                      android:text="红色"
21                      android:textSize="25sp"
22                      android:background="#F00"/>  →背景颜色为白色
23                  <TextView
24                      android:layout_width="wrap_content"
25                    android:layout_height="match_parent"
26                      android:layout_weight="2"            →权重比为 2
27                      android:gravity="center"             →文本编辑框内部居中
28                      android:text="绿色"
29                      android:textSize="25sp"
30                      android:background="#0F0"/>          →背景颜色为绿色
31                  <TextView
32                      android:layout_width="wrap_content"
33                      android:layout_height="match_parent"
34                      android:layout_weight="3"            →权重比为 3
35                      android:text="蓝色"
36                      android:gravity="center"             →文本编辑框内部居中
37                      android:textSize="25sp"
38                      android:background="#00F"/>          →背景颜色为蓝色
39              </LinearLayout>
40            <LinearLayout
41                android:layout_width="match_parent"
42                android:layout_height="wrap_content"
43                android:gravity="center_horizontal"        →线性布局内部水平居中
44                android:orientation="vertical">            →线性布局方向垂直
45                <ImageView
46                    android:layout_width="150dp"
47                    android:layout_height="150dp"
48                    android:scaleType="fitCenter"  →图片缩放格式保持纵横比,完全显示
49                    android:padding="3dp"          →图片内边距 3dp
50                    android:src="@drawable/a000"/→引用图片地址
```

```
51                <ImageView
52                    android:layout_width = "150dp"
53                    android:layout_height = "150dp"
54                    android:scaleType = "fitCenter"→图片缩放格式保持纵横比,完全显示
55                    android:padding = "3dp"           →图片内边距 3dp
56                    android:src = "@drawable/a001"/>→引用图片地址
57                <ImageView
58                    android:layout_width = "150dp"
59                    android:layout_height = "150dp"
60                    android:scaleType = "fitCenter"→图片缩放格式保持纵横比,完全显示
61                    android:padding = "3dp"           →图片内边距 3dp
62                    android:src = "@drawable/a002"/>→引用图片地址
63                <ImageView
64                    android:layout_width = "150dp"
65                    android:layout_height = "150dp"
66                    android:scaleType = "fitCenter"→图片缩放格式保持纵横比,完全显示
67                    android:padding = "3dp"                          →图片内边距 3dp
68                    android:src = "@drawable/a003"/>                 →引用图片地址
69                <ImageView
70                    android:layout_width = "150dp"
71                    android:layout_height = "150dp"
72                    android:scaleType = "fitCenter"→图片缩放格式保持纵横比,完全显示
73                    android:padding = "3dp"                          →图片内边距 3dp
74                    android:src = "@drawable/a004"/>                 →引用图片地址
75            </LinearLayout>
76        </LinearLayout>
77    </ScrollView>
```

3.2.2 表格布局

视频详解

表格布局是指以行和列的形式来管理界面组件,由 TableLayout 类表示,不必明确声明包含几行几列,而通过添加 TableRow 来添加行,在 TableRow 中添加组件来添加列。

TableRow 就是一个表格行,本身也是容器,可以不断地添加其他组件,每添加一个组件就是在该行中增加一列,如果直接向 TableLayout 中添加组件,而没有添加 TableRow,那么该组件将会占用一行。

在表格布局中,每列的宽度都是一样的,列的宽度由该列中最宽的那个单元决定,整个表格布局的宽度则取决于父容器的宽度,默认总是占满父容器本身。

TableLayout 继承了 LinearLayout,因此它完全支持 LinearLayout 所支持的全部 XML 属性,另外,TableLayout 还增加了自己所特有的属性。

(1) android:collapseColumns:隐藏指定的列,其值为列所在的序号,从 0 开始,如果需要隐藏多列,可用逗号隔开这些序号。

(2) android:shrinkColumns:收缩指定的列以适合屏幕,使整行能够完全显示,不会超出屏幕,用于当某一行的内容超过屏幕的宽度时,会使该列自动换行,其值为列所在的序号。如果没有该属性,则超出屏幕的部分会自动截取,不会显示。

(3) android:stretchColumns:尽量把指定的列填充空白部分。该属性用于某一行的内容不足以填充整个屏幕,这样指定某一列的内容扩张以填满整个屏幕,其他列的宽度不变。

如果某一列有多行，而每行的列数可能不相同，那么可扩展列的宽度是一致的，不会因为某一行有多余的空白而填充整行。也就是说，不管在哪一行，它的宽度都是相同的。

(4) android:layout_column：控件在 TableRow 中所处的列。如果没有设置该属性，默认情况下，控件在一行中是一列挨着一列排列的。通过设置该属性，可以指定控件所在的列，这样就可以达到中间某一个列为空的效果。

(5) android:layout_span：该控件所跨越的列数，即将多列合并为一列。

【例 3-6】 下面以“欢乐购商城”首页面商品分类图片按钮栏为例，介绍表格使用方法，界面布局代码如下，程序运行效果如图 3-11 所示。

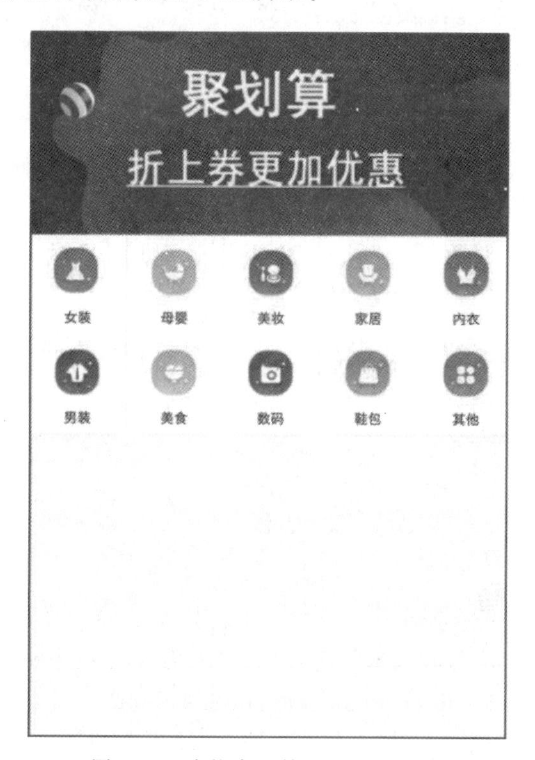

图 3-11　表格布局管理器实现效果

程序清单 3-10：char0307\app\src\main\res\layout\activity_main. xml

```
1    <LinearLayout xmlns:android = "http://schemas.android.com/apk/res/android"
2        xmlns:app = "http://schemas.android.com/apk/res - auto"
3        xmlns:tools = "http://schemas.android.com/tools"
4        android:layout_width = "match_parent"
5        android:layout_height = "match_parent"
6        android:orientation = "vertical"              →线性布局垂直方向
7        tools:context = ".MainActivity">
8        <ImageView
9            android:layout_width = "match_parent"
10             android:layout_height = "150dp"
11             android:scaleType = "centerCrop"        →图片缩放格式保持纵横比,覆盖控件
12             android:src = "@drawable/jhs" />        →引用图片地址
13        <TableLayout
14             android:layout_gravity = "center_horizontal"  →表格布局对外水平居中摆放
```

```
15            android:layout_width = "wrap_content"
16            android:layout_height = "wrap_content"
17            android:gravity = "center">         →组件内部居中显示
18            <TableRow>                          →添加一行
19                <ImageView
20                    android:layout_width = "73dp"
21                    android:layout_height = "73dp"
22                    android:clickable = "true"             →可单击
23                    android:src = "@drawable/a" />         →引用图片地址
24                <ImageView
25                    android:layout_width = "73dp"
26                    android:layout_height = "73dp"
27                    android:src = "@drawable/b" />         →引用图片地址
28                <ImageView
29                    android:layout_width = "73dp"
30                    android:layout_height = "73dp"
31                    android:src = "@drawable/c" />         →引用图片地址
32                <ImageView
33                      android:layout_width = "73dp"
34                    android:layout_height = "73dp"
35                    android:src = "@drawable/d" />         →引用图片地址
36                <ImageView
37                    android:layout_width = "73dp"
38                    android:layout_height = "73dp"
39                    android:src = "@drawable/e" />         →引用图片地址
40            </TableRow>
41            <TableRow>                                     →添加一行
42                <ImageView
43                    android:layout_width = "73dp"
44                    android:layout_height = "73dp"
45                    android:src = "@drawable/f" />         →引用图片地址
46                <ImageView
47                    android:layout_width = "73dp"
48                    android:layout_height = "73dp"
49                    android:src = "@drawable/g" />         →引用图片地址
50                <ImageView
51                    android:layout_width = "73dp"
52                    android:layout_height = "73dp"
53                    android:src = "@drawable/h" />         →引用图片地址
54                <ImageView
55                    android:layout_width = "73dp"
56                    android:layout_height = "73dp"
57                    android:src = "@drawable/i" />         →引用图片地址
58                <ImageView
59                    android:layout_width = "73dp"
60                    android:layout_height = "73dp"
61                    android:src = "@drawable/k" />         →引用图片地址
62            </TableRow>
63        </TableLayout>
64    </LinearLayout>
```

3.2.3 相对布局

视频详解

相对布局，顾名思义就是相对于某个组件的位置，由 RelativeLayout 类表示，这种布局的关键是找到一个合适的参照物，如果甲组件的位置需要根据乙组件的位置来确定，那么要求先定义乙组件，再定义甲组件。

在相对布局中，每个组件的位置可通过它相对于某个组件的方位以及对齐方式来确定，因此相对布局中常见的属性如表 3-3 所示。由于父容器是确定的，所以与父容器方位与对齐的关系取值为 true 或 false。

表 3-3　相对布局中常用属性设置

属　　性	说　　明
android：layout_centerHorizontal	设置该组件是否位于父容器的水平居中位置
android：layout_centerVertical	设置该组件是否位于父容器的垂直居中位置
android：layout_centerInParent	设置该组件是否位于父容器的正中央位置
android：layout_alignParentTop	设置该组件是否与父容器顶端对齐
android：layout_alignParentBottom	设置该组件是否与父容器底端对齐
android：layout_ alignParentLeft	设置该组件是否与父容器左边对齐
android：layout_ alignParentRight	设置该组件是否与父容器右边对齐
android：layout_toRightOf	指定该组件位于给定的 ID 组件的右侧
android：layout_toLeftOf	指定该组件位于给定的 ID 组件的左侧
android：layout_above	指定该组件位于给定的 ID 组件的上方
android：layout_below	指定该组件位于给定的 ID 组件的下方
android：layout_alignTop	指定该组件与给定的 ID 组件的上边界对齐
android：layout_ alignBottom	指定该组件与给定的 ID 组件的下边界对齐
android：layout_ alignLeft	指定该组件与给定的 ID 组件的左边界对齐
android：layout_ alignRight	指定该组件与给定的 ID 组件的右边界对齐

【例 3-7】 在智能手机中，当系统中有软件更新时，经常会显示提示软件更新页面。本实例使用相对布局实现一个显示软件更新提示的界面，提示文字相对于父容器居中，“以后再说”按钮相对于提示文字右对齐、底部方向，“现在更新”按钮相对于“以后再说”按钮底部对齐、左部方向。界面布局代码如下，程序运行效果如图 3-12 所示。

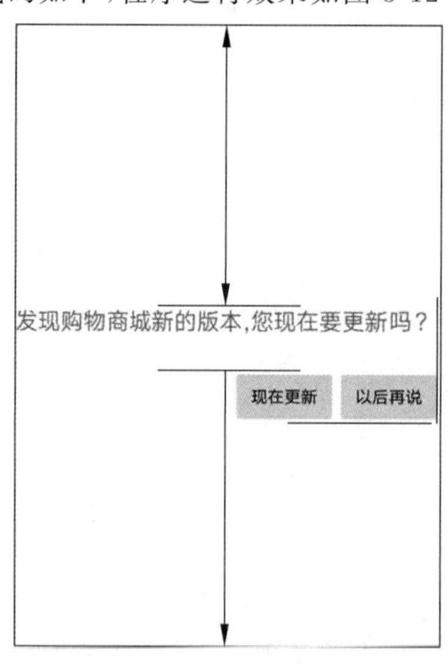

图 3-12　相对布局管理器实现效果

程序清单 3-11：char0308\app\src\main\res\layout\activity_main.xml

```
1   <RelativeLayout xmlns:android = "http://schemas.android.com/apk/res/android"
2       xmlns:app = "http://schemas.android.com/apk/res-auto"
3       xmlns:tools = "http://schemas.android.com/tools"
4       android:layout_width = "match_parent"
5       android:layout_height = "match_parent"
6       tools:context = ".MainActivity">                    →线性布局垂直方向
7       <TextView
8           android:id = "@+id/text"
9           android:layout_width = "wrap_content"
10           android:layout_height = "wrap_content"
11           android:textSize = "20sp"      →图片缩放格式保持纵横比,覆盖控件
12           android:text = "发现购物商城新的版本,您现在要更新吗?"      →引用图片地址
13           android:layout_centerInParent = "true"/>
14       <Button                                           →表格布局对外水平居中摆放
15           android:text = "现在更新"
16           android:id = "@+id/button1"
17           android:layout_width = "wrap_content"          →组件内部居中显示
18           android:layout_height = "wrap_content"         →添加一行
19           android:layout_below = "@id/text"
20           android:layout_toLeftOf = "@id/button2"/>
21       <Button
22           android:text = "以后再说"                      →可单击
23           android:id = "@+id/button2"                   →引用图片地址
24           android:layout_width = "wrap_content"
25           android:layout_height = "wrap_content"
26           android:layout_alignRight = "@id/text"
27           android:layout_below = "@id/text"/>            →引用图片地址
28   </RelativeLayout>
```

3.2.4　层布局

视频详解

层布局也叫帧布局,由 FrameLayout 类表示。其每个组件占据一层,后面添加的层会覆盖前面的层,后面的组件会叠放在先前的组件之上。如果后面组件大于前面的组件,那么前面的组件将会完全被覆盖,不可见;如果后面组件无法完全覆盖前面的组件,则未覆盖部分显示先前的组件。该布局在开发中设计地图时经常用到,因为是按层次方式布局,需要实现层面显示的样式时就可以采用这种布局方式,比如要实现一个类似百度地图的布局,我们移动的标志是在一个图层的上面。

【例 3-8】　通过实例讲解 Android 程序中使用层布局管理器,界面布局代码如下。

程序清单 3-12：char0309\app\src\main\res\layout\activity_main.xml

```
1   <FrameLayout xmlns:android = "http://schemas.android.com/apk/res/android"
2       xmlns:app = "http://schemas.android.com/apk/res-auto"
3       xmlns:tools = "http://schemas.android.com/tools"
4       android:layout_width = "match_parent"
5       android:layout_height = "match_parent"
```

```
6         tools:context = ".MainActivity">
7         <TextView
8             android:layout_width = "300dp"        →控件宽度为 300dp
9             android:layout_height = "300dp"       →控件高度为 300dp
10            android:background = "#f00"           →背景颜色为红色
11            android:layout_gravity = "center"     →控件对外水平居中摆放
12            />
13        <TextView
14            android:layout_width = "200dp"        →控件宽度为 200dp
15            android:layout_height = "200dp"       →控件高度为 200dp
16            android:background = "#0f0"           →背景颜色为绿色
17            android:layout_gravity = "center"     →控件对外水平居中摆放
18            />
19        <TextView
20            android:layout_width = "100dp"        →控件宽度为 100dp
21            android:layout_height = "100dp"       →控件高度为 100dp
22            android:background = "#00f"           →背景颜色为蓝色
23            android:layout_gravity = "center"     →控件对外水平居中摆放
24            />
25        <TextView
26            android:layout_width = "50dp"         →控件宽度为 50dp
27            android:layout_height = "50dp"        →控件高度为 50dp
28            android:background = "#fff"           →背景颜色为白色
29            android:layout_gravity = "center"     →控件对外水平居中摆放
30            />
31        <TextView
32            android:layout_width = "20dp"         →控件宽度为 20dp
33            android:layout_height = "20dp"        →控件高度为 20dp
34            android:background = "#000"           →背景颜色为黑色
35            android:layout_gravity = "center"     →控件对外水平居中摆放
36            />
37    </FrameLayout>
```

程序运行效果如图 3-13 所示。

图 3-13　层布局管理器实现效果

3.2.5　网格布局

视频详解

网格布局由 GridLayout 代表，是 Android 4.0 新增的布局管理器，因此需要在 Android 4.0 之后的版本中才能使用该布局管理器。如果希望在更早的 Android 平台上使用该布局管理器，则需要导入相应的支撑库。GridLayout 的作用类似于 HTML 中的 table 标签，它把整个容器划分成若干行和若干列个网格，每个网格可以放置一个组件。除此之外，也可以设置一个组件横跨多个列、一个组件纵跨多个行。网格布局和 TableLayout（表格布局）有点儿类似，不过它功能更多，使用更加方便，具有以下优势：

（1）可以自己设置布局中组件的排列方式。

(2) 可以自定义网格布局有多少行,多少列。

(3) 可以直接设置组件位于某行某列。

(4) 可以设置组件横跨几行或者几列。

3.2.6　项目实战——首页布局

在讲解表格布局时,以"欢乐购商城"首页面商品分类图片按钮栏为例进行实现,现对表格布局实现案例使用网格布局进行优化,把表格布局替换成网格布局,界面布局代码如下,程序运行效果如图 3-11 所示。

程序清单 3-13: char0310\app\src\main\res\layout\activity_main.xml

```
1   <LinearLayout xmlns:android = "http://schemas.android.com/apk/res/android"
2       xmlns:app = "http://schemas.android.com/apk/res-auto"
3       xmlns:tools = "http://schemas.android.com/tools"
4       android:layout_width = "match_parent"
5       android:layout_height = "match_parent"
6       android:orientation = "vertical"              →线性布局垂直方向
7       tools:context = ".MainActivity">
8       <ImageView
9           android:layout_width = "match_parent"
10           android:layout_height = "150dp"
11           android:scaleType = "centerCrop"          →图片缩放格式保持纵横比,覆盖控件
12           android:src = "@drawable/jhs" />          →引用图片地址
13       <GridLayout
14           android:layout_gravity = "center_horizontal"→网格布局对外水平居中摆放
15           android:layout_width = "wrap_content"
16           android:layout_height = "wrap_content"
17           android:gravity = "center"                →组件内部居中显示
18           android:columnCount = "5"                 →网格布局设置 5 列
19           android:orientation = "horizontal"        →控件水平摆放
20           android:rowCount = "2">                   →网格布局设置 2 行
21               <ImageView
22                   android:layout_width = "73dp"     →可单击
23                   android:layout_height = "73dp"    →引用图片地址
24                   android:clickable = "true"
25                   android:src = "@drawable/a" />
26               <ImageView
27                   android:layout_width = "73dp"     →引用图片地址
28                   android:layout_height = "73dp"
29                   android:src = "@drawable/b" />
30               <ImageView
```

```
31                      android:layout_width = "73dp"           →引用图片地址
32                      android:layout_height = "73dp"
33                      android:src = "@drawable/c" />
34                  < ImageView
35                      android:layout_width = "73dp"           →引用图片地址
36                      android:layout_height = "73dp"
37                      android:src = "@drawable/d" />
38                  < ImageView
39                      android:layout_width = "73dp"           →引用图片地址
40                      android:layout_height = "73dp"
41                      android:src = "@drawable/e" />          →添加一行
42                  < ImageView
43                      android:layout_width = "73dp"
44                      android:layout_height = "73dp"
45                      android:src = "@drawable/f" />          →引用图片地址
46                  < ImageView
47                      android:layout_width = "73dp"
48                      android:layout_height = "73dp"
49                      android:src = "@drawable/g" />          →引用图片地址
50                  < ImageView
51                      android:layout_width = "73dp"
52                      android:layout_height = "73dp"
53                      android:src = "@drawable/h" />          →引用图片地址
54                  < ImageView
55                      android:layout_width = "73dp"
56                      android:layout_height = "73dp"
57                      android:src = "@drawable/i" />          →引用图片地址
58                  < ImageView
59                      android:layout_width = "73dp"
60                      android:layout_height = "73dp"
61                      android:src = "@drawable/k" />          →引用图片地址
62          </GridLayout >
63      </LinearLayout >
```

本章小结

本章围绕“欢乐购商城”项目登录页面、注册页面和首页面商品分类栏引入 Android 基础界面控件和布局管理器讲解。详细介绍了几种最基本的界面组件的功能和常用属性，包括文本显示框、文本编辑框和按钮等，并通过实现登录页面案例演示了基本界面控件具体使用用法。为了使这些组件排列美观，继续学习了 Android 中几种常见的布局管理器，包括线性布局、表格布局、相对布局、层布局和网格布局，它们各有优缺点。线性布局方便，需使用的属性较少，但不够灵活；表格布局中通过 TableRow 添加行，每列的宽度一致；相对布局则通过提供一个参照物来准确定义各个控件的具体位置，通常在一个实例中会用到多种布局，把各种布局结合起来达到所要的界面效果。网格布局相对表格布局使用更加方便，只需要设置行、列、摆放方向就可以控制控件摆放。

自测习题

1. 下列(　　)可作 EditText 编辑框的提示信息。

A. android:inputType　　B. android:text

C. android:digits　　D. android:hint

2. 为下面控件添加 android:text="Hello"属性，运行时无法显示文字的控件是(　　)。

A. Button　　B. EditText　　C. ImageButton　　D. TextView

3. 下列选项中，前后两个类不存在继承关系的是(　　)。

A. TextView、EditText　　B. TextView、Button

C. Button、ImageButton　　D. ImageView、ImageButton

4. 假设手机屏幕宽度为 400px，现采取水平线性布局放置 5 个按钮，设定每个按钮的宽度为 100px，那么该程序运行时，界面显示效果为(　　)。

A. 自动添加水平滚动条，拖动滚动条可查看 5 个按钮

B. 只可以看到 4 个按钮，超出屏幕宽度部分无法显示

C. 按钮宽度自动缩小，可看到 5 个按钮

D. 程序运行出错，无法显示

5. 表格布局中，设置某一列是可扩展的正确的做法是(　　)。

A. 设置 TableLayout 的属性：android:stretchColumns="x"，x 表示列的序号

B. 设置 TableLayout 的属性：android:shrinkColumns="x"，x 表示列的序号

C. 设置具体列的属性：android:stretchable="true"；

D. 设置具体列的属性：android:shrinkable="true"；

6. 相对布局中，设置以下属性时，属性值只能为 true 或 false 的是(　　)。

A. android:layout_below　　B. android:layout_alignParentLeft

C. android:layout_alignBottom　　D. android:layout_toRightOf

7. 布局文件中有一个按钮(Button)，如果要让该按钮在其父容器中居中显示，正确的设置是(　　)。

A. 设置按钮的属性：android: layout_gravity="center"

B. 设置按钮的属性：android: gravity="center"

C. 设置按钮父容器的属性：android: layout_gravity="center"

D. 设置按钮父容器的属性：android: gravity="center"

8. 根据所学的相对布局的知识，设计出如图 3-14 所示界面，要求在文本编辑框内只能输入数字，并且输入的内容会以“密码隐藏”的形式显示。

9. 运用所学知识，实现“欢乐购商城”注册页面。

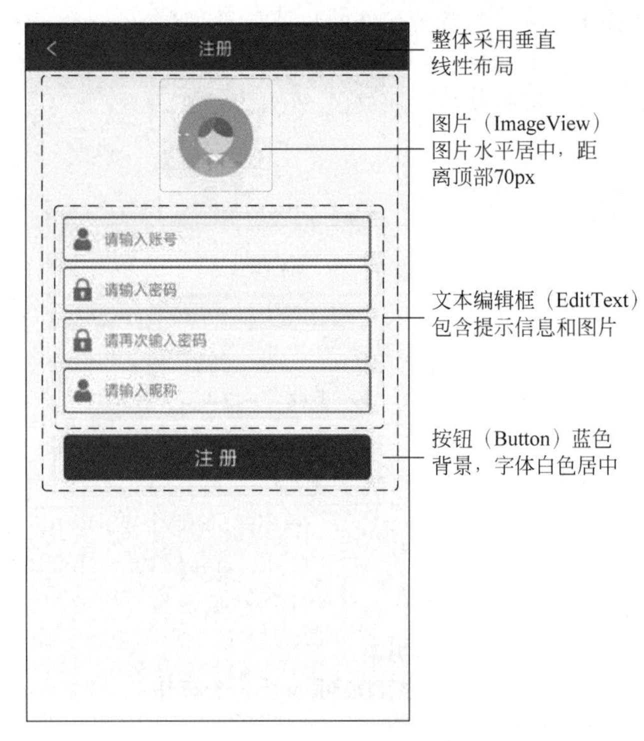

图 3-14　相对布局实现输入框

第4章 Android高级界面控件

在第3章中学习了Android中一些简单界面组件以及常用布局管理，使这些组件按我们的需求排列在界面上，能够设计出一些简单的界面效果。可以根据需求设计界面控件，然而要想设计出一些界面复杂、功能强大的控件，让界面美观还是存在一些困难。Android提供了一些常用的、功能强大的高级组件，如图片组件、图片按钮组件、事件监听机制，本章将集中讲解。

高级界面控件提供了强大的功能和界面优化机制，所以要开发设计功能强大和美观的项目必须要掌握高级界面控件。本章对“欢乐购商城”高级组件进行分析，学习图片、图片按钮、事件处理机制、对话框等，综合运用高级界面控件实现个人中心、登录注册页面优化、购物车支付对话框等。

学习目标

本章要点

(1) 掌握图片按钮的功能和用法。

(2) 掌握自定义XML图片功能和用法。

(3) 掌握基于监听事件处理。

(4) 掌握绑定标签事件处理。

(5) 掌握Handler消息传递机制。

(6) 掌握警告框的功能和用法。

(7) 掌握自定义对话框的功能和用法。

视频详解

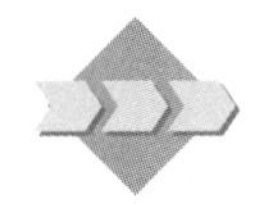

4.1 布局美化

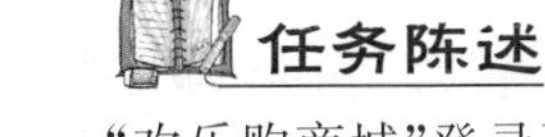

任务陈述

“欢乐购商城”登录页面在第3章中已经初步实现，但是效果和项目中登录页面存在不一样之处，比如编辑框无边框、“确定”按钮为矩形边框无弯角，如图4-1所示。在单击“登录”按钮时按钮颜色发生变化。在个人中心页面中单击头像跳转进入登录页面，该图片其实是一个图片按钮，如图 4-2 所示。

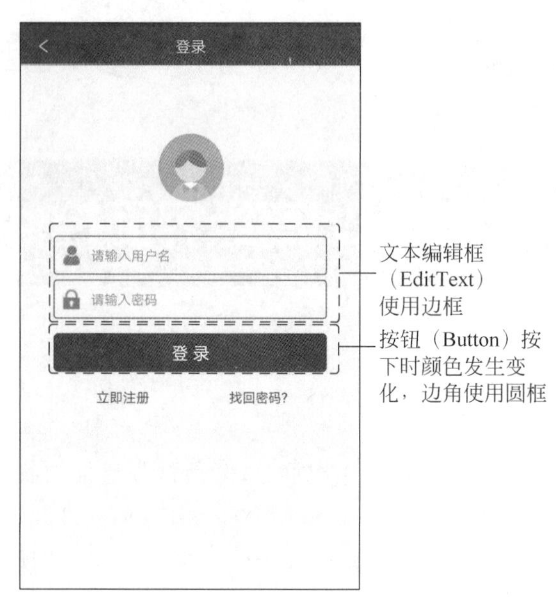

图 4-1　登录页面

图 4-2　个人中心页面

分析：在个人中心页面单击头像能够产生动作，该处使用图片按钮(ImageView)。

登录页面用户名和密码框的边框使用 shape 自定义 XML 类型图片作为背景。“登录”按钮实现边角同样使用 shape 标签自定义 XML 类型图片作为背景，但不同的是，当单击“登录”按钮时按钮颜色发生变化，该功能使用 selector 标签自定义 XML 图片根据按压状态显示不同的图片。

视频详解

4.1.1　图片按钮

ImageButton 的作用与 Button 的作用类似，主要是用于添加单击事件处理。Button 类从 TextView 继承而来，相应的 ImageButton 从 ImageView 继承而来，主要区别是，Button 按钮上显示的是文字，而 ImageButton 按钮上显示的是图片。需要注意的是，在 ImageView、ImageButton 上是无法显示文字的，即使在 XML 文件中为 ImageButton 添加 android: text 属性，虽然程序运行时不会报错，但运行结果仍无法显示文字。

如果想在按钮上既显示文字又显示图片，应该怎么办呢？一种方法是直接将图片和文字设计成一张图片，然后将其作为 ImageButton 的 src 属性的值，但这种方法不够灵活，当需要改变文字或图片时，需重新设计整张图片；另一种方式是直接将图片作为 Button 的背景，并为 Button 按钮添加 android: text 属性，在这种情况下，图片和文字是分离的，可以单独进行设置，灵活性较好，但缺点就是图片作为背景时可能会变形，以适应 Button 的大小。

在 ImageButton 中，既可以设置 background 属性也可以设置 src 属性。这两个属性的值都可以指向一张图片，那么这两个属性有什么区别呢？

src 属性表示的是图标，background 属性表示的是背景。图标是中间的一块区域，而背景是我们所能看到的控件范围。简单来说，一个是前景图（src），一个是背景图（background）。这两个属性最大的区别是：用 src 属性时是原图显示，不会改变图片的大小；用 background 属性时，会按照 ImageButton 的大小来放大或者缩小图片。举例来说，如果 ImageButton 的宽和高是 100×100，而原图片的大小是 80×80。如果用 src 属性来引用该图片，则图片会按 80×80 的大小居中显示在 ImageButton 上。如果用 background 属性来引用该图片，则图片会被拉伸成 100×100。

【例 4-1】 使用 ImageButton 实现个人中心"登录"图标，单击"登录"图标提示"您单击了登录图标"。

程序清单 4-1：chart0401\app\src\main\res\layout\activity_main.xml

```
1   <LinearLayout xmlns:android = "http://schemas.android.com/apk/res/android"
2       xmlns:app = "http://schemas.android.com/apk/res-auto"
3       xmlns:tools = "http://schemas.android.com/tools"
4       android:layout_width = "fill_parent"
5       android:layout_height = "200dp"
6       android:background = "#3F51B5"                →背景颜色设置为蓝色
7       android:orientation = "vertical"               →方向为垂直
8       tools:context = ".MainActivity">
9       <ImageButton                                   →图片按钮控件
10          android:id = "@+id/head_icons"             →为图片按钮控件添加 id
11          android:layout_width = "70dp"
12          android:layout_height = "70dp"
13          android:layout_gravity = "center_horizontal"       →图片按钮控件相对于外部
                                                               →容器水平居中
14          android:layout_marginTop = "40dp"          →外边距离顶部 40dp
15          android:background = "#0000"               →背景设置为透明
16          android:scaleType = "centerCrop"           →图片缩放设置为居中覆盖
17          android:src = "@drawable/head" />          →引用图片地址
18      <TextView
19          android:id = "@+id/tv_user_name"
20          android:layout_width = "wrap_content"
21          android:layout_height = "wrap_content"
22          android:layout_gravity = "center_horizontal"       →图片按钮控件相对于外部
                                                               →容器水平居中
23          android:layout_marginTop = "10dp"                  →外边距离顶部 10dp
24          android:text = "单击登录"
25          android:textColor = "@android:color/white"         →字体颜色为白色
26          android:textSize = "16sp" />
27  </LinearLayout>
```

程序清单 4-2：chart0401\app\src\main\res\layout\activity_main.xml

```
1   public class MainActivity extends AppCompatActivity {
2       @Override                                              →标识重写父类方法
3       protected void onCreate(Bundle savedInstanceState) {
4           super.onCreate(savedInstanceState);
5           setContentView(R.layout.activity_main);            →指定布局文件
```

```
6            ImageButton head_icons = (ImageButton)findViewById(R.id.head_icons);
                                                    →获取图片按钮 id
7            head_icons.setOnClickListener(new View.OnClickListener() {
                                                    →对图片按钮设置监听
8                public void onClick(View view) {
9                    Toast.makeText(MainActivity.this,"您单击了登录图标!
                 ",Toast.LENGTH_LONG).show();       →单击按钮时执行 Toast 提示内容
10               }
11           });
12       }
13   }
```

单击图片按钮产生动作需要对图片按钮对象事件监听，当单击图片按钮时就会执行onclick()方法内代码，其中包含一个 Toast 提示块。

Toast 是为了给当前视图显示一个浮动的显示块，它永远不会获得焦点。一般用于提示一些不那么引人注目，但是又希望用户看见的消息，无须用户自己维护它的消失。如果只是提示简单的信息，使用 Android 为 Toast 提供的两个静态方法最为方便，它们会返回一个 Toast 对象，如果需要显示，只需要调用 show()方法显示即可，下面是这两个方法的签名。

(1) static Toast makeText(Context context,int resId,int duration)

(2) static Toast makeText(Context context,CharSequence text,int duration)

上面两个方法，参数大致相同，一个上下文对象 context，一个显示消息来源，一个设置持续时间。消息来源可以指定 String 资源，最后的 duration 参数设置了 Toast 的持续时间，一般使用 Toast 自带的两个整型常量：LENGTH_LONG (1，时间稍长)，LENGTH_SHORT(0，时间稍短)。针对不同的使用场景，选择不用的持续时间。

图 4-3　单击"登录"按钮效果

单击登录图片按钮效果如图 4-3 所示。

视频详解

4.1.2　自定义 XML 图片

1. shape XML 图片

有时候为了满足特定功能需求，要用 shape 标签去定义一些背景，shape 的用法跟图片一样，可以给 View 设置 Android:background="@drawable/shape"，定义的 shape 文件放在 res\shape 目录下。通常可以用 shape 作 button 的背景选择器、编辑框背景，也可以作切换 Tab 时底部的下画线。通过 shape 绘制不同的形状，如矩形、圆形、直线等。下面介绍 shape 的属性。

(1) android：shape 属性值如表 4-1 所示。

表 4-1 android：shape 属性值

属　　性	说　　明
rectangle	矩形，默认的形状，可以画出直角矩形、圆角矩形、弧形等
oval	椭圆形，用得比较多的是画正圆
line	线形，可以画实线和虚线
ring	环形，可以画环形进度条

(2) 包含的元素。

① solid：设置形状填充的颜色，只有 android：color 一个属性，如表 4-2 所示。

表 4-2 android：Solid

XML 属性	说　　明
android：color	填充的颜色

② padding：设置内容与形状边界的内间距，可分别设置左右上下的距离，如表 4-3 所示。

表 4-3 android：padding

XML 属性	说　　明
android：left	左内间距
android：right	右内间距
android：top	上内间距
android：bottom	下内间距

③ gradient：设置形状的渐变颜色，可以是线性渐变、辐射渐变、扫描性渐变，如表 4-4 所示。

表 4-4 android：gradient

XML 属性	说　　明
android：type	渐变的类型
android：startColor	渐变开始的颜色
android：endColor	渐变结束的颜色
android：centerColor	渐变中间的颜色
android：angle	渐变的角度，线性渐变时才有效，必须是 45 的倍数，0 表示从左到右，90 表示从下到上。
android：centerX	渐变中心的相对 X 坐标，放射渐变时才有效，为 0.0～1.0，默认为 0.5，表示正中间
android：centerY	渐变中心的相对 X 坐标，放射渐变时才有效，在 0.0～1.0，默认为 0.5，表示在正中间

④ corners：设置圆角，只适用于 rectangle 类型，可分别设置四个角不同半径的圆角，当设置的圆角半径很大时，如 200dp，就变成弧形边了，如表 4-5 所示。

表 4-5 android：corners

XML 属性	说 明
android：radius	圆角半径，会被下面每个特定的圆角属性重写
android：topLeftRadius	左上角的半径
android：topRightRadius	右上角的半径
android：bottomLeftRadius	左下角的半径
android：bottomRightRadius	右下角的半径

⑤ stroke：设置描边，可描成实线或虚线，如表 4-6 所示。

表 4-6 android：stroke

XML 属性	说 明
android：color	描边的颜色
android：width	描边的宽度
android：dashWidth	设置虚线时的横线长度
android：dashGap	设置虚线时的横线之间的距离

2. selector XML 图片

给控件不同状态设置背景色时，经常起不到效果，这就需要使用 selector 标签根据不同的状态加载不同的背景图片。selector 集合定义在 XML 文件里面，View 对象背景颜色的选择取决于自身当前的状态，共有 pressed、focused、selected、checkable、checked、enabled、window_focused 和 default 八种状态，View 对象的每一种状态可以设置不同的背景色。

Android 的 selector 要在 drawable 下配置，其中，selector 可以设置的属性如表 4-7 所示。

表 4-7 selector 属性

XML 属性	True	False
android：state_pressed	当被单击时显示该图片	没被按下时显示默
android：state_focused	获得焦点时显示	没获得焦点显示
android：state_selected	当被选择时显示该图片	当不被选择时显示该图片
android：state_checkable	当 CheckBox 能使用时显示该图片	当 CheckBox 不能使用时显示该图片
android：state_checked	当 CheckBox 选中时显示该图片	当 CheckBox 不选中时显示该图片
android：state_enabled	当该组件能使用时显示该图片	当该组件不能使用时显示该图片
android：state_window_focused	当此 activity 获得焦点在最前面时显示该图片	当复制没在最前面时显示该图片

注意：

（1）因为匹配的第一项作为当前 View 对象的背景颜色，结束匹配。如果第一项每次都能被匹配到，就不会往下匹配，那么之后的状态也就失效了。然而每个 View 对象都有默认状态，所以这也就是为什么默认状态不能放在开头，而要放到最后的原因。

（2）不是所有的 View 子类都拥有八种状态，比如 Button 对象，拥有 pressed、enabled、

window_focused、default 四种状态，如果设置 selected 是无效的，要根据组件的具体情况设置。

(3) 状态的顺序排放。

如果控件设置了多种状态。排放顺序依次为 pressed→focused→selected→checkable→checked→enabled→window_focused→default，具体排放顺序还要根据当前 View 对象拥有的状态来排列。注意，默认的权重是最大的，要放在最后面。

4.1.3 实战演练——登录页面布局美化

视频详解

在第 2 章中已经初步实现“欢乐购商城”登录页面，但是具体布局实现效果和项目实现页面不一样，这就需要借助 shape、selector 标签制作自定义 XML 图片作为用户名框、密码框、“登录”按钮的背景。

使用 shape 标签制作用户名框、密码框背景的 editer_borders. xml 代码如下。

程序清单 4-3：chart0402\app\src\main\res\drawable\editer_borders. xml

```
1   <shape xmlns:android="http://schemas.android.com/apk/res/android"
2       android:shape="rectangle">                        →形状为矩形
3       <corners android:radius="3dp" />                  →圆角 3dp
4       <stroke
5           android:width="1dp"                           →线宽度 1dp
6           android:color="#3F51B5" />                    →线的颜色蓝色
7   </shape>
```

因为“登录”按钮在按下和抬起时背景是两种状态，就需要创建 btnback. xml 背景图片，采用 selector 标签设置不同状态对应的图片，不同状态的图片使用 shape 标签实现：btnback. xml 和 btnback_press. xml。

btnback_press. xml 自定义图片代码如下。

程序清单 4-4：chart0402\app\src\main\res\drawable\btnback_press. xml

```
1   <shape xmlns:android="http://schemas.android.com/apk/res/android"
2       android:shape="rectangle">                              →形状为矩形
3       <solid
4           android:color="#6673CA" />                          →填充颜色为蓝色
5       <stroke android:color="#6673CA"></stroke>               →边线颜色为蓝色
6       <corners android:radius="5dp" />                        →圆角为 5dp
7   </shape>
```

btnback_unpress. xml 自定义图片代码如下。

程序清单 4-5：chart0402\app\src\main\res\drawable\btnback_unpress. xml

```
1   <shape xmlns:android="http://schemas.android.com/apk/res/android"
2       android:shape="rectangle">                              →形状为矩形
3       <solid
4           android:color="#4050B5" />                          →填充颜色为蓝色
```

```
5       < stroke android:color = "#4050B5"></stroke >          →边线颜色为蓝色
6       < corners android:radius = "5dp" />                     →圆角为 5dp
7   </shape >
```

btnback.xml 自定义图片代码如下。

程序清单 4-6：chart0402\app\src\main\res\drawable\btnback.xml

```
1   < selector xmlns:android = "http://schemas.android.com/apk/res/android">
2       < item
3           android:state_pressed = "true"                      →被按压的状态
4           android:drawable = "@drawable/btnback_press">
5       </item >
6       < item
7           android:state_pressed = "false"                     →未被按压的状态
8           android:drawable = "@drawable/btnback_unpress">
9       </item >
```

“欢乐购商城”登录页面整体布局代码如下。

程序清单 4-7：chart0402\app\src\main\res\layout\activity_main.xml

```
1   < LinearLayout
2       xmlns:android = "http://schemas.android.com/apk/res/android"
3       xmlns:app = "http://schemas.android.com/apk/res - auto"
4       xmlns:tools = "http://schemas.android.com/tools"
5       android:layout_width = "match_parent"
6       android:layout_height = "match_parent"
7       android:orientation = "vertical"
8       tools:context = ".MainActivity">
9       < ImageView
10          android:id = "@ + id/iv_head"                       →对图片添加 id
11          android:layout_width = "70dp"
12          android:layout_height = "70dp"
13          android:layout_gravity = "center_horizontal"        →相对外部水平居中
14          android:layout_marginTop = "70dp"                   →顶部外边距 70dp
15          android:background = "@drawable/head" />            →加载背景图片
16       < EditText
17          android:id = "@ + id/et_user_name"                  →对图片添加 id
18          android:layout_width = "fill_parent"
19          android:layout_height = "40dp"
20          android:layout_gravity = "center_horizontal"        →相对外部水平居中
21          android:layout_marginLeft = "35dp"                  →左边外边距 35dp
22          android:layout_marginRight = "35dp"                 →右边外边距 35dp
23          android:layout_marginTop = "35dp"                   →顶部外边距 35dp
24          android:background = "@drawable/editer_borders"     →加载边框背景图片
25          android:drawableLeft = "@drawable/user_name_icon"   →加载左侧图标
26          android:drawablePadding = "10dp"                    →图标外边距为 10dp
27          android:gravity = "center_vertical"                 →内部对齐方式垂直居中
```

```
28            android:hint = "请输入用户名"
29            android:paddingLeft = "8dp"
30            android:singleLine = "true"                →单行显示
31            android:textColor = "#000000"              →字体颜色为白色
32            android:textColorHint = "#a3a3a3"          →提示文字颜色为灰色
33            android:textSize = "14sp" />               →字体大小为 14sp
34        <EditText
35            android:id = "@+id/et_psw"
36            android:layout_width = "fill_parent"
37            android:layout_height = "40dp"
38            android:layout_gravity = "center_horizontal"        →相对外部水平居中
39            android:layout_marginLeft = "35dp"
40            android:layout_marginRight = "35dp"
41            android:background = "@drawable/editer_borders"    →加载边框背景图片
42            android:layout_marginTop = "5dp"
43            android:drawableLeft = "@drawable/psw_icon"
44            android:drawablePadding = "10dp"
45            android:hint = "请输入密码"
46            android:inputType = "textPassword"         →文字输入类型文本密码形式
47            android:paddingLeft = "8dp"                →左内边距 8dp
48            android:singleLine = "true"                →单行显示
49            android:textColor = "#000000"
50            android:textColorHint = "#a3a3a3"          →提示文字颜色为灰色
51            android:textSize = "14sp" />
52        <Button
53            android:id = "@+id/btn_login"
54            android:layout_width = "fill_parent"
55            android:layout_height = "40dp"
56            android:layout_gravity = "center_horizontal"      →相对外部水平居中
57            android:layout_marginLeft = "35dp"
58            android:layout_marginRight = "35dp"
59            android:layout_marginTop = "15dp"
60            android:background = "@drawable/btnback"         →添加背景颜色
61            android:text = "登 录"
62            android:textColor = "@android:color/white"       →字体文字为白色
63            android:textSize = "18sp" />
64        <LinearLayout
65            android:layout_width = "fill_parent"
66            android:layout_height = "fill_parent"
67            android:layout_marginLeft = "35dp"
68            android:layout_marginRight = "35dp"
69            android:layout_marginTop = "8dp"
70            android:gravity = "center_horizontal"
71            android:orientation = "horizontal">              →线性布局方向为水平
72            <TextView
73                android:id = "@+id/tv_register"
74                android:layout_width = "0dp"                 →宽度设置为 0
75                android:layout_height = "wrap_content"
76                android:layout_weight = "1"                  →权重设置为 1
77                android:gravity = "center_horizontal"        →对齐方式水平居中
```

```
78                  android:padding = "8dp"
79                  android:text = "立即注册"
80                  android:textColor = "#3F51B5"
81                  android:textSize = "14sp" />
82              <TextView
83                  android:id = "@+id/tv_find_psw"
84                  android:layout_width = "0dp"                    →宽度设置为 0
85                  android:layout_height = "wrap_content"
86                  android:layout_weight = "1"                     →权重设置为 1
87                  android:gravity = "center_horizontal"           →对齐方式水平居中
88                  android:padding = "8dp"
89                  android:text = "找回密码?"
90                  android:textColor = "#3F51B5"
91                  android:textSize = "14sp" />
92          </LinearLayout>
93      </LinearLayout>
```

“欢乐购商城”登录页面实现效果如图 4-4 所示。

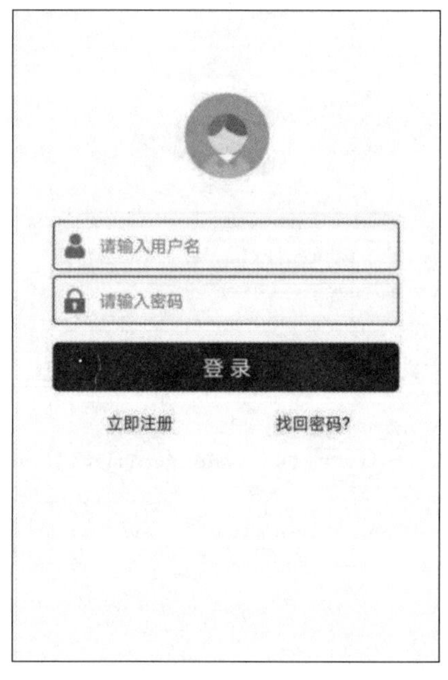

图 4-4　登录页面实现效果

视频详解

4.2　对话框

任务陈述

“欢乐购商城”项目中多处出现对话框,比如在未登录状态下添加商品进入购物车时,提示“你还没有登录,请先登录”,如图 4-5 所示;在结算时提示付款详情,如图 4-6 所示;在单击“确认付支”按钮时，提示“恭喜您支付成功!”，如图 4-7 所示。

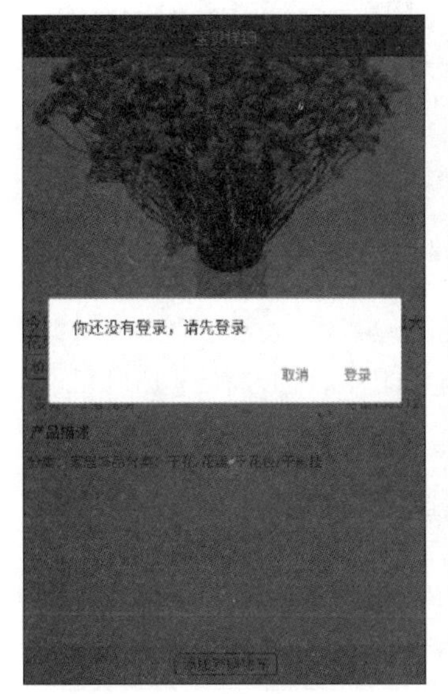

图 4-5 登录提示框

图 4-6 付款详情

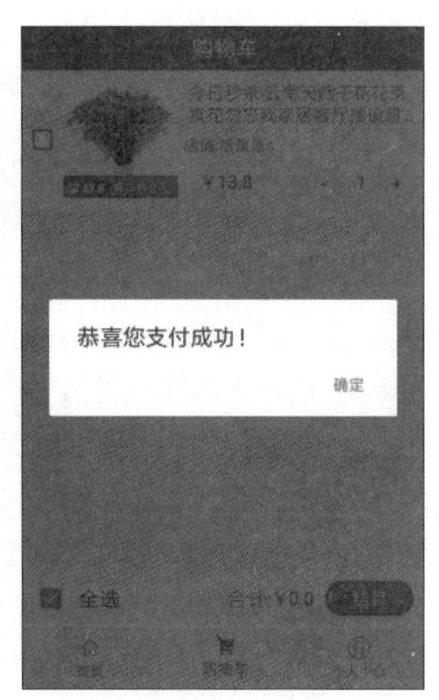

图 4-7 支付成功

分析：提示“你还没有登录，请先登录”和“恭喜您支付成功!”使用对话框中警告框实现，对用户主起到要提示交互作用。而支付详情提示采用的是自定义对话框实现。

对话框是一个漂浮在 Activity 之上的小窗口，此时，Activity 会失去焦点，对话框获取用户的所有交互。对话框通常用于通知，它会临时打断用户，执行一些与应用程序相关的小任务，例如，任务执行进度或登录提示等。在 Android 中，提供了丰富的对话框支持，主要分为以下四种。

AlertDialog：警示框，功能最丰富、应用最广的对话框，该对话框可以包含 0～3 个按钮，或者是包含复选框或单选按钮的列表。

ProgressDialog：进度对话框，主要用于显示进度信息，以进度环或进度条的形式显示任务执行进度，该类继承于 AlertDialog，也可添加按钮。

DatePickerDialog：日期选择对话框，允许用户选择日期。

TimePickerDialog：时间选择对话框，允许用户选择时间。

除此之外，Android 也支持用户创建自定义的对话框，只需要继承 Dialog 基类，或者是 Dialog 的子类，然后定义一个新的布局就可以了。下面着重讲解 AlertDialog 和自定义 Dialog 的使用。

4.2.1 普通对话框

视频详解

AlertDialog 是 Dialog 的子类，它能创建大部分用户交互的对话框，也是系统推荐的对话框类型。常见的 AlertDialog 的类型主要有普通对话框、单选对话框、多选对话框和进度对话框。

创建 AlertDialog 对话框的方式有两种：一种是通过 AlertDialog 的内部类 Builder 对象创建；另一种是通过 Activity 的 onCreateDialog()方法进行创建，通过 showDialog()进行显示，但该方法在 4.1 版本中已经被废弃了，不推荐使用。

使用 AlertDialog 创建对话框，大致步骤如下。

(1) 创建 AlertDialog. Builder 对象，该对象是 AlertDialog 的创建器。

(2) 调用 AlertDialog. Builder 的方法，为对话框设置图标、标题、内容等。

(3) 调用 AlertDialog. Builder 的 create()方法，创建 AlertDialog 对话框。

(4) 调用 AlertDialog. Builder 的 show()方法，显示对话框。

在上述步骤中，主要是 AlertDialog 的内部类 Builder 在起作用，下面来看看 Builder 类提供了哪些方法。Builder 内部类的主要方法如表 4-8 所示。

表 4-8　Builder 类中主要的方法及其作用

方法名	作　　用
public BuildersetTitle	设置对话框标题
public Builder setMessage	设置对话框内容
public BuildersetIcon	设置对话框图标
public BuildersetPositiveButton	添加肯定按钮(Yes)
public BuildersetNegativeButton	添加否定按钮(No)
public BuildersetNeutralButton	添加普通按钮
public BuildersetOnCancelListener	添加取消监听器
public BuildersetCancelable	设置对话框是否可取消
public BuildersetItems	添加列表
public BuildersetMultiChoiceItems	添加多选列表
public Builder setSingleChoiceItems	添加单选列表
publicAlertDialog create()	创建对话框
publicAlertDialog show()	显示对话框

注意：*表中很多方法的返回类型都是 Builder 类型，也就是说，调用 Builder 对象的这些方法后，返回的是该对象本身。Builder 对象每调用一个方法就是为对话框添加一些内容，是对对话框的不断完善，调用方法就是构造对话框的过程，每次返回的都是构建好的对话框。*

【例 4-2】 显示提示信息对话框。本程序的界面非常简单，程序界面上只有一个简单的文本框和一个按钮，当用户单击按钮时将会显示普通对话框。由于用户界面十分简单，故此处不再给出界面布局文件。显示提示信息对话框在 MainActivity 中实现代码如下。

程序清单 4-8：chart0403\app\src\main\java\com\jxcia\chart0403\MainActivity.java

```
1    Button generalDialog = (Button)findViewById(R.id.generalDialog);
2    generalDialog.setOnClickListener(new View.OnClickListener() {
3        @Override
4        public void onClick(View v) {
5            AlertDialog.Builder builder = new AlertDialog.Builder(MainActivity.this);
6            builder.setTitle("普通对话框");      //设置标题
```

```
7              //builder.setIcon(R.drawable.ic_launcher);       //设置图标
8              builder.setMessage("恭喜您支付成功");              //为对话框设置内容
9              //为对话框设置取消按钮
10              builder.setNegativeButton("取消", new DialogInterface.OnClickListener() {
11                  public void onClick(DialogInterface dialog, int which) {
12                      Toast.makeText(MainActivity.this, "您单击了取消按钮",Toast.LENGTH
   _LONG).show();
13                  }
14              });
15              builder.setPositiveButton("确定", new DialogInterface.OnClickListener() {
16                  public void onClick(DialogInterface dialog, int which) {
17                      Toast.makeText(MainActivity.this, "您单击了确定按钮",Toast.LENGTH
   _LONG).show();
18                  }
19              });
20              builder.create().show();      //使用 show()方法显示对话框
21          }
22      });
```

程序运行效果如图 4-8 所示。

图 4-8　普通对话框

4.2.2　列表对话框

视频详解

AlertDialog. Builder 除了提供了 setMessage()方法来设置对话框所显示的消息之外，还提供了如下方法来设置对话框显示列表内容。

setItems(int itemsId, DialogInterface. OnClickListener listener)：创建普通列表对话框。

setMultiChoiceltems(CharSequence[]items, boolean[]checkedltems, DialogInterface.OnMultiChoiceClickListener listener)：创建多选列表对话框。

setSingleChoiceltems（CharSequence[] items，int checkedltem，DialogInterface.OnClickListener listener)：创建单选列表对话框。

【例 4-3】 下面通过演示单选列表对话框实现选择性别，单选对话框在 MainActivity 中实现代码如下。由于用户界面十分简单，故此处不再给出界面布局文件。

程序清单 4-9：chart0403\app\src\main\java\com\jxcia\chart0403\activity_main. xml

```
1    Button singleDialog = (Button)findViewById(R. id. singleDialog);
2    singleDialog. setOnClickListener(new View. OnClickListener() {
3        @Override
4        public void onClick(View v) {
5            AlertDialog. Builder builder  =  new AlertDialog. Builder(MainActivity. this);
6            builder. setTitle("单选对话框");           //设置对话框标题
7            final String[] items  =  { "男", "女" };    //创建一个存放选项的数组
8            final boolean[] checkedItems  =  { true, false };
9            //存放选中状态,true 为选中,false 为未选中,和 setSingleChoiceItems 中第二个
参数对应
10           //为对话框添加单选列表项
11           //第一个参数存放选项的数组,第二个参数存放默认被选中的项,第三个参数单击
事件
12            builder. setSingleChoiceItems(items, 0, new DialogInterface. OnClickListener
 () {
13                @Override
14                public void onClick(DialogInterface dialog, int which) {
15                    //TODO Auto - generated method stub
16                    for (int i  =  0; i < checkedItems. length; i++) {
17                        checkedItems[i]  =  false;
18                    }
19                    checkedItems[which]  =  true;
20                }
21            });
22            builder. setNegativeButton("取消", new DialogInterface. OnClickListener() {
23                @Override
24                public void onClick(DialogInterface dialog, int which) {
25                    dialog. dismiss();                //取消显示对话框
26                }
27            });
28            builder. setPositiveButton("确定", new DialogInterface. OnClickListener() {
29                @Override
30                public void onClick(DialogInterface dialog, int which) {
31                    String str  =  "";
32                    for (int i  =  0; i < checkedItems. length; i++) {
33                        if (checkedItems[i]) {
34                            str  =  items[i];          //获取被单击的列表项内容
35                        }
36                    }
```

```
37                     Toast.makeText(MainActivity.this, "您选择了" + str, Toast.LENGTH_
  LONG).show();
38                 }
39             });
40             builder.create().show();
41         }
42     });
```

程序运行效果如图 4-9 所示。

只要调用 AlertDialog Builder 的 selMultiChoiceItems()方法即可创建一个多选列表的对话框。由于创建多选列表的对话框与创建单选列表对话框十分相似,故下面的程序只给出关键的代码片段。复选对话框在 MainActivity 中实现代码如下,运行效果图如 4-10 所示。

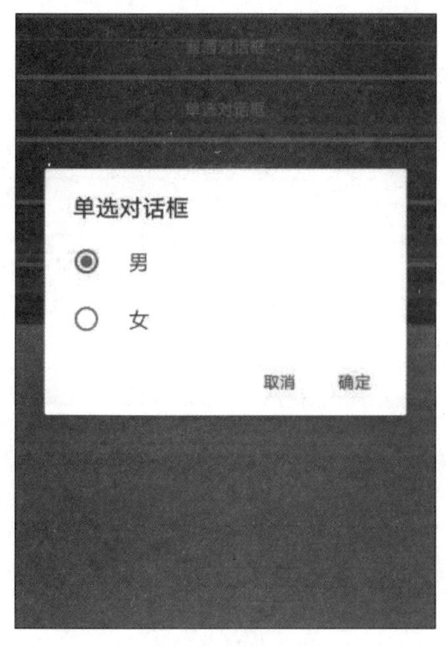

图 4-9 单选对话框

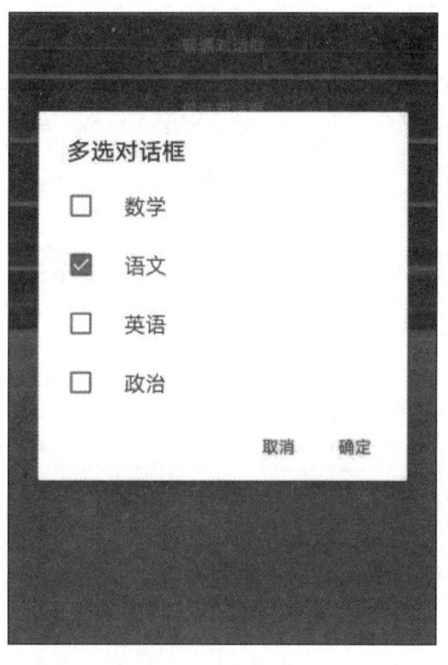

图 4-10 多选对话框

程序清单 4-10:chart0403\app\src\main\java\com\jxcia\chart0403\activity_main.xml

```
1     Button multiDialog = (Button)findViewById(R.id.multiDialog);
2     multiDialog.setOnClickListener(new View.OnClickListener() {
3         @Override
4         public void onClick(View v) {
5             AlertDialog.Builder builder = new AlertDialog.Builder(MainActivity.this);
6             builder.setTitle("多选对话框");          //设置对话框标题
7             final String[] items = { "数学", "语文", "英语", "政治" }; //存放选项的数组
8             final boolean[] checkedItems = { false, true, false, false };
```

```
9               //第一个参数选项,第二个参数选项的状态,第三个单击事件
10              builder.setMultiChoiceItems(items, checkedItems,
11                      new DialogInterface.OnMultiChoiceClickListener() {
12                          public void onClick(DialogInterface dialog, int which,
13                                                  boolean isChecked) {
14                              if (isChecked) {
15                                  checkedItems[which] = true;
16                              } else {
17                                  checkedItems[which] = false;
18                                    Toast.makeText(MainActivity.this, "您取消了" items
  [which],Toast.LENGTH_LONG).show();
19                              }
20                          }
21                      });
22              builder.setNegativeButton("取消", new DialogInterface.OnClickListener() {
23                  @Override
24                  public void onClick(DialogInterface dialog, int which) {
25                      dialog.dismiss();     //取消对话框
26                  }
27              });
28              builder.setPositiveButton("确定", new DialogInterface.OnClickListener() {
29                  @Override
30                  public void onClick(DialogInterface dialog, int which) {
31                      String str = "";
32                      for (int i = 0; i < checkedItems.length; i++) {
33                          if (checkedItems[i]) {
34                              str = str + items[i];
35                          }
36                      }
37                      Toast.makeText(MainActivity.this, "您选择了" + str,Toast.LENGTH_
  LONG).show();
38                  }
39              });
40              builder.create().show();
41          }
42      });
```

视频详解

4.2.3 进度对话框

ProgressDialog 本身就代表了进度对话框，程序只要创建 ProgressDialog 实例，并将它显示出来就是一个进度对话框。当然，开发者也可以设置进度对话框里进度条的方法，ProgressDialog 包含如下常用的方法。

setIndeterminate(boolean indeterminate)：设置对话框里的进度条不显示进度值。

setMax(int max)：设置对话框里进度条的最大值。

setMessage(CharSequence message)：设置对话框里显示的消息。

setProgress(int value)：设置对话框里进度条的进度值。

setProgressStyle(int style)：设置对话框里进度条的风格。

【例 4-4】 不带进度条对话框在 MainActivity 中实现代码如下，运行效果如图 4-11 所示。

程序清单 4-11：chart0403\app\src\main\java\com\jxcia\chart0403\activity_main. xml

```
1    Button progressDialog1 = (Button)findViewById(R. id. ProgressDialog1 );
2    progressDialog1. setOnClickListener(new View. OnClickListener() {
3        @Override
4        public void onClick(View v) {
5            ProgressDialog dialog = new ProgressDialog(MainActivity. this);
6            dialog. setTitle("不带进度条的对话框");      //设置对话框标题
7            dialog. setMessage("正在加载中～");      //设置对话框内容
8            dialog. show();
9        }
10   });
```

【例 4-5】 带进度条对话框在 MainActivity 中实现代码如下，运行效果如图 4-12 所示。

图 4-11　不带进度条的对话框

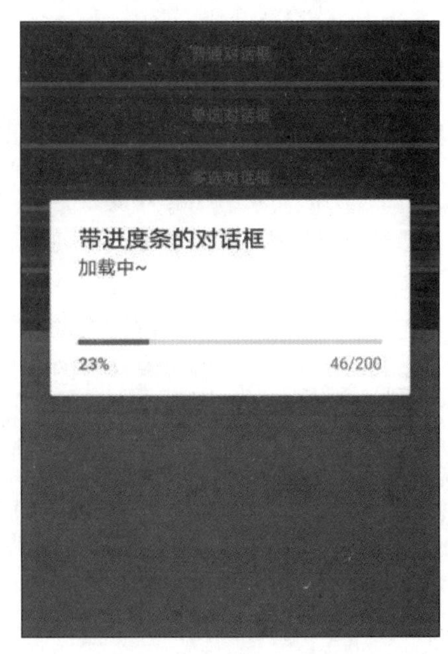

图 4-12　带进度条的对话框

程序清单 4-12：chart0403\app\src\main\java\com\jxcia\chart0403\activity_main. xml

```
1    Button progressDialog2 = (Button)findViewById(R. id. ProgressDialog2 );
2    progressDialog2. setOnClickListener(new View. OnClickListener() {
3        @Override
4        public void onClick(View v) {
5            final ProgressDialog dialog = new ProgressDialog(MainActivity. this);
6            dialog. setProgressStyle(ProgressDialog. STYLE_HORIZONTAL); //设置水平样式
7            dialog. setTitle("带进度条的对话框");      //设置对话框标题
```

```
8            dialog.setMessage("加载中～");              //设置对话框内容
9            dialog.setMax(200);                        //设置对话框内容
10           dialog.show();
11           new Thread(new Runnable() {//开启线程
12               @Override
13               public void run() {
14                   for (int j = 0; j <= 200; j++) {
15                       try {
16                           Thread.sleep(100);         //休眠 0.1s
17                       } catch (InterruptedException e) {
18                           e.printStackTrace();
19                       }
20                       dialog.setProgress(j);         //设置对话框进度值
21                   }
22               }
23           }).start();
24       }
25   });
```

视频详解

4.2.4 实战演练——支付对话框

前面创建了 AlertDialog 并设置自定义布局实现常见的对话框，我们可以根据自己的需求添加已经设定好的模块，这样很方便但也暴露出不能灵活设置自己想要的对话框样式的问题，这时就用到了自定义 Dialog。

自定义对话框，主要是对对话框的显示进行自定义，Builder 对象提供了一个 setView()和 setAdapter()方法，只需将定义好的布局控件传递进去即可达到想要的效果。在“欢乐购商城”项目支付详情页面就使用了自定义对话框，下面进行任务实施完成该部分功能。

在“欢乐购商城”中购物车结算时，使用了对话框，该处自定义对话框和普通对话框嵌套使用，在此从项目中拿出部分代码演示其功能。其中，单击“结算”按钮弹出付款详情对话框，由于“结算”按钮较为简单，此处不提供代码。

要实现自定义对话框需要创建对话框布局文件，通过 Builder 类中的 setView()方法传入才能实现，对话框布局文件为 dialoglayout.xml，代码如下。

程序清单 4-13：chart0404\app\src\main\res\layout\dialoglayout.xml

```
1    <LinearLayout
2        xmlns:android="http://schemas.android.com/apk/res/android"
3        android:layout_width="match_parent"
4        android:layout_height="match_parent"
5        android:orientation="vertical">
6        <TextView
7            android:layout_width="match_parent"
8            android:layout_height="wrap_content"
9            android:layout_marginVertical="15dp"
10           android:layout_marginBottom="8dp"
11           android:gravity="center"
```

```
12            android:text = "付款详情"
13            android:textSize = "30sp" />
14        < View
15            android:layout_width = "match_parent"
16            android:layout_height = "0.5dp"
17            android:background = "#A8A8A8" />
18        < LinearLayout
19            android:layout_width = "match_parent"
20            android:layout_height = "60dp"
21            android:layout_marginTop = "10dp"
22            android:layout_marginBottom = "10dp"
23            android:orientation = "horizontal"
24            android:paddingLeft = "10dp">
25            < ImageView
26                android:layout_width = "60dp"
27                android:layout_height = "60dp"
28                android:src = "@drawable/money" />
29            < TextView
30                android:layout_width = "wrap_content"
31                android:layout_height = "match_parent"
32                android:gravity = "center_vertical"
33                android:paddingLeft = "20dp"
34                android:text = "钱包"
35                android:textColor = "#A6A6A6"
36                android:textSize = "25sp" />
37        </LinearLayout >
38        < View
39            android:layout_width = "match_parent"
40            android:layout_height = "0.5dp"
41            android:background = "#A8A8A8" />
42        < LinearLayout
43            android:layout_width = "match_parent"
44            android:layout_height = "60dp"
45            android:layout_marginTop = "10dp"
46            android:layout_marginBottom = "10dp"
47            android:orientation = "horizontal"
48            android:paddingLeft = "10dp">
49            < ImageView
50                android:layout_width = "60dp"
51                android:layout_height = "60dp"
52                android:src = "@drawable/zhifubao" />
53            < TextView
54                android:layout_width = "wrap_content"
55                android:layout_height = "match_parent"
56                android:gravity = "center_vertical"
57                android:paddingLeft = "20dp"
58                android:text = "支付宝"
59                android:textColor = "#A6A6A6"
60                android:textSize = "25sp" />
61        </LinearLayout >
```

```
62        <View
63            android:layout_width="match_parent"
64            android:layout_height="0.5dp"
65            android:background="#A8A8A8" />
66        <LinearLayout
67            android:layout_width="match_parent"
68            android:layout_height="60dp"
69            android:layout_marginTop="10dp"
70            android:layout_marginBottom="10dp"
71            android:orientation="horizontal"
72            android:paddingLeft="10dp">
73            <ImageView
74                android:layout_width="60dp"
75                android:layout_height="60dp"
76                android:src="@drawable/wechat" />
77            <TextView
78                android:layout_width="wrap_content"
79                android:layout_height="match_parent"
80                android:gravity="center_vertical"
81                android:paddingLeft="20dp"
82                android:text="微信"
83                android:textColor="#A6A6A6"
84                android:textSize="25sp" />
85        </LinearLayout>
86        <View
87            android:layout_width="match_parent"
88            android:layout_height="0.5dp"
89            android:background="#A8A8A8" />
90        <LinearLayout
91            android:layout_width="match_parent"
92            android:layout_height="60dp"
93            android:layout_marginTop="10dp"
94            android:layout_marginBottom="10dp"
95            android:orientation="horizontal"
96            android:paddingLeft="10dp">
97
98            <ImageView
99                android:layout_width="60dp"
100               android:layout_height="60dp"
101               android:src="@drawable/bankcard" />
102           <TextView
103               android:layout_width="wrap_content"
104               android:layout_height="match_parent"
105               android:gravity="center_vertical"
106               android:paddingLeft="20dp"
107               android:text="银行卡"
108               android:textColor="#A6A6A6"
109               android:textSize="25sp" />
110       </LinearLayout>
111       <View
```

```
112              android:layout_width = "match_parent"
113              android:layout_height = "0.5dp"
114              android:background = "#A8A8A8" />
115         < TextView
116              android:id = "@ + id/sum"
117              android:layout_width = "match_parent"
118              android:layout_height = "40dp"
119              android:gravity = "center_vertical|right"
120              android:text = "需付款? 50 元"
121              android:textColor = "#A8A8A8"
122              android:paddingRight = "10dp"
123              android:textSize = "20sp" />
124         < View
125              android:layout_width = "match_parent"
126              android:layout_height = "0.5dp"
127              android:background = "#A8A8A8" />
128     </LinearLayout >
```

接下来在应用程序中调用 AlertDialog. Builder 类中的 setView(View view)方法，让对话框显示该输入界面即可。该程序与前面介绍的列表对话框程序比较相似，只是将原来的调用 setItems()设置列表项，改为现在的调用 setView()来设置自定义视图。下面给出该程序的关键代码。

程序清单 4-14：chart0404\app\src\main\java\com\jxcia\chart0404\activity_main. xml

```
1     Button dialog = (Button)findViewById(R.id.dialog);
2     dialog.setOnClickListener(new View.OnClickListener() {
3         @Override
4         public void onClick(View v) {
5             final AlertDialog.Builder paybuilder  = new AlertDialog.Builder(MainActivity.
  this);
6             View view = getLayoutInflater().inflate(R.layout.dialoglayout,null);
7
8             TextView text = view.findViewById(R.id.sum);
9             text.setText("需付款?100 元");
10             paybuilder.setView(view);
11                  paybuilder. setPositiveButton ( " 确 认 支 付 ", new DialogInterface.
  OnClickListener() {
12                 @Override
13                 public void onClick(DialogInterface dialog, int which) {
14                         AlertDialog. Builder sucbuilder  = new AlertDialog. Builder
  (MainActivity.this);
15                     sucbuilder.setTitle("恭喜您支付成功!");
16                         sucbuilder. setPositiveButton ( " 确 定 ", new DialogInterface.
  OnClickListener() {
17                         @Override
18                         public void onClick(DialogInterface dialog, int which) {
19                         }
```

```
20                      });
21                      sucbuilder.create().show();
22                  }
23              });
24              paybuilder.create().show();
25          }
26      });
```

“欢乐购商城”付款详情和付款成功对话框运行效果如图4-13和图4-14所示。

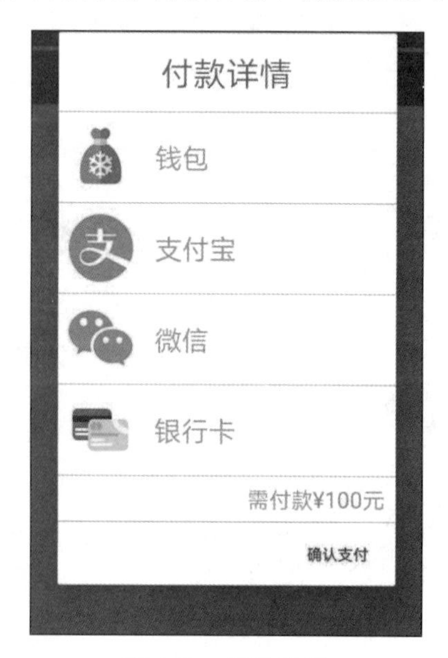

图4-13　付款详情

图4-14　付款成功

视频详解

4.3　事件监听

任务陈述

在“欢乐购商城”中很多页面使用了事件监听，比如对按钮、图片按钮和列表项事件监听，程序通过户单击监听控件进行交互。在项目中还使用了消息传递机制，客户端通过HTTP访问服务端，获取商品数据。

分析：不管是什么手机应用，都离不开与用户的交互，只有通过用户的操作，才能知道用户的需求，从而实现具体的业务功能。因此，应用中经常需要处理的就是用户的操作，也就是需要为用户的操作提供响应，这种为用户操作提供响应的机制就是事件处理。

Android提供了强大的事件处理机制，包括三种事件处理机制：基于监听的事件处理、基于回调的事件处理、直接绑定到标签。本节主要通过个人中心页面讲解基于监听的事件处理、直接绑定到标签事件处理。

Android平台不允许Activity新启动的线程访问该Activity里的界面组件，也不允许将运行状态外送出去，这样就会导致新启动的线程无法动态改变界面组件的属性值，与

Activity 进行交互。比如客户端访问服务器是一个耗时的过程，就必须使用多线程，Android 提供 Handler 消息传递机制解决这一问题。

事件处理在项目中多处使用，本节通过实现个人中心页面学习事件监听，如图 4-15 所示。

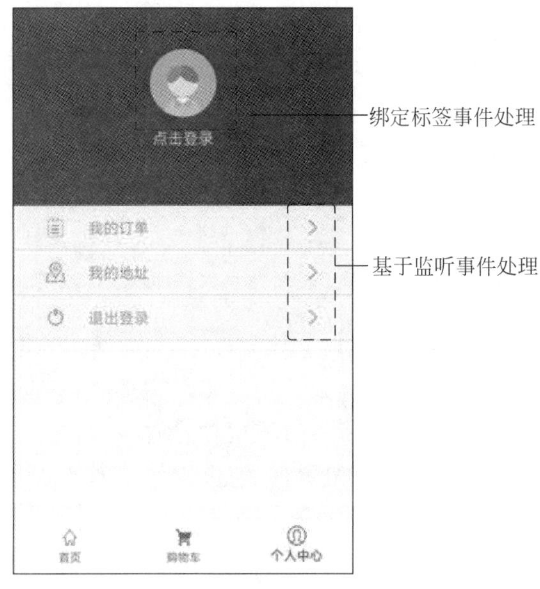

图 4-15　个人中心页面

视频详解

4.3.1　基于监听的事件处理

Android 的基于监听的事件处理模型与 Java 的 AWT、Swing 的处理方式几乎完全一样，只是相应的事件监听器和事件处理方法名有所不同。在基于监听的事件处理模型中，主要涉及以下三类对象。

(1) EventSource(事件源)：产生事件的组件即事件发生的源头，如按钮、菜单等。

(2) Event(事件)：具体某一操作的详细描述，事件封装了该操作的相关信息，如果程序需要获得事件源上所发生事件的相关信息，一般通过 Event 对象来取得。例如，在按键事件中按下的是哪个键、触摸事件发生的位置等。

(3) EventListener(事件监听器)：负责监听用户在事件源上的操作，并对用户的各种操作做出相应的响应。事件监听器中可包含多个事件处理器，一个事件处理器实际上就是一个事件处理方法。

那么在基于监听的事件处理中，这三类对象又是如何协作的呢？实际上，基于监听的事件处理是一种委托式事件处理。普通组件(事件源)将整个事件处理委托给特定的对象(事件监听器)；当该事件源发生指定的事情时，系统自动生成事件对象，并通知所委托的事件监听器，由事件监听器相应的事件处理器来处理这个事件。具体的事件处理模型如图 4-16 所示。当用户在 Android 组件上进行操作时，系统会自动生成事件对象，并将这个事件对象以参数的形式传给注册到事件源上的事件监听器，事件监听器调用相应的事件处理器来处理。

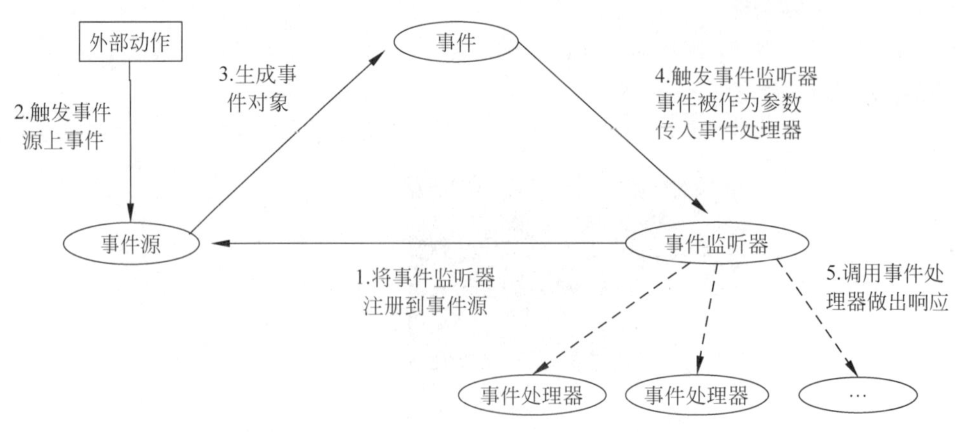

图 4-16　基于监听的事件处理模型

委托式事件处理非常好理解，就类似于生活中每个人能力都有限，当碰到一些自己处理不了的事情时，就委托给某个机构或公司来处理。我们需要把所遇到的事情和要求描述清楚，这样，其他人才能比较好地解决问题，然后该机构会选派具体的员工来处理这件事。其中，我们自己就是事件源，遇到的事情就是事件，该机构就是事件监听器，具体解决事情的员工就是事件处理器。

基于监听的事件处理模型的编程步骤主要如下。

（1）获取普通界面组件（事件源），也就是被监听的对象。

（2）实现事件监听器类，该监听器类是一个特殊的 Java 类，必须实现一个 XxxListerner 接口，并实现接口里的所有方法，每个方法用于处理一种事件。

（3）调用事件源的 setXxxListener()方法将事件监听器对象注册给普通组件（事件源），即将事件源与事件监听器关联起来，这样，当事件发生时就可以自动调用相应的方法。

在上述步骤中，事件源比较容易获取，一般就是界面组件，根据 findViewById()方法即可得到；调用事件源的 setXxxListener()方法是由系统定义好的，只需要传入一个具体的事件监听器。所以，我们所要做的就是实现事件监听器。

在 Android 应用开发过程中，常用监听事件方法如表 4-9 所示。

表 4-9　常用事件监听方法

事件监听	方　法	说　明
ListView	setOnItemClickListener	单击时触发
EditText	setOnKeyListener	获取焦点时触发
RadioGroup	setOnCheckedChangeListener	单击时触发
CheckBox	setOnCheckedChangeListener	单击时触发
Spinner	setOnItemSelectedListener	单击时触发
DatePicker	onDateChangedListener	日期改变时触发
DatePickerDialog	onDateSetListener	设置日期时触发
TimePicker	onTimeChangedListener	时间改变时触发
TimePickerDialog	onTimeSetListener	设置时间时触发
Button, ImageButton	setOnClickListener	单击时触发
Menu	onOptionsItemSelected	单击时触发
Gallery	setOnItemClickListener	单击时触发
GridView	setOnItemClickListener	单击时触发

所谓事件监听器，其实就是实现了特定接口的Java类的实例。在程序中实现事件监听器，通常有如下几种形式。

（1）内部类形式：将事件监听器类定义为当前类的内部类。

（2）外部类形式：将事件监听器类定义成一个外部类。

（3）类自身作为事件监听器类：让Activity本身实现监听器接口，并实现事件处理方法。

（4）匿名内部类形式：使用匿名内部类创建事件监听器对象。

下面依次介绍四种事件监听器。

1. 匿名内部类作为事件监听器类

大部分时候事件处理器都没有什么利用价值，因此大部分事件监听器只是临时使用一次，所以使用匿名内部类形式的事件监听器更合适。实际上，这种形式是目前最广泛的事件监听器形式。

【例4-6】 匿名内部类创建事件监听器。

```
1    Button button = (Button)findViewById(R.id.button);
2    button.setOnClickListener(new View.OnClickListener() {
3        @Override
4        public void onClick(View v) {
5            Toast.makeText(MainActivity.this,"匿名内部类作为事件监听器",Toast.LENGTH_
   LONG).show();
6        }
7    });
```

2. 内部类作为事件监听器

将事件监听器类定义成当前类的内部类。使用内部类可以在当前类中复用监听器类，因为监听器类是外部类的内部类，所以可以自由访问外部类的所有界面组件，这也是内部类的两个优势。

【例4-7】 内部类的形式。

```
1    public class ButtonTest extends AppCompatActivity {
2        protected void onCreate(Bundle savedInstanceState) {
3            super.onCreate(savedInstanceState);
4            setContentView(R.layout.activity_main);
5            Button button = (Button)findViewById(R.id.button);
6            MyButton myButton = new MyButton();
7            button.setOnClickListener(myButton);      //绑定事件监听器
8        }
9        class MyButton implements View.OnClickListener {//定义内部类事件监听器
10            public void onClick(View view){
11                Toast.makeText(MainActivity.this,"内部类作为事件监听器",Toast.LENGTH
   _LONG).show();
12            }
13        }
14   }
```

3. Activity 本身作为事件监听器

使用 Activity 本身作为监听器类，可以直接在 Activity 类中定义事件处理器方法，这种形式非常简洁。但这种做法有两个缺点：①这种形式可能造成程序结构混乱。Activity 的主要职责应该是完成界面初始化；但此时还需包含事件处理器方法，从而引起混乱。②如果 Activity 界面类需要实现监听器接口，会让人感觉比较怪异。

让 Activity 类实现 OnClickListener 事件监听接口，从而可以在该 Activity 类中直接定义事件处理器方法 onClick(view v)，当为某个组件添加该事件监听器对象时，直接使用 this 作为事件监听器对象即可。

【例 4-8】 类自身的形式。

```
1     public class ButtonTest extends AppCompatActivity implements View.OnClickListener{//类自身实现接口
2         protected void onCreate(Bundle savedInstanceState) {
3             super.onCreate(savedInstanceState);
4             setContentView(R.layout.activity_main);
5             Button button = (Button)findViewById(R.id.button);
6             button.setOnClickListener(this);       //控件绑定事件监听器
7         }
8         public void onClick(View view){//重写父类方法
9                 Toast.makeText(MainActivity.this,"类自身作为事件监听器",Toast.LENGTH_LONG).show();
10
11         }
12    }
```

4. 外部类作为事件监听器

使用顶级类定义事件监听器类的形式比较少见，主要有如下两个原因。

(1) 事件监听器通常属于特定的 GUI 界面，定义成外部类不能提高程序的内聚性。

(2) 外部类形式的事件监听器不能自由访问创建 GUI 界面的类中的组件，编程不够简洁。但如果某个事件监听器确实需要被多个 GUI 界面所共享，而且主要是完成某种业务逻辑的实现，则可以考虑使用外部类的形式来定义事件监听器类。

【例 4-9】 外部类的形式。

```
1     public class ButtonTest extends AppCompatActivity {
2         protected void onCreate(Bundle savedInstanceState) {
3             super.onCreate(savedInstanceState);
4             setContentView(R.layout.activity_main);
5             Button button = (Button)findViewById(R.id.button);
6             button.setOnClickListener(new MyButtonListener( "类自身作为事件监听器"));
7         }
8     }
9     class MyButtonListener implements View.OnClickListener{//定义外部类事件监听器
10         private String str;
```

```
11        public MyButtonListener(String str){
12            super();
13            this.str = str;
14        }
15        public void onClick(View view){
16            Toast.makeText(this,str,Toast.LENGTH_LONG).show();
17        }
18    }
```

4.3.2 直接绑定到标签

视频详解

Android 还有一种简单的绑定事件的方式，即直接在界面布局文件中为指定标签绑定事件处理方法。对于很多 Android 界面组件标签而言，它们都支持如 onClick、onLongClick 等属性，这种属性的属性值就是一个形如 xxx(View view)的方法的方法名。

【例 4-10】 在布局文件中为组件添加单击事件的处理方法，布局文件如下。

程序清单 4-15：chart0405\app\src\main\res\layout\activity_main.xml

```
1    <LinearLayout xmlns:android = "http://schemas.android.com/apk/res/android"
2        xmlns:app = "http://schemas.android.com/apk/res - auto"
3        xmlns:tools = "http://schemas.android.com/tools"
4        android:layout_width = "match_parent"
5        android:layout_height = "match_parent"
6        android:orientation = "vertical"
7        tools:context = ".MainActivity">
8        <TextView                                          →用于显示按钮被单击后内容
9            android:layout_width = "wrap_content"
10           android:layout_height = "wrap_content"
11           android:layout_marginLeft = "5dp"
12           android:text = "测试"
13           android:id = "@ + id/textView"                 →添加 id
14           android:textSize = "20sp" />
15       <Button
16           android:layout_width = "wrap_content"
17           android:layout_height = "wrap_content"
18           android:layout_marginLeft = "5dp"
19           android:background = "@color/colorPrimary"     →引用程序主颜色
20           android:textColor = "#fff"                     →字体为白色
21           android:text = "单击我"
22           android:id = "@ + id/button"
23           android:onClick = "clickButton"/>              →绑定一个事件处理方法
24   </LinearLayout>
```

然后在该界面布局对应的 Activity 中定义一个 void clickButton(View view)方法，该方法将会负责处理该按钮上的单击事件。详细代码如下运行效果如图 4-17 和图 4-18 所示。

图 4-17　按钮单击前

图 4-18　按钮单击后

程序清单 4-16：chart0405\app\src\main\java\com\jxcia\chart0405\MainActivity .java

```
1    public class MainActivity extends AppCompatActivity {
2        TextView tView;
3        protected void onCreate(Bundle savedInstanceState) {
4            super.onCreate(savedInstanceState);
5            setContentView(R.layout.activity_main);
6        }
7        /*实现 clickButton 方法*/
8        public void clickButton(View view) {     //定义 clickButton 方法和布局中属性对应
9            tView = (TextView) findViewById(R.id.textView);
10            tView.setText("按钮被单击了");
11         }
12   }
```

视频详解

4.3.3　Handler 消息传递机制

出于性能优化的考虑，Android 的 UI 操作并不是线程安全的，这意味着如果有多个线程并发操作 UI，可能导致线程安全问题。为了解决这个问题，Android 制定了一条简单的规则：只允许 UI 线程修改 Activity 的 UI 组件。

当一个程序第一次启动时，Activity 会同时启动一条主线程，主线程主要负责处理与 UI 相关的事件，如用户的按键操作、用户触摸屏幕的事件及屏幕绘制事件，并把相关的事件分发到对应的组件进行处理。所以，主线程通常又被叫作 UI 线程。

Android 的消息传递机制是另一种形式的"事件处理"，这种机制主要为解决 Android 应用的多线程问题——Android 平台只允许 UI 线程修改 Activity 里的 UI 组件，这就会导致新启动的线程无法动态改变界面组件的属性值。但在实际开发中，尤其是涉及动画的游戏开发中，需要让新启动的线程周期性地改变界面组件的属性值，这就需要借助于 Handler 的消息传递机制。Handler 类的常用方法如表 4-10 所示。

表 4-10　Handler 类的常用方法

方法签名	描　　述
public voidhandleMessage(Message msg)	通过该方法获取、处理信息
public final booleansendEmptyMessage(int what)	发送一个只含有 what 值的消息
public final booleansendMessage(Message msg)	发送消息到 Handler，通过 handleMessage( )方法接收
public final booleanhasMessages(int what)	监测消息队列中是否有 what 值的消息
public final boolean post(Runnable r)	将一个线程添加到消息队列

Handler 类主要有以下两个作用。

(1) 在新启动的线程中发送消息。

(2) 在主线程中获取、处理消息。

上面的说法看上去很简单,似乎只分成两步:在新线程中发送消息,然后在主线程中获取并处理消息即可。但过程中涉及一些问题,新启动的线程何时发送消息?主线程又何时处理消息?时机如何控制?

为了解决处理消息问题,只能通过回调的方式来实现,重写 Handler 类的 handleMessage( )方法。当新启动的线程发送消息时,消息会发送到与之关联的 MessageQueue,而 Handler 会不断从 MessageQueue 中获取并处理消息,这将导致 Handler 中处理消息的方法被回调。

开发带有 Handler 类的程序步骤如下。

(1) 创建 Handler 类对象,并重写 handleMessage()方法。

(2) 在新启动的线程中,调用 Handler 对象的发送消息方法。

(3) 利用 Handler 对象的 handleMessage( )方法接收消息,然后根据不同的消息执行不同的操作。

【例 4-11】 运用多线程和 Handler 实现数字输入,罗列其质数。

程序布局包含一个编辑框、一个按钮和一个文本视图,编辑框用于输入要求的质数,按钮用于与用户产生交互,文本视图用于显示求出的质数,运行效果如图 4-19 所示。

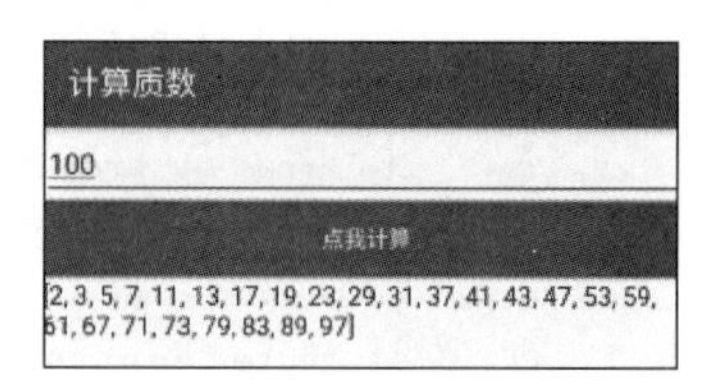

图 4-19 计算质数

程序清单 4-17:chart0406\app\src\main\res\layout\activity_main.xml

```
1    <LinearLayout
2        xmlns:android = "http://schemas.android.com/apk/res/android"
3        xmlns:app = "http://schemas.android.com/apk/res-auto"
4        xmlns:tools = "http://schemas.android.com/tools"
5        android:layout_width = "match_parent"
6        android:layout_height = "match_parent"
7        android:orientation = "vertical"              →线性布局方向垂直
8        tools:context = ".MainActivity">
9        <EditText                                      →用于用户输入要求的质数
10           android:id = "@+id/edit"
11           android:layout_width = "match_parent"
12           android:layout_height = "wrap_content"
13           android:hint = "请输入要计算的质数"
14           />
15        <Button
16           android:onClick = "cal"                     →绑定标签事件监听
17           android:text = "点我计算"
18           android:background = "@color/colorPrimary"
19           android:textColor = "#fff"
20           android:layout_width = "match_parent"
21           android:layout_height = "wrap_content"/>
```

```
22        <TextView                                    →用于显示结果
23            android:id="@+id/result"
24            android:textSize="16sp"
25            android:textColor="#00f"
26            android:layout_width="wrap_content"
27            android:layout_height="wrap_content"/>
28    </LinearLayout>
```

然后在该界面布局对应的 Activity 中定义一个 void cal(View view)方法，该方法将会负责处理该按钮上的单击事件，当被单击时会启动线程 Handler 发送消息，handleMessage(Message msg)方法接收消息更改页面，详细代码如下。

程序清单 4-18：chart0406\app\src\main\java\com\jxcia\chart0406\MainActivity .java

```
1     public class MainActivity extends AppCompatActivity {
2         private static final String DATA_NUM = "num";  //用户存储用户输入数据,关键字
3         private static final String RESULT_KEY = "result_key";  //线程发送的消息标记
4         EditText editText;
5         MyThread myThread;                              //自定义线程内部类
6         TextView result;                                //用户显示结果
7         @Override
8         protected void onCreate(Bundle savedInstanceState) {
9             super.onCreate(savedInstanceState);
10             setContentView(R.layout.activity_main);
11             editText = findViewById(R.id.edit);
12             result = findViewById(R.id.result);
13             myThread = new MyThread();                 //实例化线程
14             myThread.start();                          //启动线程
15         }
16         public void cal(View view) {                   //用户监听按钮
17             Message message = new Message();
18             message.what = 0x110;                      //定义消息标记
19             Bundle bundle = new Bundle();
20             bundle.putInt(DATA_NUM, Integer.parseInt(editText.getText().toString()));
21             message.setData(bundle);
22             myThread.mHandler.sendMessage(message);
23         }
24         class MyThread extends Thread {
25             public Handler mHandler;
26             @Override
27             public void run() {
28                 Looper.prepare();                      //为当前线程准备消息队列
29                 mHandler = new Handler() {
30                     @Override
31                     public void handleMessage(Message msg) {
32                         super.handleMessage(msg);
33                         if (msg.what == 0x110) {
34                             int num = msg.getData().getInt(DATA_NUM);     //通过关键
   字获取数据
```

```
35                              List < Integer > nums = new ArrayList <>();   //用户存储结
果数据
36                              //计算从 2 开始到 num 的所有质数
37                              for (int i = 2; i <= num; i++) {
38                                  //用 i 除以从 2 开始,到 i 的平方根的所有数
39                                  boolean isadd = true;
40                                  for (int j = 2; j <= Math.sqrt(i); j++) {
41                                      //如果可以整除,则不是质数
42                                      if (i != 2 && i % j == 0) {
43                                          isadd = false;
44                                          continue;
45                                      }
46                                  }
47                                  if (isadd) {
48                                      nums.add(i);     //把求得数据结果存入集合 nums
49                                  }
50                              }
51                              //Toast 显示
52                              Toast.makeText(MainActivity.this, nums.toString(),
                                  Toast.LENGTH_LONG).show();
53                              Message message = new Message();
54                              message.what = 0x111; //存入由子线程发送给主线程标记
55                              Bundle bundle = new Bundle();
56                              bundle.putString(RESULT_KEY, nums.toString());
                                                                        //把数据存入 bundle
57                              message.setData(bundle);
58                              mUiHandler.sendMessage(message);
59                          }
60                      }
61                  };
62                  Looper.loop();      //让 Looper 开始工作,从消息队列里取消息,处理消息
63              }
64          }
65          Handler mUiHandler = new Handler() {//用户接收子线程发出的消息
66              @Override
67              public void handleMessage(Message msg) {
68                  super.handleMessage(msg);
69                  if (msg.what == 0x111) {//判断子线程发出的消息标记
70                      result.setText(msg.getData().getString(RESULT_KEY));
                                                                        //在主线程中显示结果
71                  }
72              }
73          };
74      }
```

该程序重写了 Handler 类的 handleMessage(Message msg)方法，该方法用于处理消息，当新线程发送消息时，该方法会被自动回调，然后根据消息的标记，对不同的消息进行不同的业务逻辑处理，由于 handleMessage(Message msg)方法依然位于主线程，所以可以动

态地修改 TextView 组件的文本。

注意：发送消息和处理消息的是同一个 Handler 对象。

视频详解

4.3.4 实战演练——个人中心页面

在“欢乐购商城”项目中多处使用事件监听，下面以个人中心页面作为任务完成任务实施，其中包括个人头像绑定标签事件监听，我的订单、我的地址、退出登录使用内部类事件监听器实现完成，如图 4-20 和图 4-21 所示。

图 4-20 绑定标签事件监听

图 4-21 内部类事件监听

个人中心事件监听主布局 activity_main.xml 如下。

程序清单 4-19：chart0407\app\src\main\res\layout\activity_main.xml

```
1   <LinearLayout xmlns:android = "http://schemas.android.com/apk/res/android"
2       xmlns:app = "http://schemas.android.com/apk/res - auto"
3       xmlns:tools = "http://schemas.android.com/tools"
4       android:layout_width = "match_parent"
5       android:layout_height = "match_parent"
6       android:orientation = "vertical"                 →整体采用垂直线性布局
7       tools:context = ".MainActivity">
8       <LinearLayout
9           android:id = "@ + id/ll_head"
10          android:layout_width = "fill_parent"
11          android:layout_height = "200dp"
12          android:background = "#3F51B5"               →头像登录栏背景为蓝色
13          android:orientation = "vertical">            →方向为垂直方向
14          <ImageButton
15              android:onClick = "login"                →标签绑定事件监听器
```

```
16                  android:layout_width = "70dp"
17                  android:layout_height = "70dp"
18                  android:layout_gravity = "center_horizontal"
19                  android:scaleType = "centerCrop"
20                  android:layout_marginTop = "40dp"                →距离顶部外边距为 40dp
21                  android:background = "#0000"
22                  android:src = "@drawable/head" />
23              <TextView                                            →提示文字
24                  android:id = "@ + id/tv_user_name"
25                  android:layout_width = "wrap_content"
26                  android:layout_height = "wrap_content"
27                  android:layout_gravity = "center_horizontal"
28                  android:layout_marginTop = "10dp"                →距离顶部 10dp
29                  android:text = "单击登录"
30                  android:textColor = "@android:color/white"
31                  android:textSize = "16sp" />
32          </LinearLayout>
33          <RelativeLayout
34              android:id = "@ + id/rl_my_order"                    →添加 id
35              android:layout_width = "fill_parent"
36              android:layout_height = "45dp"
37              android:background = "#F7F8F8"                       →背景颜色为灰色
38              android:gravity = "center_vertical"
39              android:paddingLeft = "10dp"
40              android:paddingRight = "10dp">
41              <ImageView
42                  android:id = "@ + id/iv_my_order"
43                  android:layout_width = "20dp"
44                  android:layout_height = "20dp"
45                  android:layout_centerVertical = "true"           →相对于外部垂直居中
46                  android:layout_marginLeft = "25dp"               →距离顶部 25dp
47                  android:src = "@drawable/order" />
48              <TextView
49                  android:layout_width = "wrap_content"
50                  android:layout_height = "wrap_content"
51                  android:layout_centerVertical = "true"           →相对于外部垂直居中
52                  android:layout_marginLeft = "25dp"               →距离顶部 25dp
53                  android:layout_toRightOf = "@id/iv_my_order"
54                  android:text = "我的订单"
55                  android:textColor = "#A3A3A3"
56                  android:textSize = "16sp" />
57              <ImageView
58                  android:layout_width = "15dp"
59                  android:layout_height = "15dp"
60                  android:layout_alignParentRight = "true"         →在父容器中靠右
61                  android:layout_centerVertical = "true"           →相对于外部垂直居中
62                  android:layout_marginRight = "25dp"              →距离顶部 25dp
```

```
63                  android:src = "@drawable/iv_right_arrow" />
64          </RelativeLayout>
65          <View                                               →横线
66              android:layout_width = "fill_parent"
67              android:layout_height = "1dp"
68              android:background = "#E3E3E3" />
69
70          <RelativeLayout
71              android:id = "@ + id/rl_my_address"             →添加 id
72              android:layout_width = "fill_parent"
73              android:layout_height = "45dp"
74              android:background = "#F7F8F8"                  →背景颜色为灰色
75              android:gravity = "center_vertical"
76              android:paddingLeft = "10dp"
77              android:paddingRight = "10dp">
78
79              <ImageView
80                  android:id = "@ + id/iv_my_websit"
81                  android:layout_width = "20dp"
82                  android:layout_height = "20dp"
83                  android:layout_centerVertical = "true"      →相对于外部垂直居中
84                  android:layout_marginLeft = "25dp"          →距离顶部 25dp
85                  android:src = "@drawable/address" />
86
87              <TextView
88                  android:layout_width = "wrap_content"
89                  android:layout_height = "wrap_content"
90                  android:layout_centerVertical = "true"      →相对于外部垂直居中
91                  android:layout_marginLeft = "25dp"          →距离顶部 25dp
92                  android:layout_toRightOf = "@id/iv_my_websit"→在 iv_my_websit 的右边
93                  android:text = "我的地址"
94                  android:textColor = "#A3A3A3"
95                  android:textSize = "16sp" />
96              <ImageView
97                  android:layout_width = "15dp"
98                  android:layout_height = "15dp"
99                  android:layout_alignParentRight = "true"    →在父容器中靠右
100                 android:layout_centerVertical = "true"      →相对于外部垂直居中
101                 android:layout_marginRight = "25dp"         →距离顶部 25dp
102                 android:src = "@drawable/iv_right_arrow" />
103         </RelativeLayout>
104         <View                                               →横线
105             android:layout_width = "fill_parent"
106             android:layout_height = "1dp"
107             android:background = "#E3E3E3" />
108         <RelativeLayout
109             android:id = "@ + id/rl_loginout"               →添加 id
110             android:layout_width = "fill_parent"
111             android:layout_height = "45dp"
```

```
112            android:background = "#F7F8F8"                →背景颜色为灰色
113            android:gravity = "center_vertical"
114            android:paddingLeft = "10dp"
115            android:paddingRight = "10dp">
116            < ImageView
117                android:id = "@ + id/iv_loginout_icon"
118                android:layout_width = "20dp"
119                android:layout_height = "20dp"
120                android:layout_centerVertical = "true"   →相对于外部垂直居中
121                android:layout_marginLeft = "25dp"       →距离顶部 25dp
122                android:src = "@drawable/exit" />
123            < TextView
124                android:layout_width = "wrap_content"
125                android:layout_height = "wrap_content"
126                android:layout_centerVertical = "true"   →相对于外部垂直居中
127                android:layout_marginLeft = "25dp"       →距离顶部 25dp
128                  android: layout _ toRightOf = "@ + id/iv _ loginout _ icon"     → 在 iv_
       loginout_icon 的右边
129                android:text = "退出登录"
130                android:textColor = "#A3A3A3"
131                android:textSize = "16sp" />
132            < ImageView
133                android:layout_width = "15dp"
134                android:layout_height = "15dp"
135                android:layout_alignParentRight = "true"→在父容器中靠右
136                android:layout_centerVertical = "true"   →相对于外部垂直居中
137                android:layout_marginRight = "25dp"      →距离顶部 25dp
138                android:src = "@drawable/iv_right_arrow" />
139        </RelativeLayout >
140        < View                                           →横线
141            android:layout_width = "fill_parent"
142            android:layout_height = "1dp"
143            android:background = "#E3E3E3" />
144    </LinearLayout >
```

然后在该界面布局对应的 Activity 中定义一个 void login(View view)方法，该方法将会负责处理单击头像进入登录页面的单击事件。我的订单、我的地址、退出登录监听绑定内部类事件监听器 relativeListener 中，详细代码如下，运行效果如图 4-20 和图 4-21 所示。

程序清单 4-20：chart0407\app\src\main\java\com\jxcia\chart0407\MainActivity .java

```
1    public class MainActivity extends AppCompatActivity {
2        private RelativeLayout rl_my_order,rl_my_address,rl_loginout;
3        @Override
4        protected void onCreate(Bundle savedInstanceState) {
5            super.onCreate(savedInstanceState);
6            setContentView(R.layout.activity_main);
7            rl_my_order = findViewById(R.id.rl_my_order);
8            rl_my_address = findViewById(R.id.rl_my_address);
```

```
9            rl_loginout = findViewById(R.id.rl_loginout);
10           RelativeListener relativeListener = new RelativeListener();
11           rl_my_order.setOnClickListener(relativeListener);     //绑定事件监听器
12           rl_my_address.setOnClickListener(relativeListener);   //绑定事件监听器
13           rl_loginout.setOnClickListener(relativeListener);     //绑定事件监听器
14       }
15       public void login(View view) {//绑定标签事件监听器
16           dialog("下一步将跳转到登录页面!");
17       }
18       private class RelativeListener implements View.OnClickListener {//定义内部类事件
                                                                         //监听器
19           @Override
20           public void onClick(View v) {
21               switch (v.getId()){
22                   case R.id.rl_my_order://判断当单击我的订单栏
23                       dialog("下一步将查询个人订单!");
24                       break;
25                   case R.id.rl_my_address://判断当单击地址栏
26                       dialog("下一步将编辑个人地址!");
27                       break;
28                   case R.id.rl_loginout://判断当单击退出栏
29                       dialog("下一步将退出登录!");
30                       break;
31               }
32           }
33       }
34       public void dialog(String str){//定义对话框方法可多次使用
35           AlertDialog.Builder builder  = new AlertDialog.Builder(this);
36           builder.setMessage(str);
37           builder.setPositiveButton("确定", new DialogInterface.OnClickListener() {
38               @Override
39               public void onClick(DialogInterface dialog, int which) {
40               }
41           });
42           builder.create().show();
43       }
44   }
```

本章小结

本章是对前面章节的补充，通过学习图片按钮和 shape、selector 标签自定义图片对前面布局进行美化。对话框为人机交互提供了比较好的用户体验，能够时刻提示用户进行操作，以避免不必要的失误。Android 中使用最为广泛的就是 AlertDialog，本章详细讲解了几种 AlertDialog 的创建与使用，以及如何创建自定义的 AlertDialog。

Android 开发图形界面编程肯定需要与事件处理相结合，设计了界面友好的应用之后，必须为界面上的相应组件提供响应，从而当用户操作时，能执行相应的功能，这种响应动作就是由事件处理来完成的。本章的重点就是掌握 Android 的事件处理机制：基于监听的事件处理、直接绑定到标签的事件处理。此外，还着重讲解了动态改变界面组件的显示。需要注意的是，Android 不允许在子线程中更新主线程的界面组件。因此，我们讲解了 Handler 消息处理机制，当子线程需要更改界面显示时，就向主线程发送一条消息，主线程接收到消息后，自己对界面显示进行修改。最后讲解了异步任务处理，异步任务主要处理一些比较耗时的操作，是对消息处理机制的一种补充。

自测习题

1. 简单描述 ImageButton 的 src 属性与 background 属性的区别。
2. 如何将 ImageButton 默认的背景去除？
3. 简述 AlertDialog 创建的一般步骤。
4. Android 中事件处理方式主要有哪三种？
5. 基于监听的事件处理模型中，主要包含的三类对象是什么？
6. 简单描述基于监听的事件处理的过程。
7. 实现事件监听器的方式有________、________、________和________。
8. 简要描述 Handler 消息传递机制的步骤。

列表控件

在现实生活中，人们经常会使用QQ、微信等应用程序。在这些应用程序中通常会有一个页面展示多个条目，并且每个条目的布局都是一样的。如果利用前面所学习的知识实现这种布局，需要创建大量相同的布局，使用这种方式并不利于程序维护和扩展。针对上述情况，Android系统提供了功能强大的列表控件，通过列表控件可轻松实现上述需求。

列表控件是Android系统开发中使用最广泛的控件之一，常见的列表控件包括Spinner（下拉列表）、ListView（普通列表）、GridView（网格列表）、RecyclerView（增强列表）等。通过列表控件可展示多项数据，并且开发者可动态配置数据源，列表控件可根据所适配的数据源不同展示不同的内容。Android中主要采用适配器模式帮助建立列表控件和数据源之间的联系。因此在使用列表控件的时候，还需要创建适配器对象并为适配器提供数据源。Adapter（适配器）对象用来指明数据源中的每一项数据在列表控件中如何显示。列表控件调用setAdapter()方法把Adapter对象传递进来，即可将数据显示在列表中。

本章通过对"欢乐购商城"分析，帮助读者学会使用列表控件。通过对不同种类的列表控件综合运用，实现"欢乐购商城"中首页、商品列表、订单列表等页面的功能。

学习目标

本章要点

(1) 掌握下拉列表Spinner的功能和用法。

(2) 掌握普通列表ListView的功能和用法。

(3) 掌握网格列表GridView的功能和用法。

(4) 掌握增强列表RecyclerView的功能和用法。

视频详解

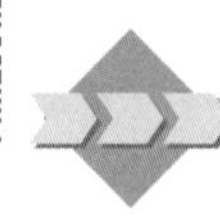

5.1 下拉列表Spinner

任务陈述

在"欢乐购商城"项目中修改收货地址页面使用到了Spinner控件，下面对修改收货地址页面做概要分析。

分析：在修改用户收货地址页中，需要用户填写地址。在填写地址时需要选择不同的省份市区。由于省份等信息是固定不变的，用户只需从一组数据中选择，因此在这里采用Spinner控件实现，当用户单击此控件时，下拉列表中显示出所有省份信息选项供用户选择，如图5-1所示。

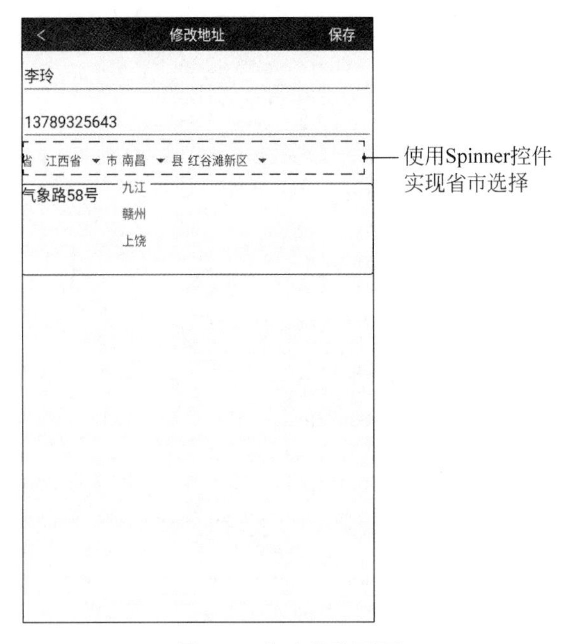

图5-1　修改地址页面

5.1.1　Spinner控件

视频详解

下拉列表Spinner类似于下拉菜单，默认情况下展示列表项中的第一项，单击Spinner控件时会弹出一个包含所有数据的下拉列表。Spinner比较节省空间，常用于固定值选择或者条件筛选等。例如，在用户输入地址信息的时候，在选择省份或者地区时通常可以采用Spinner列表控件提供给用户，让用户从中选择，这样可减少用户的输入和避免用户输入错误信息。

使用列表控件的关键步骤如下。

(1) 在布局文件中添加列表控件，在Java代码文件中通过id属性获取到列表控件。

(2) 准备数据源，数据源可以是数组或者集合。

(3) 构建Adapter对象，指定列表中每一项数据的显示样式，并将数据源数据通过构造函数等方式传递给Adapter对象。

(4) 列表控件调用setAdapter()方法，关联创建好的Adapter对象，展示数据源中的数据。

下面就以选择编程语言为例，讲解Spinner控件的具体用法。

(1) 创建名字为chart0501的应用程序，指定包名为com. example. administrator.

chart0501。

（2）设置布局文件，在 res/layout/activity_main.xml 文件中添加一个 TextView 用于显示标题，再添加一个 Spinner 控件用于显示下拉列表。布局文件的核心代码如下。

程序清单 5-1：chart0501\app\src\main\res\layout\activity_main.xml

```
1   <?xml version = "1.0" encoding = "utf - 8"?>
2   <LinearLayout xmlns:android = "http://schemas.android.com/apk/res/android"
3   xmlns:tools = "http://schemas.android.com/tools"
4       android:layout_width = "match_parent"
5       android:layout_height = "wrap_content"
6       android:orientation = "horizontal"
7       android:gravity = "center_vertical"
8       tools:context = ".MainActivity">
9       <TextView
10          android:layout_width = "wrap_content"
11          android:layout_height = "wrap_content"
12          android:text = "你学习的编程语言:"
13          android:textSize = "16sp"/>
14      <Spinner
15          android:id = "@ + id/spinner"                     →Spinner 添加 id 属性
16          android:layout_width = "wrap_content"             →为 Spinner 设置宽度
17        android:layout_height = "wrap_content"/>            →为 Spinner 设置高度
18   </LinearLayout>
```

（3）Spinner 列表由若干 Item 组成，每个 Item 显示编程语言的名称，因此需要设置 Item 选项的布局。在 res/layout 文件夹中创建一个 Item 界面的布局文件 simple_list_item.xml，在该文件中添加一个 TextView 用来展示每项的内容。注意：不需要在 TextView 的外层添加布局，完整的布局文件代码如下。

程序清单 5-2：chart0501\app\src\main\res\layout\simple_list_item.xml

```
1   <?xml version = "1.0" encoding = "utf - 8"?>
2   <TextView xmlns:android = "http://schemas.android.com/apk/res/android"
3       android:layout_width = "wrap_content"
4       android:layout_height = "wrap_content"
5       android:padding = "5dp"
6       android:textSize = "16sp"/>
```

（4）编写界面交互代码，在 MainActivity 中定义一个数组 programLang，存储 Spinner 下拉列表中显示的编程名称，并创建 ArrayAdapter 对象，调用 Spinner 控件的 setAdapter() 方法，将 Adapter 与列表控件关联起来，实现数据适配，具体代码如下。

程序清单 5-3：chart0501\java\com\example\administrator\chart0401\MainActivity.java

```
1       public class MainActivity extends AppCompatActivity {
2           private String[] programLang = {"Java", "C#",        →定义数组保存编程语言
3               "Python", "Javascript", "C++", "C"};
4           private Spinner spinner;
```

```
5          @Override
6          protected void onCreate(Bundle savedInstanceState) {
7              super.onCreate(savedInstanceState);
8              setContentView(R.layout.activity_main);
9              spinner = (Spinner) findViewById(R.id.spinner);      →获取 Spinner 对象
10         ArrayAdapter<String> stringArrayAdapter = new ArrayAdapter<String>
11                 (this, R.layout.simple_list_item, programLang);
12                                             →创建适配器对象
13              spinner.setAdapter(stringArrayAdapter);
                                               →为 Spinner 对象绑定适配器对象
14          }
15      }
```

上述代码中，第 2～3 行代码定义一个数组 programLang，存储 Spinner 下拉列表中显示的编程名称。第 10 行代码定义一个 ArrayAdapter 对象，并传入三个参数，第一个参数表示当前对象，第二个参数表示每个 Item 的布局文件，第三个参数表示数据源。第 13 行代码通过 setAdapter()方法为 Spinner 控件设置适配器。

上述代码的运行效果如图 5-2 所示。

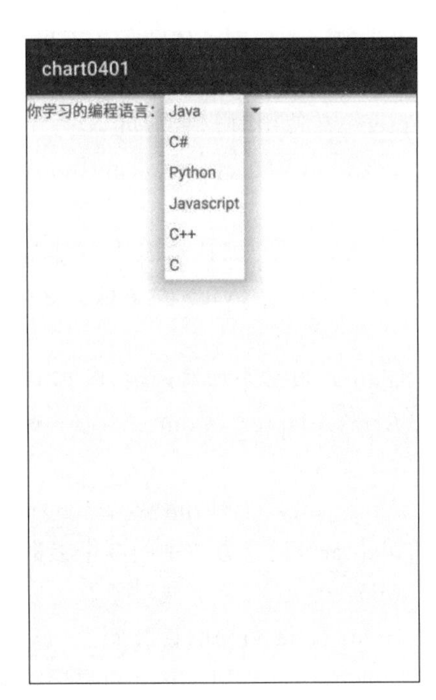

图 5-2　程序运行效果

5.1.2　Adapter 适配器

视频详解

在为 Spinner 控件添加数据的时候用到了数据适配器(Adapter)，数据适配器是数据和视图之间的桥梁，类似于一个转换器，将复杂的数据转换成用户设定的样式呈现出来，在 Android 系统中提供了多种适配器，它们之间的关系如图 5-3 所示。

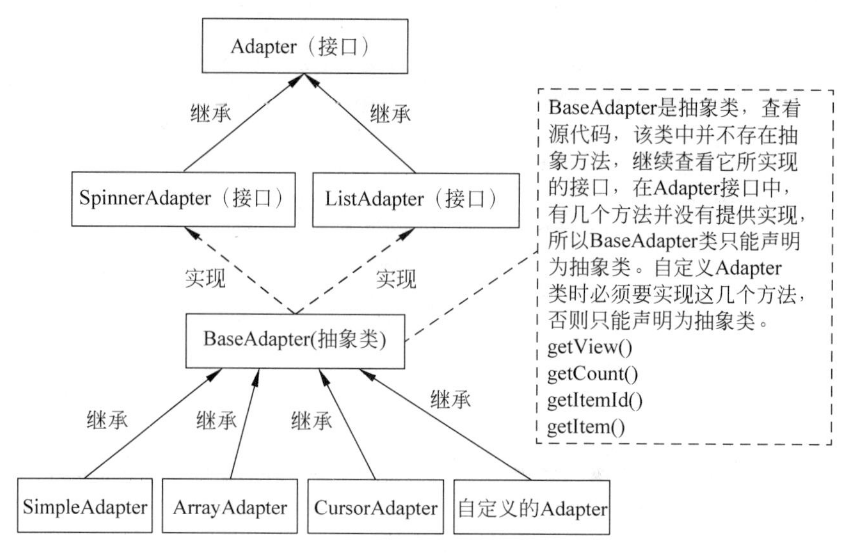

图 5-3 Adapter 相关的接口

从图 5-3 可知，实现 Adapter 相关接口的基类是 BaseAdapter，它实现了按口里的大部分方法，对于少数需要根据具体情景才能确定的方法则没有实现，因此它被声明为不能实例化。针对一些常用情景系统提供了三个子类：ArrayAdapter、SimpleAdapter、CursorAdapter。在使用时需要自定义一个 Adapter 继承自 BaseAdapter，并重写里面的 getView()、getCount()、geItem()、getItemId()方法。自定义 Adapter 的优点是可以使数据按照设定的形式显示，非常灵活，缺点是代码相对较多，需要用户自己重写各个方法。

(1) ArrayAdapter：默认情况下只能显示文本，如果想显示其他的 View 控件，比如 ImageView，需要重写 getView()方法。通常，ArrayAdapter 的数据源是数组或者集合的形式。

ArrayAdapter 有多个构造函数，以其中两个举例说明：

public ArrayAdapter (Context context, int resource, int textViewResourceId, T[] objects);

public ArrayAdapter (Context context, int resource, List<T> objects)。

ArrayAdapter()构造方法中的 5 个参数含义如下。

① Context：context 上下文对象。

② resource：Item 布局的资源 id。

③ textViewResourceId：Item 布局中相应 TextView 的 id。

④ T[] objects：需要适配的数组类型的数据。

⑤ List<T> objects：需要适配的 List 类型的数据。

(2) SimpleAdapter：可用于显示列表项相对复杂的列表，要求所有的列表项结构相同，显示样式相同。SimpleAdapter 继承自 BaseAdapter 并实现 BaseAdapter 的四个抽象方法并对其进行封装。因此在使用 SimpleAdapter 进行数据适配时，只需要在构造方法中传入相应的参数即可。

SimpleAdapter 的构造方法的具体信息如下。

public SimpleAdapter (Context context, List <? extends Map < String, ? >> data,int resource, string[] from, int[] to)

SimpleAdapter()构造方法中的 5 个参数的含义如下。

① context：表示上下文对象。

② data：数据集合，data 中的每一项对应 ListView 控件中的条目的数据。

③ resource：Item 布局的资源 id。

④ from：Map 集合中的 key 值。

⑤ to：Item 布局中对应的控件。

(3) BaseAdapter：本质上是一个抽象类，通常在自定义适配器时会继承 BaseAdapter。该类有四个抽象方法，在使用自定义适配器时需要实现这几个抽象方法，如表 5-1 所示。

表 5-1 BaseAdapter 类抽象方法

方 法 名	功 能 描 述
public int getCount()	获取 Item 条目的总数
public object getItem(int position)	根据 position(位置)获取某个 Item 对象
public long getItemId(int position)	根据 position(位置)获取某一个 Item 的 id
public View getView (int position, View convertView, ViewGroup parent)	获取相应 position 对应的 Item 视图

5.1.3 实战演练——收货地址

视频详解

通过对 Spinner 的学习，我们可以对“欢乐购商城”中修改地址页面进行实现。布局页面的代码如下。

程序清单 5-4：chart0502\app\src\main\res\layout\activity_main.xml

```
1    <?xml version = "1.0" encoding = "utf - 8"?>
2    < RelativeLayout xmlns:android = "http://schemas.android.com/apk/res/android"
3        xmlns:tools = "http://schemas.android.com/tools"
4        android:layout_width = "match_parent"
5        android:layout_height = "match_parent"
6        tools:context = ".MainActivity">
7        < EditText
8            android:id = "@ + id/edit_name"
9            android:layout_width = "match_parent"
10            android:layout_height = "wrap_content"
11            android:layout_marginTop = "5dp"
12            android:hint = "姓名" />
13        < EditText
14            android:id = "@ + id/edit_phone"
15            android:layout_width = "match_parent"
16            android:layout_height = "wrap_content"
17            android:layout_below = "@ + id/edit_name"
18            android:layout_marginTop = "5dp"
19            android:hint = "电话" />
```

```
20      <TextView
21          android:id="@+id/tv_proName"
22          android:layout_width="wrap_content"
23          android:layout_height="30dp"
24          android:layout_below="@+id/edit_phone"
25          android:layout_marginTop="5dp"
26          android:gravity="center"
27          android:text="省" />
28      <Spinner
29          android:id="@+id/proviceSpinner"          →Spinner 添加 id 属性
30          android:layout_width="wrap_content"       →为 Spinner 设置宽度
31          android:layout_height="30dp"              →为 Spinner 设置高度
32          android:layout_alignTop="@+id/tv_proName"
33          android:layout_marginLeft="10dp"
34          android:layout_toRightOf="@+id/tv_proName">
        </Spinner>
35      <TextView
36          android:id="@+id/tv_cityName"
37          android:layout_width="wrap_content"
38          android:layout_height="30dp"
39          android:layout_alignBottom="@+id/proviceSpinner"
40          android:layout_marginTop="5dp"
41          android:layout_toRightOf="@+id/proviceSpinner"
42          android:text="市"
43          android:gravity="center"
44          android:textSize="13sp" />
45      <Spinner
46          android:id="@+id/citySpinner"
47          android:layout_width="wrap_content"
48          android:layout_height="30dp"
49          android:layout_alignBottom="@+id/tv_cityName"
50          android:layout_toRightOf="@+id/tv_cityName"/>
51      <TextView
52          android:id="@+id/tv_countryName"
53          android:layout_width="wrap_content"
54          android:layout_height="30dp"
55          android:layout_alignBottom="@+id/citySpinner"
56          android:layout_marginTop="5dp"
57          android:layout_toRightOf="@+id/citySpinner"
58          android:gravity="center"
59          android:text="县" />
60      <Spinner
61          android:id="@+id/countrySpinner"
62          android:layout_width="wrap_content"
63          android:layout_height="30dp"
64          android:layout_alignBottom="@+id/tv_countryName"
65          android:layout_toRightOf="@+id/tv_countryName"/>
66      <EditText
67          android:id="@+id/edit_detailAddress"
68          android:layout_width="match_parent"
```

```
69              android:layout_height = "100dp"
70              android:layout_below = "@ + id/tv_proName"
71              android:layout_marginTop = "10dp"
72              android:background = "@drawable/textview_borders"
73              android:gravity = "top"
74              android:minLines = "5"
75              android:textSize = "16sp"
76              android:text = "中华路 58 号"/>
77      </RelativeLayout>
```

spinner_item. xml 的布局文件如下。

程序清单 5-5：char0502\app\src\main\res\layout\spinner_item. xml

```
1     <?xml version = "1.0" encoding = "utf - 8"?>
2     < TextView xmlns:android = "http://schemas.android.com/apk/res/android"
3         android:layout_width = "match_parent"
4         android:layout_height = "wrap_content"
5         android:padding = "5dp"
6         android:textSize = "14sp" />
```

MainActivity 页面的代码如下。

程序清单 5-6：char0502app\src\main\\res\layout\activity_main. xml

```
1     public class MainActivity extends AppCompatActivity {
2         private EditText edit_name;
3         private EditText edit_phone;
4         private Spinner proPinner;
5         private EditText detailAddress;
6         private Spinner citySpinner;
7         private Spinner countrySpinner;
8         private String provinceName;
9         private String cityName;
10        private String[ ] partCitys;
11        private String[ ] partCountry;
12        private String[ ] province  =  {"江西省", "河南省", "湖北省"};
13        @Override
14        protected void onCreate(Bundle savedInstanceState) {
15            super.onCreate(savedInstanceState);
16            setContentView(R.layout.activity_main);
17            initView();
18            initData();
19        }
20        private void initView() {        →创建方法初始化控件
21            edit_name  =  (EditText) findViewById(R.id.edit_name);
22            edit_phone  =  (EditText) findViewById(R.id.edit_phone);
23            proPinner  =  (Spinner) findViewById(R.id.proviceSpinner);
24            citySpinner  =  (Spinner) findViewById(R.id.citySpinner);
25            countrySpinner  =  (Spinner) findViewById(R.id.countrySpinner);
```

```
26            detailAddress = (EditText) findViewById(R.id.edit_detailAddress);
27        }
28        private void initData() {
29            edit_name.setText("李玲");      →为 edit_name 设置数据
30            edit_phone.setText("13789325643");      →为 edit_phone 设置数据
31            detailAddress.setText("气象路 58 号");
32            proPinner.setAdapter(new ArrayAdapter<String>(this, R.layout.spinner_
                                  item, province));
                                  →创建适配器对象,并为 proPinner 绑定该对象
33            proPinner.setOnItemSelectedListener(new AdapterView.OnItemSelectedListener() {
                                  →为 proPinner 设置项被选择的事件
34                @Override
35                public void onItemSelected(AdapterView<?> parent, View
36                view, int position, long id) {
37                    provinceName = province[position];
38                    switch (position) {
                                  →根据选择不同的省份,加载不同的地区
39                    case 0:
40                        partCitys = new String[]{"南昌", "九江", "赣州", "上饶"};
41                cityySpinner.setAdapter(new ArrayAdapter<String>(MainActivity.this, R.
                layout.spinner_item, partCitys));

                                  →创建地区的适配器对象,并为 citySpinner 绑定该对象
42
43                            break;
44                    case 1:
45                        partCitys = new String[]{"郑州", "南阳", "洛阳", "安阳"};
46                        citySpinner. setAdapter ( new ArrayAdapter < String >
(MainActivity.this,
47                            R.layout.spinner_item, partCitys));
48                            break;
49                      case 2:
50                        partCitys = new String[]{"武汉", "黄石", "阳新", "鄂州"};
51                            citySpinner.setAdapter(new ArrayAdapter<String>
                                  ( MainActivity. this, R. layout. spinner _ item,
                            partCitys));
52
53                            break;
54                    }
55                }
56                @Override
57                public void onNothingSelected(AdapterView<?> parent) {
58                }
59            });
60
61
62
```

```
63                          citySpinner. setOnItemSelectedListener ( new  AdapterView.
                                     OnItemSelectedListener() {
                                     →为 citySpinner 设置项被选择事件
64                  @Override
65                  public void onItemSelected(AdapterView<?> parent, View
66                        view, int position, long id) {
67                     cityName = partCitys[position];
68                     switch (position) {
                                           →根据选择不同的地区,加载不同的县
69                        case 0:
70                           partCountry = new String[]{"红谷滩新区", "东湖区", "西
                                                 湖区"};
71                           countrySpinner.setAdapter(new
                             →创建县的适配器对象,并为 countrySpinner 绑定该对象
72                             ArrayAdapter < String >(MainActivity.this, R.layout.
                                                 spinner_item,
73                               partCountry));
74                           break;
75                        case 1:
76                            partCountry  =  new String[]{"修水", "永修", "都昌", "德
  安"};
77                            countrySpinner. setAdapter ( new  ArrayAdapter  <  String  >
  (MainActivity.this,
                                R.layout.spinner_item, partCountry));
                          break;
78
79
80                        case 2:
81                            partCountry  =  new String[]{"信丰", "大余", "兴国", "于
都"};
82                           countrySpinner.setAdapter(new ArrayAdapter<String>(
83                           MainActivity.this, R .layout.spinner_item, partCountry));
84                           break;
85                     }
86                  }
87                  @Override
88                  public void onNothingSelected(AdapterView<?> parent) {
89                  }
90               });
91           }
92     }
```

上述程序运行效果如图 5-4 所示。

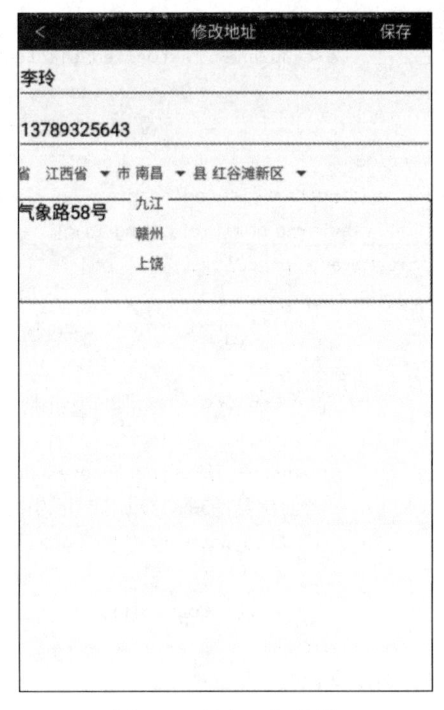

图 5-4　程序运行效果

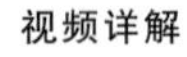
视频详解

5.2　普通列表 ListView

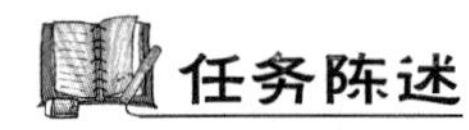

“欢乐购商城”项目中订单页面使用到了 ListView 控件，下面对修改订单页面做概要分析。

分析：在订单页面需要展示用户的订单信息，用户的订单信息展示的内容的格式是一致的，有订单商品标题、订单状态、总价和创建时间。对于列表展示并且每一项的格式一致适合采用 ListView 控件实现，如图 5-5 所示。

视频详解

5.2.1　ListView 控件

ListView 是使用非常广泛的一种列表控件，它以垂直列表的形式展示所有的数据项，并且能够根据列表的高度自适应屏幕显示，也是早期 Android 开发中用于列表界面开发最多的控件。早期的开发者为了拓展 ListView，对其进行继承和改写，也为其丰富了许多独特的功能，例如，经典的上拉加载新的列表数据，下拉刷新原有的列表数据等。

下面就以展示植物信息为例，讲解 ListView 控件的具体用法。

(1) 创建程序，创建名字为 chart0502 的应用程序。

(2) 导入图片，在 Android Studio 中切换到 Project 选项卡，选中程序中的 res 文件夹，

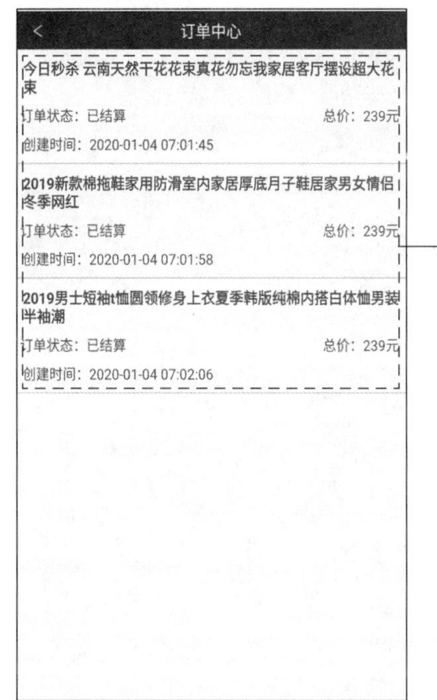

图 5-5 订单中心

右击选择 New→Directory 选项，创建一个名为 drawable-hdpi 的文件夹。将程序所需要的图片导入到文件夹。

(3) 设置布局文件，在 res/layout/activity_main. xml 文件中添加一个 ListView 控件，并对其相应属性进行设置，布局文件的核心代码如下。

程序清单 5-7：char0503\app\src\main\res\layout\activity_main. xml

```
1    <ListView
2        android:id = "@ + id/listView"
3        android:layout_width = "match_parent"
4        android:layout_height = "match_parent"
5        android:listSelector = "#A9A9A9"></ListView>
```

(4) 设置每个 Item 的布局，在 res/layout 文件夹中创建一个 Item 界面的布局文件 list_item. xml，在该文件中添加一个 ImageView 用来显示植物的图片，添加两个 TextView 分别用于展示植物的名称和植物的描述。完整的布局文件代码如下。

程序清单 5-8：chart0503\app\src\main\res\layout\list_item. xml

```
1    <?xml version = "1.0" encoding = "utf - 8"?>
2    <RelativeLayout xmlns:android = "http://schemas.android.com/apk/res/android"
3        android:layout_width = "match_parent"
4        android:layout_height = "match_parent">
5        <ImageView
6            android:id = "@ + id/iv"
7            android:layout_width = "80dp"
8            android:layout_height = "80dp"
9            android:layout_margin = "5dp"
```

```
10              android:background = "@drawable/b"
11              android:scaleType = "centerCrop" />
12          < TextView
13              android:id = "@ + id/tv_title"
14              android:layout_width = "match_parent"
15              android:layout_height = "wrap_content"
16              android:layout_alignTop = "@id/iv"
17              android:layout_toRightOf = "@id/iv"
18              android:text = "植物名称"
19              android:textSize = "18sp"
20              android:textStyle = "bold" />
21          < TextView
22              android:id = "@ + id/tv_depict"
23              android:layout_width = "match_parent"
24              android:layout_height = "wrap_content"
25              android:layout_below = "@id/tv_title"
26              android:layout_marginTop = "10dp"
27              android:layout_toRightOf = "@id/iv"
28              android:text = "植物的描述"
29              android:textSize = "16sp" />
30      </RelativeLayout >
```

（5）在 MainActivity 中创建一个内部类 MyBaseAdapter 继承自 BaseAdapter，并且在 MyBaseAdapter 中实现对 ListView 控件的适配。完整的代码如下。

程序清单 5-9：chart0503\java\com\example\administrator\chart0503\MainActivity. java

```
1     public class MainActivity extends AppCompatActivity {
2         private ListView mListView;
3         private int[] icons  =  {R.drawable.a, R.drawable.b,
4                 R.drawable.c, R.drawable.d, R.drawable.e, R.drawable.f};
5     private String[] titles  =  {"白鹃梅", "五色椒", "小檗",
6                 "李叶绣线菊", "多花兰", "獐耳细辛"};
7         private String[] depicts  =
8             {"白鹃梅又名茧子花、金瓜果等,是蔷薇科白鹃梅属灌木.",
9              "五色椒又名朝天椒,五彩辣椒,为辣椒 变种,味涩",
10             "小檗春日黄花簇簇,秋日红果满枝", "蔷薇科、绣线菊属灌木
11             ,高可达 3 米.", "多花兰是兰科兰属的一种附生植物",
12             "植物獐耳细辛,分布于我国辽宁、安徽、浙江、河南"};
13          @Override
14          protected void onCreate(Bundle savedInstanceState) {
15              super.onCreate(savedInstanceState);
16              setContentView(R.layout.activity_main);
17              mListView  =  (ListView) findViewById(R.id.listView);
18              MyBaseAdapter myBaseAdapter  =  new MyBaseAdapter();
19              mListView.setAdapter(myBaseAdapter);
20          }
21
```

```
22        private class MyBaseAdapter extends BaseAdapter {
23            @Override
24            public int getCount() {
25                return titles.length;
26            }
27            @Override
28            public Object getItem(int position) {
29                return titles[position];
30            }
31            @Override
32            public long getItemId(int position) {
33                return position;
34            }
35            @Override
36            public View getView(int position, View convertView, ViewGroup parent) {
37                View view = View.inflate(MainActivity.this, R.layout.list_item, null);
38                TextView title = (TextView) view.findViewById(R.id.tv_title);
39                TextView depict = (TextView) view .findViewById(R.id.tv_depict);
40                ImageView image = (ImageView) view .findViewById(R.id.iv);
41                image.setBackgroundResource(icons[position]);
42                title.setText(titles[position]);
43                depict.setText(depicts[position]);
44                return view;
45            }
46        }
47    }
```

上述代码中，在第3～12行代码定义了三个数组：titles、depicts和icons，分别存储植物的名称、植物的描述和植物的图片，并且这三个数组的长度是一致的。

第22～47行创建了一个MyBaseAdapter继承自BaseAdapter类，并重写了BaseAdapter类中的getCount()、getItem()、getItemId()和getView()方法。在getView()方法中通过inflate()方法将layout.list_item布局转换成视图对象。并通过findViewById()获取到layout.list_item布局中的各个控件，最后通过setText()和setBackgroundResource()方式将文本信息和图片信息展示出来。

该程序的运行效果如图5-6所示。

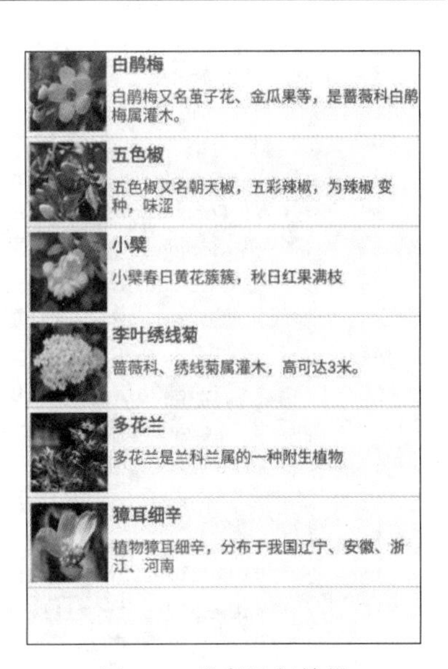

图5-6 程序运行效果

5.2.2 提升ListView运行效率

视频详解

在运行上述代码的时候，当ListView上加载的Item过多并且快速滑动ListView控件的时候，界面会出现卡顿。出现此状况的原因如下。

(1) 当不断滑动ListView控件时，就会不断创建Item对象。ListView控件在屏幕上显示多少个 Item，就会在适配器 MyBaseAdapter 中的 getView()中创建多少个 Item 对象。当滑动 ListView 控件时，滑出屏幕的 Item 对象会被销毁，新加载到屏幕上的 Item 会创建新的对象，因此在滑动的过程中就是在不断地创建和销毁 Item 对象。

(2) 在 getView()中不断执行 findViewById()方法初始化控件。每当创建一个 Item 对象就会加载一个 Item 布局，在加载布局的过程中会不断调用 findViewById()方法初始化控件。这些操作要消耗设备的内存，因此不断滑动 ListView 就是在不断地初始化控件，当占用内存过多的时候，就会出现程序内存溢出异常。

针对上述问题，需要对 ListView 控件进行优化，优化的目标是在 ListView 不断滑动过程中不再重复创建 Item 对象，减少内存的消耗。具体操作如下。

(1) 在 MainActivity 类中创建 ViewHolder 类。

程序清单 5-10：chart0503\java\com\example\administrator\chart0503\MainActivity. java

```
1    public class ViewHolder{
2        TextView tv_title;
3        TextView tv_depict;
4        ImageView image;
5    }
```

(2) 在 MyBaseAdapter 的 getView()方法中第二个参数 convertView 代表滑出屏幕的 Item 的缓存对象，当第一次加载 ListView 的时候创建 Item 对象，当滑动 ListView 控件的时候，在加载新的 Item 对象的时候可以复用缓存的 convertView 对象。在 getView()中进行优化，具体代码如下。

程序清单 5-11：chart0503\java\com\example\administrator\chart0503\MainActivity. java

```
1        public View getView(int position, View convertView, ViewGroup parent) {
2
3
4        ViewHolder viewHolder;
5        if (convertView == null) {
6            convertView = View.inflate(MainActivity.this, R .layout.list_item, null);
7            viewHolder = new ViewHolder();
8            viewHolder.tv_title = (TextView) convertView.findViewById(R.id.tv_title);
9            viewHolder.tv_depict = (TextView) convertView.findViewById(R.id.tv_depict);
10           viewHolder.image = (ImageView) convertView.findViewById(R.id.iv);
11           convertView.setTag(viewHolder); }
12       else {
13           viewHolder = (ViewHolder) convertView.getTag();
14       }
15       viewHolder.image.setBackgroundResource(icons[position]);
16       viewHolder.tv_title.setText(titles[position]);
17       viewHolder.tv_depict.setText(depicts[position]);
18       return convertView;
19      }
20   }
```

在上述代码中第 5～18 行中首先判断 convertView 对象是否为 null，如果是 null 表示是第一次加载 Item 项，需要使用 inflate()创建 Item 对象并通过 findViewById()方法找到控件。创建 viewHolder 对象，并将 Item 中的界面控件对象赋值给 viewHolder 对象的属性，最后通过 setTag()和 getTag()方法获取缓存在 convertView 对象中的 ViewHolder 对象。

5.2.3　ListView 的单击事件

视频详解

在使用 ListView 时，当单击某一个 Item 的时候，可能还需要进行其他操作。例如，在展示商品信息的时候，ListView 可展示商品概要信息，当单击某一商品时，需要跳转到另外一个页面展示商品的详细信息，此时就需要触发 ListView 中 Item 的单击事件。在 ListView 中常用的单击事件分为普通的单击事件和长按单击事件，下面以代码的形式说明如何使用单击事件。

在 onCreate( )方法中添加以下代码。

程序清单 5-12：chart0503\java\com\example\administrator\chart0503\MainActivity. java

```
1    mListView.setOnItemClickListener(new AdapterView.OnItemClickListener() {
2        @Override
3        public void onItemClick(AdapterView<?> parent, View view, int position,
4            long id) {
5            String content = "植物名称:" + titles[position] + ".描述:"
6                    + depicts[position];
7        Toast.makeText(MainActivity.this,
8        content, Toast.LENGTH_LONG).show();
9        }
10   });
```

上述代码中第 1 行使用 setOnItemClickListener()方法为 ListView 注册一个项被单击的监听器。当用户单击 ListVew 中的任意 Item 的时候就会回调 onItemClick()方法，在这个方法中通过 position 参数判断用户单击的是哪一个子项，然后获取到相应数组对应的项。最后通过 Toast 将植物的信息展示出来。

代码运行效果如图 5-7 所示。

对于 ListView 中的每个 Item，当长按某一个 Item 的时候可以触发长按事件，下面以代码的形式说明如何使用长按事件。

在 onCreate() 方法中添加以下代码。

程序清单 5-13：chart0503\java\com\example\administrator\chart0503\MainActivity. java

```
1    mListView.setOnItemLongClickListener(new AdapterView .OnItemLongClickListener() {
2        @Override
3        public boolean onItemLongClick(AdapterView<?>
4          parent, View view, int position,long id) {
5            AlertDialog alertDialog = new AlertDialog
6            .Builder(MainActivity.this).setMessage
7               ("确定要删除吗?")
```

```
8                   .setPositiveButton("确定", null)
9                   .setNegativeButton("取消", null)
10                  .create();
11             alertDialog.show();
12             return true;
13         }
14     });
```

在上述代码中第 1 行使用 setOnItemLongClickListener()方法为 ListView 注册一个监听器。当用户长按 ListVew 中的任意 Item 的时候就会回调 onItemLongClick()方法，在这个方法中通过创建对话框，提示用户是否确定要删除。

第 5～13 行代码创建一个对话框，通过调用 AlterDialog 的静态内部类 Builder 的方法创建 AlterDialog 对象。通过调用 Builder 类的 setMessage()方法设置提示的内容，通过 setPositiveButton()和 setPositiveButton()设置对话框的"确定"和"取消"按钮。通过调用 AlertDialog 对象的 show()方法显示该对话框。

上述代码运行效果如图 5-8 所示。

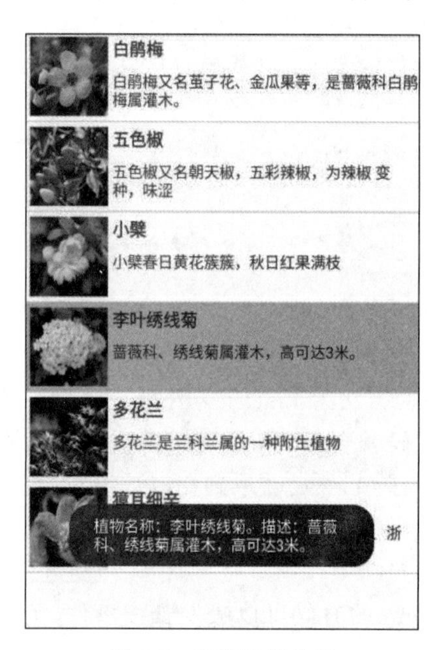

图 5-7 程序运行效果

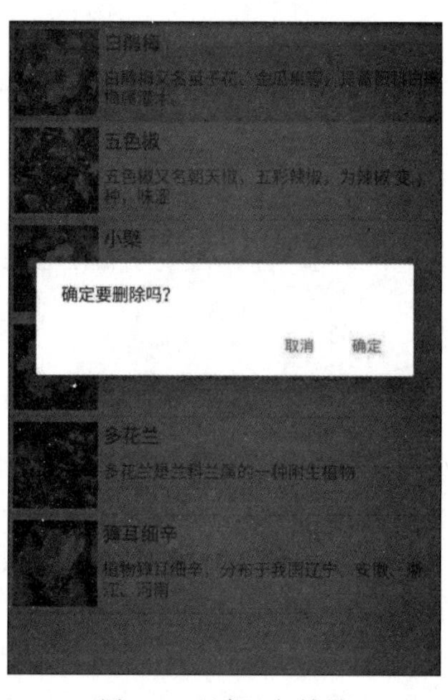

图 5-8 程序运行效果

视频详解

5.2.4 实战演练——订单中心

通过对 ListView 控件的学习，我们可以对"欢乐购商城"中的订单页面进行实现。布局页面代码如下。

程序清单 5-14：chart0504\app\src\main\res\layout\activity_main.xml

```
1    <?xml version = "1.0" encoding = "utf - 8"?>
2    < RelativeLayout xmlns:android = "http://schemas.android.com/apk/res/android"
```

```
3   xmlns:tools = "http://schemas.android.com/tools"
4       android:layout_width = "match_parent"
5       android:layout_height = "match_parent"
6       tools:context = ".MainActivity">
7       < ListView
8           android:id = "@ + id/order_ListView"
9           android:layout_width = "match_parent"
10          android:layout_height = "match_parent"/>
11   </RelativeLayout >
```

其中 Item 的布局文件如下。

程序清单 5-15：chart0504\ListView\res\layout\order_list_items.xml

```
1   <?xml version = "1.0" encoding = "utf - 8"?>
2   < RelativeLayout xmlns:android = "http://schemas.android.com/apk/res/android"
3       android:layout_width = "match_parent"
4       android:layout_height = "match_parent">
5
6       < TextView
7            android:id = "@ + id/goods_title"
8            android:layout_width = "match_parent"
9            android:layout_height = "wrap_content"
10           android:layout_margin = "8dp"
11           android:text = "云南天然干花花束真花勿忘我家居客厅摆设超大花束"
12           android:textSize = "15sp"
13           android:textStyle = "bold" />
14
15        < TextView
16           android:id = "@ + id/order_status"
17           android:layout_width = "wrap_content"
18           android:layout_height = "wrap_content"
19           android:layout_below = "@id/goods_title"
20           android:layout_marginLeft = "5dp"
21           android:text = "订单状态:已结算" />
22
23        < TextView
24           android:id = "@ + id/order_value"
25           android:layout_width = "wrap_content"
26           android:layout_height = "wrap_content"
27           android:layout_alignParentRight = "true"
28           android:layout_below = "@id/goods_title"
29           android:layout_marginRight = "8dp"
32           android:text = "总价:80 元" />
33
34        < TextView
35           android:id = "@ + id/order_time"
36           android:layout_width = "wrap_content"
37           android:layout_height = "wrap_content"
```

```
38            android:layout_below = "@id/order_status"
39            android:layout_margin = "8dp"
40            android:text = "创建时间:" />
41    </RelativeLayout>
```

MainActivity 页面的代码如下。

程序清单 5-16：chart0504\java\com\example\administrator\chart0504\MainActivity.java

```
1    public class MainActivity extends AppCompatActivity {
2        private ListView orderList;
3        private String[] orderTitle = {"云南天然干花花束真花勿忘我家居客厅摆设超大花
  束", "2019 秋装新款白色纯棉长袖 t 恤女装宽松春秋打底衫体恤大码上衣",
4                "鞋袋子装鞋子的收纳袋旅行鞋包收纳包防尘袋家用鞋罩束口鞋袋鞋套"};
5        private String[] orderValue = {"145", "213", "89"};
6        @Override
7        protected void onCreate(Bundle savedInstanceState) {
8            super.onCreate(savedInstanceState);
9            setContentView(R.layout.activity_main);
10           orderList = (ListView) findViewById(R.id.order_ListView);
11           orderList.setAdapter(new myListAdapter());
12       }
13       private class myListAdapter extends BaseAdapter {
14           private ViewHolder viewHolder;
15           @Override
16           public int getCount() {
17               return orderTitle.length;
18           }
19           @Override
20           public Object getItem(int position) {
21               return orderTitle[position];
22           }
23           @Override
24           public long getItemId(int position) {
25               return position;
26           }
27           @Override
28           public View getView(int position, View convertView, ViewGroup parent) {
29               if (convertView == null) {
30                   convertView = View.inflate(MainActivity.this,
                         R.layout.order_list_items, null);
31                   viewHolder = new ViewHolder();
32                   viewHolder.orderTitle = (TextView) convertView.
                         findViewById(R.id.goods_title);
33                   viewHolder.totalValue = (TextView)
                         convertView.findViewById(R.id.order_value);
34                   viewHolder.orderTime = (TextView)
                         convertView.findViewById(R.id.order_time);
35                   convertView.setTag(viewHolder);
```

```
36                } else {
37                    viewHolder = (ViewHolder) convertView.getTag();
38                }
39                SimpleDateFormat df = new SimpleDateFormat("yyyy-MM-dd HH:mm:ss");
40                viewHolder.orderTitle.setText(orderTitle[position]);
41                viewHolder.totalValue.setText("总价:" + orderValue[position] + "元");
42                viewHolder.orderTime.setText("创建时间:" + df.format(new Date()));
43                return convertView;
44            }
45
46            class ViewHolder {
47                TextView orderTitle;
48                TextView totalValue;
49                TextView orderTime;
50            }
51        }
52    }
```

上述程序运行结果如图 5-9 所示。

图 5-9　程序运行效果

5.3　网格列表 GridView

视频详解

任务陈述

“欢乐购商城”首页使用 GridView 控件使推荐商品按照网格的形式展示，下面对首页中的推荐商品列表做概要分析。

分析：在首页中需要展示推荐商品，展示的商品只需要展示概要信息并尽可能多展示一些商品，如果采用列表的形式展示，每行只能展示一个列表项。为了多展示一些商品信息，在项目中采用 GridView 控件展示商品，使每行可展示两个列表项，如图 5-10 所示。

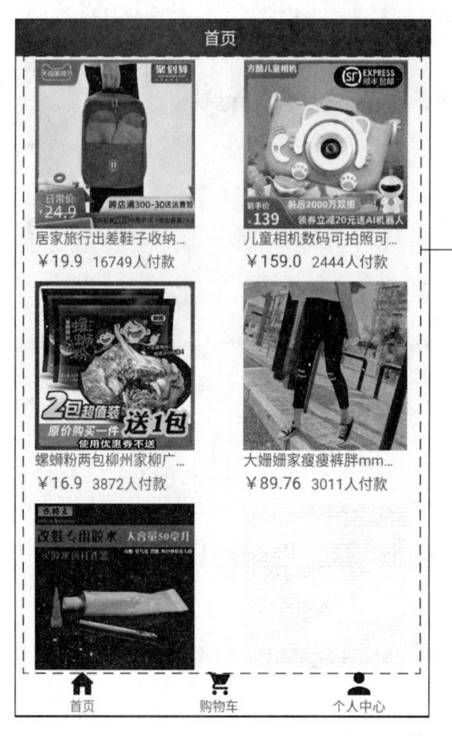

图 5-10　列表展示商品

相关知识

视频详解

5.3.1　GridView 控件

前面所学的 Spinner 和 ListView 显示列表数据时，都是以垂直方向显示，每行只显示一个列表项。无法实现一行显示多个列表项的效果。如果要实现多行多列的展示效果，可使用 GridView 控件。GridView 可将界面划分为若干个网格，可以设置每一行所能显示的列表项数量，然后根据总的列表项数来计算一共有多少行。例如，总共有 15 个列表项，每行存放 4 个列表项，则包含 4 行，如果每行存放 3 个列表项，则包含 5 行。使用 GridView 时，关键属性如表 5-2 所示。

表 5-2　GridView 关键属性

属　性　名	功 能 描 述
android：numColumns	每行中列的数量
android：horizontalSpacing	设置两个列表项之间的水平间距
android：verticalSpacing	设置两个列表项之间的垂直间距

下面以展示植物信息的例子讲解 GridView 的用法。

（1）创建程序，创建名字为 chart0503 的应用程序。

（2）导入图片，在 Android Studio 中切换到 Project 选项卡，选中程序中 res 文件夹，右击选择 New→Directory 选项，创建一个名为 drawable-hdpi 的文件夹。将程序所需要的图片导入到文件夹。

（3）设置布局文件，在 res/layout/activity_main. xml 文件中添加一个 ListView 控件，并对相应属性进行设置。布局文件的核心代码如下。

程序清单 5-17：chart0505\app\src\main\res\layout\activity_main. xml

```
1    <GridView
2        android:id = "@ + id/gridView"
3        android:layout_width = "match_parent"
4        android:layout_height = "wrap_content"
5        android:horizontalSpacing = "3dp"
6        android:numColumns = "4"
7        android:verticalSpacing = "3dp"></GridView>
```

（4）设置每个 Item 的布局，在 res/layout 文件夹中创建一个 Item 界面的布局文件 grid_item. xml，在该文件中添加一个 ImageView 用来展示植物的图片，添加一个 TextView 用于展示植物的名称。完整的布局文件代码如下。

程序清单 5-18：chart0505\app\src\main\res\layout\grid_item. xml

```
1    <?xml version = "1.0" encoding = "utf - 8"?>
2    <LinearLayout xmlns:android = "http://schemas.android.com/apk/res/android"
3        android:layout_width = "wrap_content"
4        android:layout_height = "wrap_content"
5        android:background = "#cccccc"
6        android:gravity = "center_horizontal"
7        android:orientation = "vertical"
8        android:padding = "3dp">
9        <ImageView
10           android:id = "@ + id/image"
11           android:layout_width = "100dp"
12           android:layout_height = "90dp"
13           android:scaleType = "fitXY"
14           android:src = "@drawable/a" />
15        <TextView
16           android:id = "@ + id/tv_title"
17           android:layout_width = "wrap_content"
18           android:layout_height = "wrap_content"
19           android:layout_below = "@id/image"
20           android:text = "铁茉莉"
21           android:textSize = "15sp" />
22    </LinearLayout>
```

（5）在 MainActivity 中创建一个 SimpleAdapter 对象，gridView 对象通过 setAdapter()方法将 SimpleAdapter 对象与 gridView 控件关联起来，实现数据适配。完整的代码如下。

程序清单 5-19：chart0505\java\com\example\administrator\chart0505\MainActivity. java

```
1    public class MainActivity extends AppCompatActivity {
2        private GridView gridView;
3        private List < Map < String, Object >> datas = new ArrayList < Map < String, Object >>();
4
5        private int[ ] icons = {R.drawable.a, R.drawable.b, R .drawable.c,
6                R.drawable.d, R.drawable.e,R.drawable.f, R.drawable.g,
7                R.drawable.h, R.drawable.i, R.drawable.j, R.drawable.k,
8                R.drawable.l};
9    private String[ ] titles = {"白鹃梅", "五色椒", "小檗",
10               "李叶绣菊", "多花兰", "獐耳细辛", "铁茉莉", "三色狸藻", "东洋菊",
11               "大花皇冠", "苹果花", "三弄芙蓉"};
12        @Override
13        protected void onCreate(Bundle savedInstanceState) {
14            super.onCreate(savedInstanceState);
15            setContentView(R.layout.activity_main);
16            gridView = (GridView) findViewById(R.id.gridView);
17            initData();
18            SimpleAdapter simpleAdapter = new SimpleAdapter
19                (this, datas, R.layout.grid_item, new
20            String[ ]{"image", "titles"}, new int[ ]{R
21                            .id.image, R.id.tv_title});
22            gridView.setAdapter(simpleAdapter);
23        }
24        private void initData() {
25            for (int i = 0; i < titles.length; i++) {
26                Map < String, Object > item = new
27                        HashMap < String, Object >();
28                item.put("titles", titles[i]);
29                item.put("image", icons[i]);
30                datas.add(item);
31            }
32        }
33    }
```

上述代码中，在第 3 行定义一个 List 集合对象，用于存储植物标题和植物图片的信息。第 5 行和第 9 行分别定义两个数组用于保存植物的标题和图片。

第 24～32 行定义一个方法 initData()将数组中的数据以键值对的形式保存在 Map 对象中，将每个 Map 对象保存到 List 集合对象中。

第 18 行定义 SimpleAdapter 对象，调用其构造函数，并传入 5 个相应的参数。第一个参数表示上下文对象，第二个参数表示数据源，第三个参数表示 Item 布局资源的 id，第四个参数表示 Map 集合中的 key 值，第五个参数表示 Item 布局中 key 值所代表的 value 所绑定的控件。

第 22 行通过 setAdapter()将 SimpleAdapter 对象设置给 gridView 实现数据适配。

上述程序运行效果如图 5-11 所示。

在上述案例中，GridView 控件使用的是 SimpleAdapter 对象实现了数据的适配，如果

图 5-11 程序运行效果

Item 要实现更加复杂的布局，可采用继承 BaseAdapter 抽象类的方法来实现。BaseAdapter 的使用已在 ListView 中做过相应展示，在这里不再演示。对于 GridView 中每个 Item 也可以设置其单击事件，包括普通的单击事件和长按事件，其用法已在 ListView 中做过相应的展示，在这里不再演示。

5.3.2 实战演练——首页商品列表

视频详解

通过对 GridView 的学习，我们可以对“欢乐购物商城”中首页进行实现。在首页中需要使用 ScrollView 垂直滚动条，但 GridView 在滚动条中需要重新计算高度才能在滚动条中使用。因此需要自定义一个类继承 GridView，重写 onMeasure 方法，在子布局中根据自己组件大小重新计算大小，自定义 GridView 类代码如下。

程序清单 5-20：chart0506\java\com\example\administrator\chart0506\MyGridView.java

```
1  public class MyGridView extends GridView {
2      public MyGridView(Context context) {
3          super(context);
4      }
5      public MyGridView(Context context, AttributeSet attrs) {
6          super(context, attrs);
7      }
8      public MyGridView(Context context, AttributeSet attrs, int defStyleAttr) {
9          super(context, attrs, defStyleAttr);
10     }
11     /**
12      * 重写 onMeasure 方法，重新计算高度，达到使 GridView 适应 ScrollView 的效果
```

```
13        * @param widthMeasureSpec   宽度测量规则
14        * @param heightMeasureSpec 高度测量规则
15        */
16       @Override
17       protected void onMeasure(int widthMeasureSpec, int heightMeasureSpec) {
18           //Integer.MAX_VALUE:表示 int 类型能够表示的最大值,值为(2^31) - 1
19           //>> 2:右移 N 位,相当于除以 2 的商的 N 次幂
20           //MeasureSpec.AT_MOST:子布局可以根据自己的大小选择任意大小的模式
21           int newHeightMeasureSpec = MeasureSpec.makeMeasureSpec(Integer.MAX_VALUE >> 2,
MeasureSpec.AT_MOST);
22           super.onMeasure(widthMeasureSpec, newHeightMeasureSpec);
23       }
24 }
```

其中，布局页面的代码如下。

程序清单 5-21：chart0506\app\src\main\res\layout\activity_main.xml

```
1     <?xml version = "1.0" encoding = "utf - 8"?>
2     <ScrollView xmlns:android = "http://schemas.android.com/apk/res/android"
3         android:layout_width = "match_parent"
4         android:layout_height = "match_parent"
5         android:gravity = "center">
6         <LinearLayout
7             android:layout_width = "match_parent"
8             android:layout_height = "match_parent"
9             android:orientation = "vertical">
10            <ImageView
11                android:layout_width = "match_parent"
12                android:layout_height = "150dp"
13                android:scaleType = "fitXY"
14                android:src = "@drawable/jhs" />
15           <TableLayout
16                android:id = "@ + id/table_smallIcon"
17                android:layout_width = "match_parent"
18                android:layout_height = "wrap_content"
19                android:gravity = "center">
20                <TableRow>
21                    <ImageView
22                        android:id = "@ + id/a"
23                        android:layout_width = "73dp"
24                        android:layout_height = "73dp"
25                        android:clickable = "true"
26                        android:src = "@drawable/a" />
27                    <ImageView
28                        android:id = "@ + id/b"
29                        android:layout_width = "73dp"
32                        android:layout_height = "73dp"
33                        android:src = "@drawable/b" />
```

```
34                  < ImageView
35                      android:id = "@ + id/c"
36                      android:layout_width = "73dp"
37                      android:layout_height = "73dp"
38                      android:src = "@drawable/c" />
39                  < ImageView
40                      android:id = "@ + id/d"
41                      android:layout_width = "73dp"
42                      android:layout_height = "73dp"
43                      android:src = "@drawable/d" />
44                  < ImageView
45                      android:id = "@ + id/e"
46                      android:layout_width = "73dp"
47                      android:layout_height = "73dp"
48                      android:src = "@drawable/e" />
49              </TableRow >
50              < TableRow >
51                  < ImageView
52                      android:id = "@ + id/f"
53                      android:layout_width = "73dp"
54                      android:layout_height = "73dp"
55                      android:src = "@drawable/f" />
56                  < ImageView
57                      android:id = "@ + id/g"
58                      android:layout_width = "73dp"
59                      android:layout_height = "73dp"
60                      android:src = "@drawable/g" />
61                  < ImageView
62                      android:id = "@ + id/h"
63                      android:layout_width = "73dp"
64                      android:layout_height = "73dp"
65                      android:src = "@drawable/h" />
66                  < ImageView
67                      android:id = "@ + id/i"
68                      android:layout_width = "73dp"
69                      android:layout_height = "73dp"
70                      android:src = "@drawable/i" />
71                  < ImageView
72                      android:id = "@ + id/k"
73                      android:layout_width = "73dp"
74                      android:layout_height = "73dp"
75                      android:src = "@drawable/k" />
76                  </TableRow >
77              </TableLayout >
78      < com.example.administrator.taobao.adapter.MyGridView
79                  android:id = "@ + id/gridView"
80                  android:layout_width = "match_parent"
81                  android:layout_height = "wrap_content"
82                  android:layout_marginTop = "5dp"
```

```
83                  android:gravity = "center"
84                  android:horizontalSpacing = "5dp"
85                  android:numColumns = "2"
86                  android:verticalSpacing = "5dp">
87      </com.example.administrator.taobao.adapter.MyGridView>
88          </LinearLayout>
89      </ScrollView>
```

其中，Item 的布局文件如下。

程序清单 5-22：chart0506\app\src\main\res\layout\goods_detail_item.xml

```
1     <?xml version = "1.0" encoding = "utf - 8"?>
2     <RelativeLayout xmlns:android = "http://schemas.android.com/apk/res/android"
3         android:layout_width = "wrap_content"
4         android:layout_height = "wrap_content">
5         <ImageView
6             android:id = "@ + id/iv_iamge"
7             android:layout_width = "160dp"
8             android:layout_height = "160dp"
9             android:layout_marginTop = "4dp" />
10         <TextView
11             android:id = "@ + id/tv_title"
12             android:layout_width = "160dp"
13             android:layout_height = "wrap_content"
14             android:layout_below = "@ + id/iv_iamge"
15             android:ellipsize = "end"
16             android:maxLines = "1"
17             android:text = "标题测试数据"
18             android:textSize = "14dp" />
19         <TextView
20             android:id = "@ + id/tv_prices"
21             android:layout_width = "wrap_content"
22             android:layout_height = "wrap_content"
23             android:layout_below = "@id/tv_title"
24             android:text = " ¥ 10"
25             android:textColor = " # f00"
26             android:textSize = "16sp" />
27         <TextView
28             android:id = "@ + id/tv_numbers"
29             android:layout_width = "wrap_content"
32             android:layout_height = "wrap_content"
33             android:layout_alignBottom = "@id/tv_prices"
34             android:layout_marginLeft = "10dp"
35             android:layout_toRightOf = "@id/tv_prices"
36             android:text = "1000 人付款" />
37     </RelativeLayout>
```

定义一个适配器类,代码如下。

程序清单 5-23：chart0506\java\com\example\administrator\chart0506\GoodsAdapter. Java

```
1    public class GoodsAdapter extends BaseAdapter {
2        private Context context;
3        private List < Good > goodsList;
4        private ViewHolder viewHolder;
5        public GoodsAdapter(Context context, List < Good > goodsList) {
6            this.context = context;
7            this.goodsList = goodsList;
8        }
9        @Override
10        public int getCount() {
11            return goodsList.size();
12        }
13        @Override
14        public Object getItem(int position) {
15            return goodsList.get(position);
16        }
17        @Override
18        public long getItemId(int position) {
19            return position;
20        }
21        @Override
22        public View getView(int position, View convertView, ViewGroup parent) {
23            if (convertView == null) {
24                convertView = LayoutInflater.from(context).
                      inflate(R.layout.goods_detail_item, parent, false);
25                viewHolder = new ViewHolder();
26                viewHolder.image = convertView.findViewById(R.id.iv_iamge);
27                viewHolder.textTitle = convertView.findViewById(R.id.tv_title);
28                viewHolder.textPrice = convertView.findViewById(R.id.tv_prices);
29                viewHolder.textNumber = convertView.findViewById(R.id.tv_numbers);
30                convertView.setTag(viewHolder);
31            } else {
32                viewHolder = (ViewHolder) convertView.getTag();
33            }
34     viewHolder.image.setBackgroundResource(goodsList.get(position).getImage());
35     viewHolder.textTitle.setText(goodsList.get(position).getName());
36            viewHolder.textPrice.setText(" ¥ " + goodsList.get(position).getPrice());
37     viewHolder.textNumber.setText(goodsList.get(position).getVolume() + "人付款");
38            return convertView;
39          }
40        class ViewHolder {
41            ImageView image;
42            TextView textTitle;
43            TextView textPrice;
44            TextView textNumber;
45        }
46     }
```

MainActivity 页面的代码如下。

程序清单 5-24：chart0506\java\com\example\administrator\chart0506\MainActivity. java

```
1     public class MainActivity extends AppCompatActivity {
2
3         private MyGridView gridView;
4         private List< Good> goodList = new ArrayList< Good>();
5         private int[] pictrue = {R.drawable.a000, R.drawable.a001, R.drawable.a002,
6                 R.drawable.a003,R.drawable.a004, R.drawable.a005, R.drawable.a006,
7                 R.drawable.a007, R.drawable.a008,R.drawable.a009};
8         private String[] title = {"2019 秋装新款白色纯棉长袖 t 恤女装宽松春秋打底衫体恤
          大码上衣",
              "2019 初秋新款潮韩版洋气女装宽松秋季短款衬衫女秋装方领长袖上衣",
9             "纯棉长袖 T 恤女上衣 2019 新款竹节棉宽松秋衣女装薄款初秋打底衫潮",
              "半高领打底衫女装秋冬薄款洋气莫代尔堆堆领长袖 t 恤内搭针织上衣",
10            "女装 2019 新款潮超火 cec 短袖女春夏季港味宽松百搭上衣 T 恤 ins 洋气",
              "夏大大胖 mm 韩版宽松秋装新款女装大码 2019 牛仔套装外套减龄连衣裙",
11            "大姗姗家瘦瘦裤胖 mm 韩版大码女装秋季百搭外穿打底裤网红款小脚裤",
              "中国大陆','夏装紧身纯白色 T 恤女短袖修身女装女士纯棉 2019 新款上衣潮打底衫",
12            "纯棉长袖 T 恤女上衣 2019 新款竹节棉宽松秋衣女装薄款初秋打底衫潮",
              "纯棉红色 v 领 T 恤女短袖修身夏女装 2019 新款潮体恤紧身上衣黑色短款",};
13         private String[] prices = {"39.9", "29.87", "29", "56", "20", "138", "97", "15",
  "78", "30"};
14         private String[] number = {"87816", "37384", "62165", "43221", "4561", "43211",
15                 "3011","764", "7656", "4567"};
16         @Override
17         protected void onCreate(Bundle savedInstanceState) {
18             super.onCreate(savedInstanceState);
19             setContentView(R.layout.activity_main);
20             initView();
21         }
22         private void initView() {
23             for (int i = 0; i < 10; i++) {
24                 Good good = new Good();
25                 good.setImage(pictrue[i]);
26                 good.setName(title[i]);
27                 good.setPrice(Double.valueOf(prices[i]));
28                 good.setVolume(number[i]);
29                 goodList.add(good);
30             }
31             gridView = (MyGridView) findViewById(R.id.gridView);
32             gridView.setAdapter(new GoodsAdapter(MainActivity.this, goodList));
33             gridView.setOnItemClickListener(new AdapterView.OnItemClickListener() {
34                 @Override
35                 public void onItemClick(AdapterView<?> parent, View view, int position,
  long id) {
36
37                 }
38             });
39         }
40     }
```

上述程序运行效果如图 5-12 所示。

图 5-12 程序运行效果

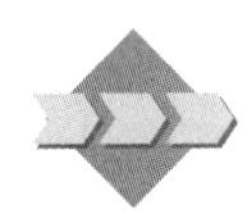

5.4 增强列表 RecyclerView

视频详解

任务陈述

“欢乐购商城”项目中商品列表页面使用了 RecyclerView 控件，下面对商品列表页面做概要分析。

分析：在商品列表页面中，需要展示分类商品的概要信息，包括商品图片、商品标题、商品产地和商品所属店铺等。每个列表项的布局较复杂，在项目中采用 RecyclerView 控件实现商品列表展示功能，如图 5-13 所示。

相关知识

5.4.1 RecyclerView 控件

视频详解

RecyclerView 是 Android 5.0 以后提供的更为强大的列表控件，不仅可以轻松实现 ListView、GridView 的效果，还优化了之前列表控件存在的各种不足之处，是目前官方推荐使用的列表控件。

RecyclerView 控件可通过 LayoutManager 类设置列表项的布局管理器，用于控制列表项的整体布局，常见的布局管理器有：LinearLayoutManager（线性布局管理器，可以控制列表项按照水平从左到右摆放，或者垂直从上到下摆放）、GridLayoutManager（网格布局管理器，按照若干行或者若干列来摆放列表项）、StaggeredGridLayoutManager（交错网格布

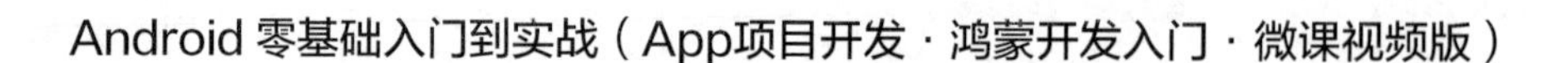

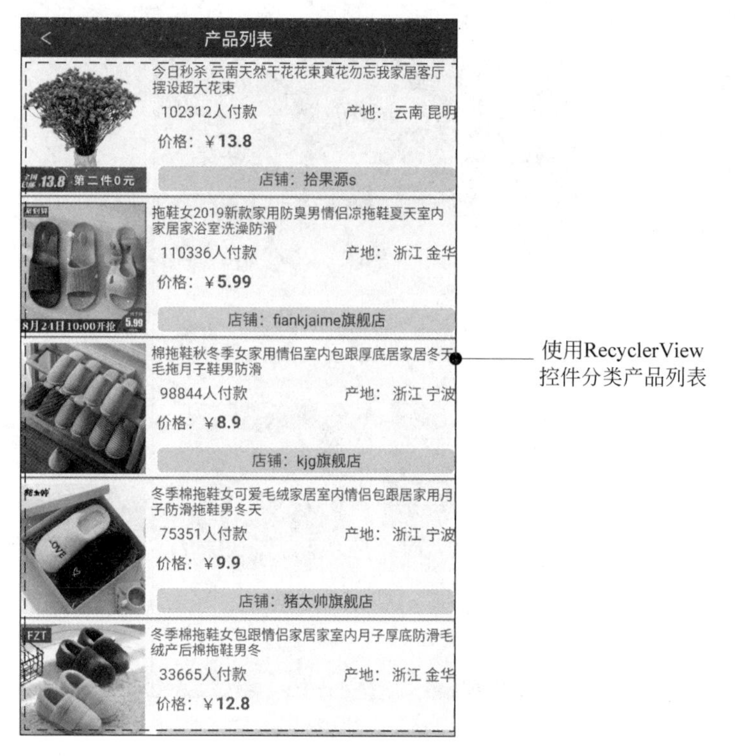

图 5-13　商品列表展示

局管理器，可以实现瀑布流效果，网格不是整齐的而是有所偏移）。

RecyclerView 控件使用 RecyclerView. Adapter 适配器，该适配器将 BaseAdapter 中的 getView()方法拆分为 onCreateViewHolder()方法和 onBindViewHolder()方法，强制使用 ViewHodler 类，使代码编写更加规范，提高性能。RecyclerView 控件复用 Item 对象是由控件自身完成，提高了灵活性。

为了使 RecyclerView 在所有的 Android 版本上都可以使用，Android 团队将其定义在兼容包中，因此如果想使用 RecyclerView 控件首先需要在项目的 build. gradle 中添加相关的依赖库。其中，RecyclerView 的版本可根据本机上安装的 Android 版本变化。

接下来通过加载植物信息的例子讲解 RecyclerView 的使用，并实现垂直列表、水平列表和网格列表效果。

(1) 创建程序，创建名字为 chart0507 的应用程序。

(2) 导入图片，在 Android Studio 中切换到 Project 选项卡，选中程序中 res 文件夹，右击，选择 New→Directory 选项，创建一个名为 drawable-hdpi 的文件夹，将程序所需要的图片导入文件夹。

(3) RecyclerView 是 Android 5.0 新增的控件，为了让 RecyclerView 在所有的 Android 版本中都能使用，RecyclerView 被定义在 support 库中，因此在使用时，需要在项目的 build. gradle 中添加相应的依赖库。打开 app\build. gradle 文件，在 dependencies{}节点中添加 com. android. support：recyclerview-v7 库。

```
dependencies {
    implementation fileTree(include: ['*.jar'], dir: 'libs')
    implementation 'com.android.support:appcompat-v7:28.0.0'
    implementation 'com.android.support:recyclerview-v7:28.0.0'
    …
}
```

注意：

① 在 Android studio 新版中 RecyclerView 类可以直接使用，只需要在布局文件中使用 androidx.recyclerview.widget.RecylerView 标签即可。如果不能使用，则需要导入 RecylerView 所在库资源。

② 添加的 com.android.support：recyclerview-v7 库的版本需要 com.android.support:appcompat 库的版本一致，否则会报错。

③ 如果在 app\build.gradle 文件中添加 recyclerview 依赖后，还是不能导入，可以在项目内 build.gradle 文件中 repositories 方法中添加 mavenCentral()方法，再进行同步导入。

(4) 设置布局文件，在 res\layout\activity_main.xml 文件中添加一个 RecyclerView 控件，并设置相应属性，布局文件的核心代码如下。

程序清单 5-25：char0507\app\src\main\res\layout\activity_main.xml

```
1    <android.support.v7.widget.RecyclerView
2        android:id="@+id/recyclerView"
3        android:layout_width="match_parent"
4        android:layout_height="match_parent">
5    </android.support.v7.widget.RecyclerView>
```

(5) 设置每个 Item 的布局，由于每个 Item 的布局效果和 GridView 中的显示效果一样，把上一个项目的图片复制过来，新建一个布局文件命名为 grid_item.xml。把上一节例子中的 Item 布局复制过来。

(6) 在 MainActivity 中获取 RecyclerView 对象并进行数据适配，最终将数据显示在列表界面中。完整的代码如下。

程序清单 5-26：chart0507\java\com\example\administrator\chart0507\MainActivity.java

```
1    public class MainActivity extends AppCompatActivity {
2
3        private int[] icons = {R.drawable.a, R.drawable.b, R
4                .drawable.c, R.drawable.d, R.drawable.e,
5                R.drawable.f, R.drawable.g, R.drawable.h, R
6                .drawable.i, R
7                .drawable.j, R.drawable.k, R.drawable.l};
8        private String[] titles = {"白鹃梅", "五色椒", "小檗",
9                "李叶绣菊", "多花兰", "獐耳细辛", "铁茉莉", "三色狸藻", "东洋菊",
10                "大花皇冠", "苹果花", "三弄芙蓉"};
11        private RecyclerView recyclerView;
12        @Override
```

```
13        protected void onCreate(Bundle savedInstanceState) {
14            super.onCreate(savedInstanceState);
15            setContentView(R.layout.activity_main);
16            recyclerView = (RecyclerView) findViewById(R.id
17                    .recyclerView);
18            LinearLayoutManager linearLayoutManager = new
19                    LinearLayoutManager(MainActivity.this);
20            recyclerView.setLayoutManager(linearLayoutManager);
21            MyAdapter myAdapter = new MyAdapter();
22            recyclerView.setAdapter(myAdapter);
23        }
24        public class MyAdapter extends RecyclerView
25                .Adapter<MyAdapter.MyViewHolder>{
26            @Override
27            public MyViewHolder onCreateViewHolder(ViewGroup viewGroup, int viewType) {
28                  MyViewHolder myViewHolder = new MyViewHolder
                    (LayoutInflater.from(MainActivity.this)
29                        .inflate(R.layout.recycler_item, viewGroup, false));
30                return myViewHolder;
31            }
32            @Override
33            public void onBindViewHolder(MyViewHolder myViewHolder, int position) {
34            myViewHolder.image.setImageResource
35                        (icons[position]);
36                myViewHolder.titleView.setText
37                        (titles[position]);
38            }
39            @Override
40            public int getItemCount() {
41                return titles.length;
42            }
43            public class MyViewHolder extends RecyclerView
44                    .ViewHolder {
45                ImageView image;
46                TextView titleView;
47          public MyViewHolder(View itemView) { super(itemView);
48            image = (ImageView) itemView.findViewById (R.id.image);
49            titleView = (TextView) itemView.findViewById(R.id.tv_title);
50                }
51            }
52        }
53    }
```

上述代码中,在第3~10行定义两个数组对象,用于存储植物标题和植物图片的信息。第16~22行首先获取recyclerView对象并为recyclerView创建布局管理器,其中linearLayoutManager对象用于指明recyclerView为线性布局方式,默认为垂直方向的线性布局。通过setLayoutManager()将linearLayoutManager对象设置给recyclerView对象。之后创建MyAdapter对象,最后通过setAdapter()方法将适配器MyAdapter对象设置到

recyclerView 控件上，这样 recyclerView 就完成了数据的适配。

第 24～42 行为 recyclerView 创建适配器对象，新建一个内部类 MyAdapter 继承自 RecyclerView. Adapter 并将 ViewHolder 泛型参数指定为 MyAdapter. MyViewHolder，其中 MyViewHolder 是 MyAdapter 中定义的一个内部类。此外，RecyclerView. Adapter 类是一个抽象类，还需要实现里面的三个抽象方法：onCreateViewHolder()、onBindViewHolder()、getItemCount()。onCreateViewHolder()中创建 MyViewHolder 对象，并将 recycler_item 布局文件转换成对象之后传入到 MyViewHolder 构造函数中，最后将 myViewHolder 对象返回。onBindViewHolder()用于对 recyclerView 每个 Item 项的数据进行赋值，会在每个子项滚动到屏幕内的时候执行，并通过 position 获取 Item 的索引。getItemCount()是获取列表总条目数。

第 43～51 行定义 MyViewHolder 类继承自 RecyclerView. ViewHolder。在该类中获取 Item 界面的控件，在 MyViewHolder 的构造函数中需要传入一个参数 itemView，这个参数通常是 recyclerView 中 Item 项的最外层布局，对于 itemView 对象通过 findViewById()获取到 Item 布局中的每一个控件对象。

图 5-14 程序运行效果

上述程序运行效果如图 5-14 所示。

5.4.2 RecyclerView 实现横向和网格布局

视频详解

在使用 RecyclerView 控件时需要注意在运行之前一定要为 RecyclerView 指定列表项的布局管理器，否则系统不知道该如何显示，就会报错并强制退出。另外，RecyclerView 不仅可以实现纵向布局，还可以实现横向布局。实现横向布局的主要代码如下。

```
LinearLayoutManager linearLayoutManager = newLinearLayoutManager(MainActivity.this,
        LinearLayoutManager.HORIZONTAL,false);
    recyclerView.setLayoutManager(linearLayoutManager);
```

上述代码中在创建线性布局管理器时传递了三个参数，第一个参数为上下文对象，通常为当前的 Activity；第二个参数用于指定方向，默认为垂直的，在此指定为水平的；第三个参数表示列表项的顺序是否反转，对于水平方向来说，列表项默认是从左到右摆放，如果需要从右到左摆放，则表示需要反转，传递 true 进去即可。

除了 LinearLayoutManager 之外，RecyclerView 还提供了 GridLayoutManager 和 StaggeredGridLayoutManager 这两种内置的布局排列方式。GridLayoutManager 可以用于实现网格布局，StaggeredGridLayoutManager 可以用于实现瀑布流布局。如果要实现网格布局，代码如下。

```
    GridLayoutManager gridLayoutManager = newGridLayoutManager
    (this, 4, GridLayoutManager.VERTICAL, false);
recyclerView.setLayoutManager(gridLayoutManager);
```

创建网格布局管理器时需传递四个参数，第一个参数为上下文对象；第二个参数为每行或者每列排列 Item 个数；第三个参数表示网格的方向，是水平摆放还是垂直摆放，GridLayoutManager.VERTICAL 表示纵向排列；第四个参数表示列表项的顺序是否反转，false 表示不反转，按照原有次序加载。上述代码运行结果如图 5-15 所示。

图 5-15 程序运行效果

视频详解

5.4.3 RecyclerView 实现单击事件

在 ListView 和 GridView 中对于每个 Item 都有单击事件，对于 RecyclerView 同样也有单击事件，不过不同于 ListView，RecyclerView 并没有提供类似于 setOnItemClickListener()这样的注册监听器方法，需要我们给 Item 中具体的 View 去注册单击事件，看似编写代码变复杂，其实是增加了控件的灵活性，这也是比 ListView 更强大的地方之一。对于 RecyclerView 控件的单击事件，需要修改自定义类 MyAdapter 中的代码，单击事件是定义在 onCreateViewHolder()方法中，具体代码如下。

程序清单 5-27：chart0507\java\com\example\administrator\chart0507\MainActivity.java

```
1   public class MyAdapter extends RecyclerView.Adapter<MyAdapter.MyViewHolder> {
2       @Override
3       public MyViewHolder onCreateViewHolder(ViewGroup viewGroup, int viewType) {
4           final MyViewHolder myViewHolder = new
5           MyViewHolder (LayoutInflater.from(MainActivity
6               .this).inflate(R.layout.recycler_item, viewGroup, false));
```

```
7           myViewHolder.view.setOnClickListener(new View
8                   .OnClickListener() {
9               @Override
10              public void onClick(View v) {
11              int adapterPosition = myViewHolder .getAdapterPosition();
12              Toast.makeText ( MainActivity.this, "你单击的植物是:" + titles
    [adapterPosition],
13              Toast.LENGTH_SHORT).show();
14              }
15          });
16          myViewHolder.image.setOnClickListener(new View.OnClickListener() {
17              @Override
18              public void onClick(View v) {
19                  int adapterPosition = myViewHolder
20                          .getAdapterPosition();
21                  Toast.makeText(MainActivity.this,
22      "你单击图片的植物名是:" + titles[adapterPosition],
23                          Toast.LENGTH_SHORT).show();
24              }
25          });
26          return myViewHolder;
27      }
28      @Override
29      public void onBindViewHolder(MyViewHolder myViewHolder, int position) {
30          myViewHolder.image.setImageResource (icons[position]);
31          myViewHolder.titleView.setText
32                  (titles[position]);
33      }
34      @Override
35      public int getItemCount() {
36          return titles.length;
37      }
38      public class MyViewHolder extends RecyclerView.ViewHolder {
39          ImageView image;
40          TextView titleView;
41          View view;
42          public MyViewHolder(View itemView) {
43              super(itemView);
44              view = itemView;
45              image = (ImageView) itemView.findViewById
46                      (R.id.image);
47              titleView = (TextView) itemView
48                      .findViewById(R.id.tv_title);
49          }
50      }
51  }
```

上述代码中,第6行声明view变量保存Item最外层布局对象。第7~15行代码为最外层布局对象view添加单击事件,第11行获取用户单击的position,第12~13行代码通过Toast弹出用户单击植物的名称。第16~27行为ImageView控件绑定单击事件,当用户单击Item项中的图片,通过Toast弹出用户单击图片所对应的植物的名称。

上述程序运行效果如图5-16所示。

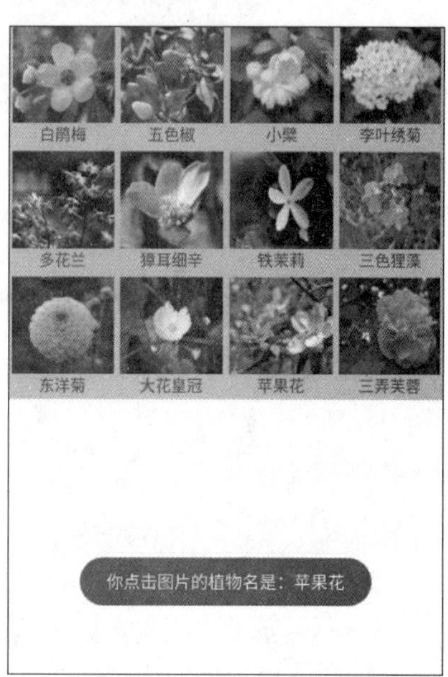

图5-16 程序运行效果

视频详解

5.4.4 实战演练——产品列表

通过对RecyclerView的学习,我们可以对"欢乐购商城"中商品列表页面进行实现。其布局页面的代码如下。

程序清单5-28:chart0508\app\src\main\res\layout\activity_main.xml

```
1    <LinearLayout xmlns:android = "http://schemas.android.com/apk/res/android"
2    xmlns:tools = "http://schemas.android.com/tools"
3    android:layout_width = "match_parent"
4    android:layout_height = "match_parent"
5    android:orientation = "vertical"
6    tools:context = ".MainActivity">
7
8        <android.support.v7.widget.RecyclerView
9          android:id = "@ + id/recycle_Viewes"
10         android:layout_width = "match_parent"
11         android:layout_height = "match_parent">
12       </android.support.v7.widget.RecyclerView>
13    </LinearLayout>
```

其中 Item 的布局文件如下。

程序清单 5-29：chart0508\app\src\main\res\layout\goods_list_items. xml

```
1    <?xml version = "1.0" encoding = "utf - 8"?>
2    < RelativeLayout xmlns:android = "http://schemas. android. com/apk/res/android"
3        android:layout_width = "match_parent"
4        android:layout_height = "match_parent">
5        < ImageView
6            android:id = "@ + id/iv_iamge"
7            android:layout_width = "120dp"
8            android:layout_height = "120dp"
9            android:layout_marginTop = "4dp"
10            android:src = "@drawable/a000" />
11        < TextView
12            android:id = "@ + id/tv_title"
13            android:layout_width = "match_parent"
14            android:layout_height = "wrap_content"
15            android:layout_alignTop = "@ + id/iv_iamge"
16            android:layout_marginLeft = "5dp"
17            android:layout_toRightOf = "@id/iv_iamge"
18            android:ellipsize = "end"
19            android:maxLines = "2"
20            android:text = "测试测试测试测试测试测试"
21            android:textSize = "13dp" />
22        < TextView
23            android:id = "@ + id/tv_count"
24            android:layout_width = "wrap_content"
25            android:layout_height = "wrap_content"
26            android:layout_below = "@id/tv_title"
27            android:layout_margin = "10dp"
28            android:layout_toRightOf = "@ + id/iv_iamge"
29            android:text = "1100" />
32        < TextView
33            android:layout_width = "wrap_content"
34            android:layout_height = "wrap_content"
35            android:layout_alignBottom = "@ + id/tv_count"
36            android:layout_marginRight = "3dp"
37            android:layout_toLeftOf = "@ + id/tv_address"
38            android:text = "产地:" />
39        < TextView
40            android:id = "@ + id/tv_address"
41            android:layout_width = "wrap_content"
42            android:layout_height = "wrap_content"
43            android:layout_alignBottom = "@ + id/tv_count"
44            android:layout_alignParentRight = "true"
45            android:text = "河南 商丘" />
46        < TextView
47            android:id = "@ + id/price_title"
48            android:layout_width = "wrap_content"
```

```
49              android:layout_height = "wrap_content"
50              android:layout_alignLeft = "@ + id/tv_count"
51              android:layout_below = "@ + id/tv_count"
52              android:layout_marginTop = "8dp"
53              android:text = "价格:¥" />
54          < TextView
55              android:id = "@ + id/tv_new_price"
56              android:layout_width = "wrap_content"
57              android:layout_height = "wrap_content"
58              android:layout_alignBottom = "@ + id/price_title"
59              android:layout_toRightOf = "@ + id/price_title"
60              android:text = "12.00"
61              android:textSize = "16sp"
62              android:textStyle = "bold" />
63          < LinearLayout
64            android:layout_width = "match_parent"
65            android:layout_height = "25dp"
66            android:layout_alignBottom = "@ + id/iv_iamge"
67            android:layout_alignLeft = "@ + id/price_title"
68            android:background = "@drawable/textview_border_style"
69            android:gravity = "center"
70            android:orientation = "horizontal">
71              < TextView
72              android:layout_width = "wrap_content"
73              android:layout_height = "wrap_content"
74              android:text = "店铺:" />
75              < TextView
76              android:id = "@ + id/tv_shopname"
77              android:layout_width = "wrap_content"
78              android:layout_height = "wrap_content"
79              android:text = "小飞的店铺"
80              android:textColor = "@color/colorPrimary" />
81          </LinearLayout >
82          < View
83              android:id = "@ + id/view"
84              android:layout_width = "match_parent"
85              android:layout_height = "1dp"
86              android:layout_below = "@id/iv_iamge"
87              android:layout_marginTop = "3dp"
88              android:background = "@color/grey" />
89      </RelativeLayout >
```

定义一个适配器类,代码如下。

程序清单 5-30:chart0508\app\src\main\java\com\example\administrator\goodsListAdapter

```
1     public class goodsListAdapter extends RecyclerView.Adapter < goodsListAdapter.
ViewHolder > {
2         private List < Good > mList;
```

```
3        private Activity mActivity;
4        private Good good;
5        public goodsListAdapter(Activity activity, List<Good> list) {
6            mList = list;
7            mActivity = activity;
8        }
9        @Override
10       public ViewHolder onCreateViewHolder(ViewGroup parent, int viewType) {
11            final ViewHolder viewHolder = new ViewHolder(View.inflate(mActivity,
                   R.layout.goods_list_items, null));
12            viewHolder.functionView.setOnClickListener(new View.OnClickListener() {
13                @Override
14                public void onClick(View v) {
15                    int adapterPosition = viewHolder.getAdapterPosition();
16                    Intent intent = new Intent(mActivity, GoodDetailActivity.class);
17                    intent.putExtra("goodInfo", mList.get(adapterPosition));
18                    mActivity.startActivity(intent);
19                }
20            });
21            return viewHolder;
22        }
23        @Override
24        public void onBindViewHolder(ViewHolder holder, int position) {
25            good = mList.get(position);
26            holder.imageView.setImageResource(good.getImage());
27            holder.title.setText(good.getName());
28            holder.price.setText("" + good.getPrice());
29            holder.count.setText(good.getVolume() + "人付款");
32            holder.address.setText(good.getProductLocation());
33            holder.shopName.setText(good.getShopName());
34        }
35        @Override
36        public int getItemCount() {
37            return mList.size();
38        }
39        class ViewHolder extends RecyclerView.ViewHolder {
40            TextView shopName;
41            TextView title;
42            TextView count;
43            TextView address;
44            TextView price;
45            ImageView imageView;
46            View functionView;
47            public ViewHolder(View itemView) {
48                super(itemView);
49                functionView = itemView;
50                imageView = itemView.findViewById(R.id.iv_iamge);
51                title = (TextView) itemView.findViewById(R.id.tv_title);
52                count = (TextView) itemView.findViewById(R.id.tv_count);
```

```
53                address = (TextView) itemView.findViewById(R.id.tv_address);
54                price = (TextView) itemView.findViewById(R.id.tv_new_price);
55                shopName = (TextView) itemView.findViewById(R.id.tv_shopname);
56            }
57
58        }
59    }
```

MainActivity 页面的代码如下。

程序清单 5-31：chart0508\app\src\main\java\com\example\administrator\MainActivity.java

```
1     public class MainActivity extends AppCompatActivity {
2
3         private RecyclerView recyclerView;
4         private LinearLayoutManager linearLayoutManager;
5         private List<Good> goodList = new ArrayList<Good>();
6         private int[] pictrue = {R.drawable.a000, R.drawable.a001, R.drawable.a002, R.
7     drawable.a003,R. drawable.a004, R.drawable.a005, R.drawable.a006,
              R.drawable.a007, S. R.drawable.a008, R.drawable.a009};
8         private String[] title = {"2019 秋装新款白色纯棉长袖 t 恤女装宽松春秋打底衫体恤
          大码上衣",
        "2019 初秋新款潮韩版洋气女装宽松秋季短款衬衫女秋装方领长袖上衣",
9           "纯棉长袖 T 恤女上衣 2019 新款竹节棉宽松秋衣女装薄款初秋打底衫潮",
             "半高领打底衫女装秋冬薄款洋气莫代尔堆堆领长袖 t 恤内搭针织上衣",
10          "女装 2019 新款潮超火 cec 短袖女春夏季港味宽松百搭上衣 T 恤 ins 洋气",
            "夏大大胖 mm 韩版宽松秋装新款女装大码 2019 牛仔套装外套减龄连衣裙",
11                 "大姗姗家瘦瘦裤胖 mm 韩版大码女装秋季百搭外穿打底裤网红款小脚裤",
                   "中国大陆",'夏装紧身纯白色 T 恤女短袖修身女装女士纯棉 2019 新款上衣潮打底衫",
12                 "纯棉长袖 T 恤女上衣 2019 新款竹节棉宽松秋衣女装薄款初秋打底衫潮",
                   "纯棉红色 v 领 T 恤女短袖修身夏女装 2019 新款潮体恤紧身上衣黑色短款",};
13          private String[] location = {"云南 昆明", "广东 广州", "江西 南昌", "广东 汕头",
             "浙江 金华", "广东 广州", "江苏 淮安",
14                 "江苏 苏州", "广东 深圳", "安徽 芜湖"};
15          private String[] shop = {"lovtis 兰蒂斯旗舰店", "lovtis 兰蒂斯旗舰店",
             "lovtis 兰蒂斯旗舰店", "lovtis 兰蒂斯旗舰店",
16                 "旗舰店", "旗舰店", "旗舰店", "旗舰店", "旗舰店", "旗舰店"};
17          private String[] number = {"87816", "37384", "62165", "43221", "4561", "43211",
18                  "3011","764", "7656", "4567"};
19          private String[] prices = {"39.9", "29.87", "29", "56", "20", "138", "97", "15",
             "78", "30"};
20          @Override
21          protected void onCreate(Bundle savedInstanceState) {
22              super.onCreate(savedInstanceState);
23              setContentView(R.layout.activity_main);
24              initView();
25          }
26          private void initView() {
27
28              for (int i = 0; i < 10; i++) {
29                  Good good = new Good();
32                  good.setImage(pictrue[i]);
```

```
33                  good.setName(title[i]);
34                  good.setProductLocation(location[i]);
35                  good.setShopName(shop[i]);
36                  good.setPrice(Double.valueOf(prices[i]));
37                  good.setVolume(number[i]);
38                  goodList.add(good);
39              }
40              recyclerView = (RecyclerView) findViewById(R.id.recycle_Viewes);
41             linearLayoutManager = new LinearLayoutManager(this);
42         recyclerView.setLayoutManager(linearLayoutManager);
43              recyclerView.setItemAnimator(new DefaultItemAnimator());
44              recyclerView.     addItemDecoration    (   new      DividerItemDecoration
   (this, DividerItemDecoration
45                      .VERTICAL));
46              recyclerView.setAdapter(new goodsListAdapter(this, goodList));
47          }
48      }
```

上述程序运行效果如图 5-17 所示。

图 5-17　程序运行效果

本章小结

本章围绕“欢乐购商城”项目首页、产品列表页面、订单列表页面和修改收货地址页面引入 Android 中列表控件的讲解。详细介绍了常用的列表控件的使用场景和基本属性，并通过实现首页、产品列表页面、订单列表页面和修改收货地址页面演示了这几种列表控件的具体使用用法。其中，Spinner 控件主要适用于具有下拉选项的应用场景，ListView 控件常用于数据量较小和简单布局的应用，GridView 主要应用于多行多列的展示，而对于要实现页面布局较复杂或者加载数据量较大的应用场景推荐使用 RecyclerView 控件，这些列表控件在以后的开发中经常使用，需要读者熟练掌握灵活运用。

自测习题

1. BaseAdapter 为什么定义为抽象类？要想实现自定义的 Adapter，必须实现哪些方法？

2. 简述 SimpleAdapter 对象创建时，各个参数的含义。

3. 根据所学的 ListView 控件，实现“欢乐购商城”的地址列表页面，如图 5-18 所示。

4. 根据所学的 RecyclerView 控件，实现“欢乐购商城”首页中的小图标加载，如图 5-19 所示。

图 5-18　地址列表页面

图 5-19　首页小图标加载

第6章 页面跳转与切换

Android 程序应用是基于四大组件进行架构的，其中，Activity 是作为负责界面显示的组件，也就是说，App 开发过程中只要用到界面，就一定离不开 Activity，它负责加载界面UI 及用户的交互流程。Fragment 原本是浮在 Activity 之上的一个块，一般只在 App 的主界面中进行使用，这些年随着 Android 开发新的架构 MVP 的流行，Fragment 也慢慢地从非主流升级到了主流技术。可以说，Android 中基于 Activity 和 Fragment 的开发是非常主要的，也是一个 App 开发中不可缺少的板块。

学习目标

本章要点

(1) 理解 Activity 的功能与作用。

(2) 创建和配置 Activity。

(3) 在程序中启动、关闭 Activity。

(4) Activity 的生命周期和使用场景。

(5) 不同 Activity 间的数据传递。

(6) 理解 Fragment 的功能与作用。

(7) 理解 Intent 的功能与作用。

(8) 学习如何编写一个 App 的主界面。

(9) 开发和编写一个合适的底部栏。

视频详解

6.1 Activity 详解

任务陈述

在“欢乐购商城”中包含多个页面，当打开应用时首先进入默认页面，然后由用户单击页面中控件跳转到其他页面。“欢乐购商城”页面跳转包含两种直接跳转和携带数据跳转。本节重点围绕页面直接跳转实现，而 Activity 携带数据跳转放在 6.2 节中重点讲解。

分析：在“欢乐购商城”登录页面中，存在很多跳转页面，比如登录跳转注册、从列表查看商品详情、添加购物车成功后页面跳转等。本节围绕单击“注册”按钮跳转进入注册页面进行讲解，如图 6-1 所示。

图 6-1　登录页面跳转注册页面

Activity 是 Android 应用的基础组成部分，如果把一个 Android 应用看成是一个网站的话，那么一个 Activity 就相当于该网站的一个具体网页。Android 应用开发的一个重要组成部分就是开发 Activity，下面将由浅入深详细地讲解 Activity 的创建、配置、启动、传值以及生命周期等相关知识。

视频详解

6.1.1 Activity 概述

Activity 是 Android 的一种应用程序组件，该组件为用户提供了一个屏幕，用户在这个屏幕上进行操作即可完成指定的功能，例如打电话、拍照、发送邮件或查看地图等。每个 Activity 都有一个用于显示用户界面的窗口。该窗口通常会充满整个屏幕，但有可能比这个屏幕更小或者是漂浮在其他窗口之上。Activity 类包含一个 setTheme()方法来设置其窗口的主题风格，例如，我们希望窗口不显示标题、以对话框形式显示窗口，都可以通过该方法来实现。

一个应用程序通常是由多个彼此之间松耦合的 Activity 组成。通常，在一个应用程序中，有一个 Activity 被指定为主 Activity。当应用程序第一次启动的时候，系统会自动运行主 Activity，前面的所有例子都只有一个 Activity，并且该 Activity 为主 Activity。在 Activity 中可以启动新的 Activity 用于执行不同的功能。当一个新的 Activity 启动后，先前的那个 Activity 就会停止，但是系统会在堆栈中保存该 Activity。新的 Activity 启动后，将会被压入栈顶，并获得用户焦点。堆栈遵循后进先出的原则。因此，当用户使用完当前的 Activity 并按 Back 键时，该 Activity 将从堆栈中取出并销毁，然后先前的那个 Activity 将恢复并获取焦点。

当一个 Activity 因为新的 Activity 的启动而停止时，系统将会调用 Activity 的生命周期的回调方法来通知这一状态的改变。Activity 类中定义了一系列的回调方法，会根据 Activity 的状态自动调用，例如创建、停止、回复、销毁等。默认情况下，重写每个回调方法内部是没有任何逻辑代码的，Activity 只是告诉开发者在哪个阶段或哪个状态下会调用哪个方法，开发者可以根据自己的用户需求在对应的地方编写相应的业务处理代码。另外，可以将 App 先放到后台中，当再次打开 App 回到该界面时，为了保障数据的时效性，需要重新请求和加载数据，这时候就可以在对应的回调方法中编写网络请求和界面数据更新的相关逻辑业务代码。

6.1.2　创建和配置 Activity

视频详解

如果想要创建自己的 Activity，一般采用的是两种方式，一种是直接选择 new→activity→自己想要的 activity 类型→命名新建的文件名，如图 6-2 所示。另一种是选择 new→class→命名 class 文件→直接或间接继承 Activity 类，如图 6-3 所示。

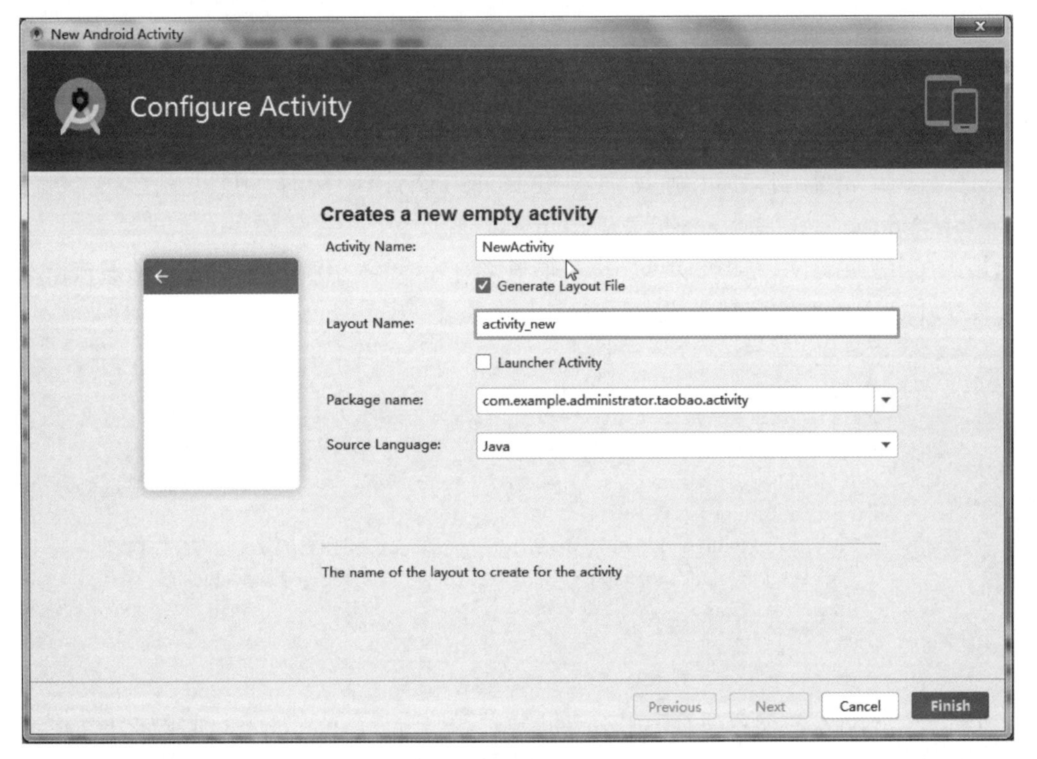

图 6-2　创建 Activity

当 Activiy 被创建时，系统将会自动回调它的 onCreate()方法，在该方法的实现中，应该初始化一些关键的界面控件，最重要的是调用 Activity 的 setContentView()方法来设置 Activity 所对应的界面布局文件。为了管理应用程序界面中的各个控件，可调用 Activity 的 findViewById(int id)方法来获取界面中的控件，然后即可修改该控件的属性和调用该控件的方法。

图 6-3　创建类并继承 AppCompatActivity

定义好自己的 Activity 后，此时系统还不能访问该 Activity，如果想让系统访问，则必须在 AndroidManiface. xml 文件中进行注册、配置。前面所写程序中，也有自己的 Activity。但并没有对它进行配置，不是也可以访问吗？这是因为在使用第一种方式创建 Activity 后，编辑器自动识别我们创建的是 Activity 文件，所以直接帮我们在 AndroidManiface. xml 注册了一条该 Activity 的声明。

其实在声明 Activity 的时候可以指定很多的内容，如 Activity 的风格、启动模式以及是否配置在开启 App 时首先启动等。

```
1   <activity
2       android:name = ".NewActivity"                                   →Activity 名称
3       android:launchMode = "singleTask"                               →启动模式
4       android:theme = "@style/AppTheme.FullScreen">                   →显示风格
5       <intent - filter>
6           <action android:name = "android.intent.action.MAIN" />      →首先启动
7           <category android:name = "android.intent.category.LAUNCHER" />
8       </intent - filter>
9   </activity>
```

此部分 Activity 的配置项看上去有点儿烦琐，其实除了 name 是必填项以外其他选项都是选填项，可以根据自己的需求进行添加，具体的属性及解释如下。

android:name：指定 Activity 实现类的类名，其中前面的点表示该类在当前应用程序所在的包下，如果该类不在当前包下，则需要用完整的包名+类名。一般情况下，编译器会找到对应名字相近的类名，故合理使用快捷键会减少类名及路径的书写错误。

android:theme：指定 Activity 的风格形式，提供了非常好看的风格，当然也可在 style. xml 文件中定义自己需要的风格，比如当前比较流行透明信号栏就可以在这里定义，但要注意手机的系统版本兼容。

android:launchMode：指定 activity 的启动模式，主要针对多次启动一个相同 Activity 进行的限制和优化，可以避免用户因操作失误多次启动同一个 Activity 导致的一些问题。

6.1.3 启动和关闭 Activity

视频详解

前面已经定义并向系统注册了 Activity，那么该 Activity 如何启动和执行呢？通常一个 Android 应用都会包含多个 Activity，但只有一个 Activity 会作为程序的入口，当该 Android 应用运行时将会自动启动并执行该 Activity。而应用中的其他 Activity，通常都由入口 Activity 来启动，或由入口 Activity 启动的 Activity 启动。Android 提供了以下两种方法来启动 Activity。

startActivity(Intent intent)：启动其他 Activity。

startActivityForResult(Intent intent,int requestCode)：程序将会得到新启动 Activity 的结果，requestCode 参数代表启动 Activity 的请求码，后面会详细讲解这一方法。

上面两个方法，都需要传入一个 Intent 类型的参数，该参数是对所需要启动的 Activity 的描述，既可以是一个确切的 Activity 类，也可以是所需要启动的 Activity 的一些特征，然后由系统查找符合该特征的 Activity，如果有多个 Activity 符合该要求时，系统将会以下拉列表的形式列出所有的 Activity，然后由用户选择具体启动哪一个，这些 Activity 既可以是本应用程序的，也可以是其他应用程序的。

Intent 的相关知识在 6.2 节会详细介绍，在此简单介绍启动一个已知的 Activity 的方法。

```
1    Intent intent = new Intent(this, OtherActivity.class);
                    →this 表示当前 Activity 的对象，OtherActivity 为一个已知的 Activity，
                    并且 OtherActivity 必须在 Manifest.xml 文件中进行了配置
2    startActivity(intent);
```

如果想从所启动的 Activity 那获取结果，则可以使用 startActivityForResult(Intent intent,int requestCode)方法启动 Activity，同时需要在自己的 Activity 中重写 onActivity-Result(…)方法，当启动的 Activity 执行结束后，它会将结果数据放入 Intent，并传给 onActivityResult(…)方法。

如果想关闭 Activity，可调用以下两个方法。

finish()：结束当前的 Activity。

finishActivity(int requestCode)：结束以 startActivityForResult(Intent intent,int requestCode)方法启动的 Activity。

注意： 大部分情况下，不建议显式调用这些方法关闭 Activity。因为 Android 系统会为我们管理 Activity 的生命周期，调用这些方法可能会影响用户的预期体验，因此，只有当你不想用户再回到当前 Activity 的时候才去关闭它。

6.1.4 Activity 生命周期

视频详解

为了方便 App 用于自我管理，Android 为每个 Activity 在创建的时候都注入了一套生命周期管理，从此 Activity 生命周期成为每个 Android 开发者管理和运用 Activity 的必备

技能,也是开发者在面试过程中常常被考到的问题。Android 是如何给每个 Activity 来设置和管理生命周期的?

Activity 具有以下四个状态。

(1) Running:当前显示在屏幕上的 Activity(位于任务栈的顶部),用户可见状态。

(2) Poused:依旧在用户可见状态,但是界面焦点已经失去,此 Activity 无法与用户进行交互。

(3) Stopped:用户看不到当前界面,也无法与用户进行交互,完全被覆盖。

(4) Killed:当前界面被销毁,等待系统被回收。

Android 系统中的 Activity 类定义了系列的回调动作,当 Activity 的状态发生变化时,相应的回调方法将会自动执行。当我们的 Activity 被启动之后,随着应用程序的运行,Activity 会不断地在各种状态之间切换,相应的方法类也就会被执行,只需要选择性地重写这些方法即可进行相应的业务处理。这些状态之间的转换就构成了 Activity 的生命周期。在 Activity 的生命周期中,主要有如下几个方法。

onCreate():Activity 被创建时自动调用。

onStart():Activity 启动时自动调用。

onRestart():Activity 重新恢复启动时自动调用。

onResume():Activity 运行在前台时自动调用。

onPause():Activity 暂停时自动调用。

onStop():Activity 停止时自动调用。

onDestroy():Activity 销毁时自动调用。

从图 6-4 中可以看出 Activity 的生命周期主要存在三个循环。

(1) 整个生命周期:从 onCreate()开始到 onDestroy()结束。Activity 在 onCreate()中执行初始化操作,例如,获取系统资源,加载布局文件,在 onDestroy()中释放所有资源。

(2) 可见生命周期:从 onStart()开始到 onStop()结束。在这段时间,可以看到 Activity 在屏幕上,尽管有可能不在前台,不能和用户交互。在这两个方法之间,需要保持显示给用户的 UI 数据和资源等。例如,可以在 onStart()中注册一个监听器来监听数据变化导致 UI 的变动,当不再需要显示时,可以在 onStop()中注销它。onStart()和 onStop()方法都可以被多次调用,因为 Activity 随时可以在可见和隐藏之间转换。

(3) 前台生命周期:从 onResume()开始到 onPause()结束。在这段时间里,该 Activity 处于所有 Activity 的最前面,和用户进行交互。Activity 可以经常性地在 resumed 和 paused 状态之间切换。例如,当设备准备休眠时,当一个 Activity 处理结果被分发时,当一个新的 Intent 被分发时。所以,在这些方法中的代码应该属于非常轻量的。

视频详解

6.1.5 实战演练——跳转到注册页面

根据本节任务分析需要实现从注册页面跳转到登录页面。本项目需要登录和注册 Activity 页面和两个 Activity 对应的布局。然后需要对登录页面的"立即注册"进行监听,实现页面跳转,整体的项目结构如图 6-5 所示。

由于登录页面在前面章节已经实现,本项目代码只展示关联控件代码,登录页面布局文件如下。

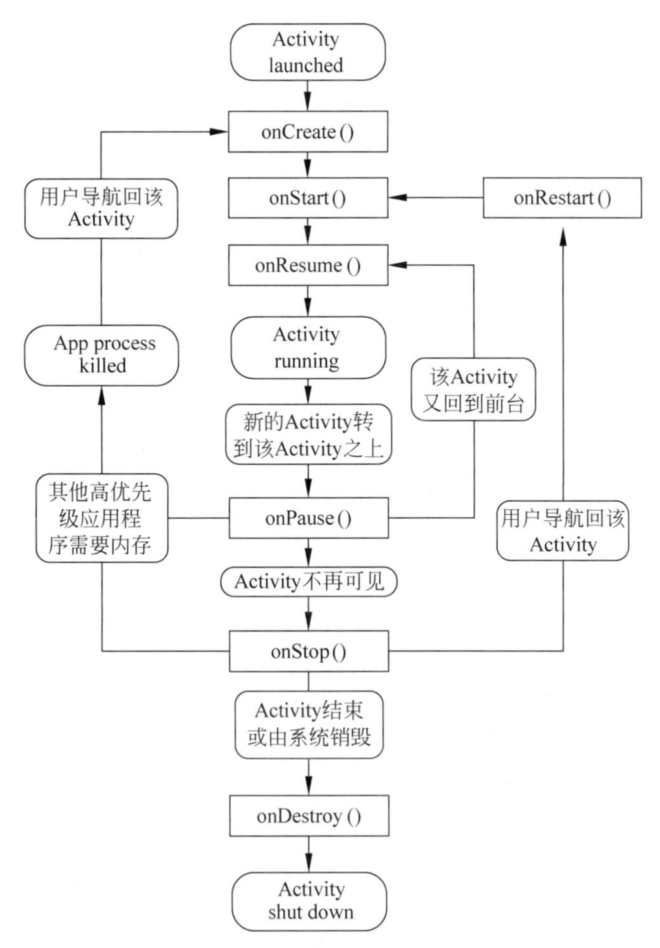

图 6-4　Activity 生命周期

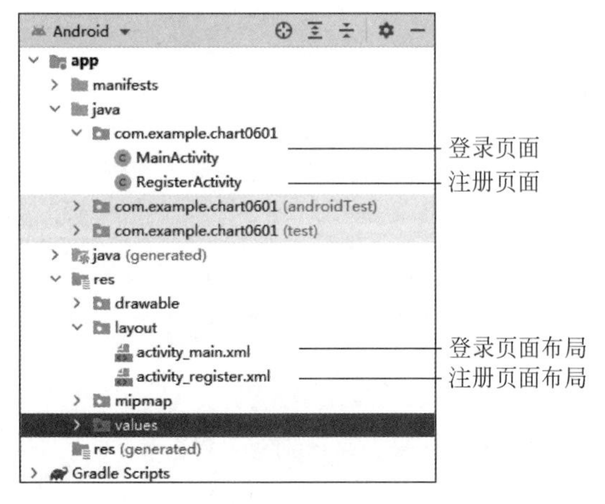

图 6-5　页面跳转结构

程序清单 6-1：chart0601\res\layout\activity_main. xml

```
1   <LinearLayout xmlns:android = "http://schemas.android.com/apk/res/android"
2       xmlns:app = "http://schemas.android.com/apk/res-auto"
3       xmlns:tools = "http://schemas.android.com/tools"
4       android:layout_width = "match_parent"
5       android:layout_height = "match_parent"
6       android:orientation = "vertical"
7       tools:context = ".MainActivity">
8        ...                                                →代码省略
9       <Button
10          android:id = "@+id/btn_login"                   →登录按钮
11          android:layout_width = "fill_parent"
12          android:layout_height = "40dp"
13          android:layout_gravity = "center_horizontal"
14          android:layout_marginLeft = "35dp"
15          android:layout_marginRight = "35dp"
16          android:layout_marginTop = "15dp"
17          android:background = "@drawable/btnback"
18          android:text = "登 录"
19          android:textColor = "@android:color/white"
20          android:textSize = "18sp" />
21       <LinearLayout
22          android:layout_width = "fill_parent"
23          android:layout_height = "fill_parent"
24          android:layout_marginLeft = "35dp"
25          android:layout_marginRight = "35dp"
26          android:layout_marginTop = "8dp"
27          android:gravity = "center_horizontal"
28          android:orientation = "horizontal">
29          <TextView                                       →注册文本
30              android:id = "@+id/tv_register"             →注册文本添加 id
31              android:layout_width = "0dp"
32              android:layout_height = "wrap_content"
33              android:layout_weight = "1"
34              android:gravity = "center_horizontal"
35              android:padding = "8dp"
36              android:text = "立即注册"
37              android:textColor = "@color/colorPrimary"
38              android:textSize = "14sp" />
39          <TextView
40              android:id = "@+id/tv_find_psw"
41              android:layout_width = "0dp"                 →找回密码文本
42              android:layout_height = "wrap_content"       →找回密码添加 id
43              android:layout_weight = "1"
44              android:gravity = "center_horizontal"
45              android:padding = "8dp"
46              android:text = "找回密码?"
47              android:textColor = "@color/colorPrimary"
48              android:textSize = "14sp" />
49       </LinearLayout>
50    </LinearLayout>
```

注册页面布局文件如下。

程序清单 6-2：chart0601\app\src\main\res\layout\activity_register. xml

```
1    <LinearLayout xmlns:android = "http://schemas.android.com/apk/res/android"
2        xmlns:tools = "http://schemas.android.com/tools"
3        android:layout_width = "match_parent"
4        android:layout_height = "match_parent"
5        android:orientation = "vertical"
6        tools:context = ".RegisterActivity">
7        < ImageView
8            android:layout_width = "70dp"
9            android:layout_height = "70dp"
10             android:layout_gravity = "center_horizontal"
11             android:layout_marginTop = "25dp"
12             android:src = "@drawable/head" />
13         < EditText
14             android:id = "@ + id/et_user_names"
15             android:layout_width = "fill_parent"
16             android:layout_height = "40dp"
17             android:layout_gravity = "center_horizontal"
18             android:layout_marginLeft = "35dp"
19             android:layout_marginRight = "35dp"
20             android:layout_marginTop = "35dp"
21             android:background = "@drawable/textview_borders"
22             android:drawableLeft = "@drawable/user_name_icon"
23             android:drawablePadding = "10dp"
24             android:gravity = "center_vertical"
25             android:hint = "请输入账号"
26             android:paddingLeft = "8dp"
27             android:singleLine = "true"
28             android:textColor = "#000000"
29             android:textColorHint = "#a3a3a3"
30             android:textSize = "14sp" />
31         < EditText
32             android:id = "@ + id/et_psw"
33             android:layout_width = "fill_parent"
34             android:layout_height = "40dp"
35             android:layout_gravity = "center_horizontal"
36             android:layout_marginLeft = "35dp"
37             android:layout_marginRight = "35dp"
38             android:layout_marginTop = "5dp"
39             android:background = "@drawable/textview_borders"
40             android:drawableLeft = "@drawable/psw_icon"
41             android:drawablePadding = "10dp"
42             android:inputType = "textPassword"
43             android:hint = "请输入密码"
44             android:paddingLeft = "8dp"
45             android:singleLine = "true"
46             android:textColor = "#000000"
47             android:textColorHint = "#a3a3a3"
48             android:textSize = "14sp" />
49         < EditText
```

```
50              android:id = "@ + id/et_psws"
51              android:layout_width = "fill_parent"
52              android:layout_height = "40dp"
53              android:layout_gravity = "center_horizontal"
54              android:layout_marginLeft = "35dp"
55              android:layout_marginRight = "35dp"
56              android:layout_marginTop = "5dp"
57              android:background = "@drawable/textview_borders"
58              android:drawableLeft = "@drawable/psw_icon"
59              android:drawablePadding = "10dp"
60              android:hint = "请再次输入密码"
61              android:inputType = "textPassword"
62              android:paddingLeft = "8dp"
63              android:singleLine = "true"
64              android:textColor = "#000000"
65              android:textColorHint = "#a3a3a3"
66              android:textSize = "14sp" />
67          <EditText
68              android:id = "@ + id/et_nick"
69              android:layout_width = "fill_parent"
70              android:layout_height = "40dp"
71              android:layout_gravity = "center_horizontal"
72              android:layout_marginLeft = "35dp"
73              android:layout_marginRight = "35dp"
74              android:layout_marginTop = "5dp"
75              android:background = "@drawable/textview_borders"
76              android:drawableLeft = "@drawable/user_name_icon"
77              android:drawablePadding = "10dp"
78              android:hint = "请输入昵称"
79              android:paddingLeft = "8dp"
80              android:singleLine = "true"
81              android:textColor = "#000000"
82              android:textColorHint = "#a3a3a3"
83              android:textSize = "14sp" />
84          <Button
85              android:id = "@ + id/btn_register"
86              android:layout_width = "fill_parent"
87              android:layout_height = "40dp"
88              android:layout_gravity = "center_horizontal"
89              android:layout_marginLeft = "35dp"
90              android:layout_marginRight = "35dp"
91              android:layout_marginTop = "20dp"
92              android:background = "@drawable/btnback"
93              android:text = "注 册"
94              android:textColor = "@android:color/white"
95              android:textSize = "18sp" />
96      </LinearLayout>
```

在登录页面 Activity 中需要实现加载布局文件，同时对“立即注册”文本进行监听，借助

Intent 对象实现页面跳转，具体代码如下。

程序清单 6-3：chart0601\app\src\main\java\com\example\chart0601\MainActivity. java

```
1   public class MainActivity extends AppCompatActivity {
2       private TextView tv_register;
3       @Override
4       protected void onCreate(Bundle savedInstanceState) {
5           super.onCreate(savedInstanceState);
6           setContentView(R.layout.activity_main);
7           tv_register = findViewById(R.id.tv_register);
8           tv_register.setOnClickListener(new View.OnClickListener() { //对"立即注册"
                                                                           文本监听
9               @Override
10              public void onClick(View v) {
11                  Intent intent = new Intent(MainActivity.this,RegisterActivity.class);
12                  startActivity(intent);      //实现页面跳转
13              }
14          });
15      }
16  }
```

注册页面 Activity 主要完成加载对应布局文件，代码详情如下。

程序清单 6-4：chart0601\app\src\main\java\com\example\chart0601\RegisterActivity. java

```
1   public class RegisterActivity extends AppCompatActivity {
2       @Override
3       protected void onCreate(Bundle savedInstanceState) {
4           super.onCreate(savedInstanceState);
5           setContentView(R.layout.activity_register);
6       }
7   }
```

因为 Activity 为 Android 系统四大组件之一，需要在清单文件 AndroidManifest. xml 中进行注册。

程序清单 6-5：chart0601\app\src\main\java\com\example\chart0601\RegisterActivity. java

```
1   <manifest xmlns:android = "http://schemas.android.com/apk/res/android"
2       package = "com.example.chart0501">
3       <application
4           android:allowBackup = "true"
5           android:icon = "@mipmap/ic_launcher"
6           android:label = "@string/app_name"
7           android:roundIcon = "@mipmap/ic_launcher_round"
8           android:supportsRtl = "true"
9           android:theme = "@style/AppTheme">
10          <activity android:name = ".MainActivity">
11              <intent-filter>
```

```
12                  <action android:name = "android.intent.action.MAIN" />
13                  <category android:name = "android.intent.category.LAUNCHER" />
14              </intent-filter>
15          </activity>
16          <activity android:name = ".RegisterActivity"/>      →注册页面
17      </application>
18  </manifest>
```

视频详解

6.2 Intent

任务陈述

Android 中四大组件包括 Activity、Service、BroadcastReceive 和 ContentProvider，它们之间互相调用和数据传递需要借助 Intent 对象。在“欢乐购商城”中多处出现 Activity 之间的跳转以及之间的数据传递。

分析：在“欢乐购商城”登录页面中，存在很多跳转页面，比如登录跳转注册、从列表查看商品详情、添加购物车成功后页面跳转等。本节重点分析项目中出现的案例，如单击商品列表子项跳转到新的页面查看商品详情，如图 6-6 所示。

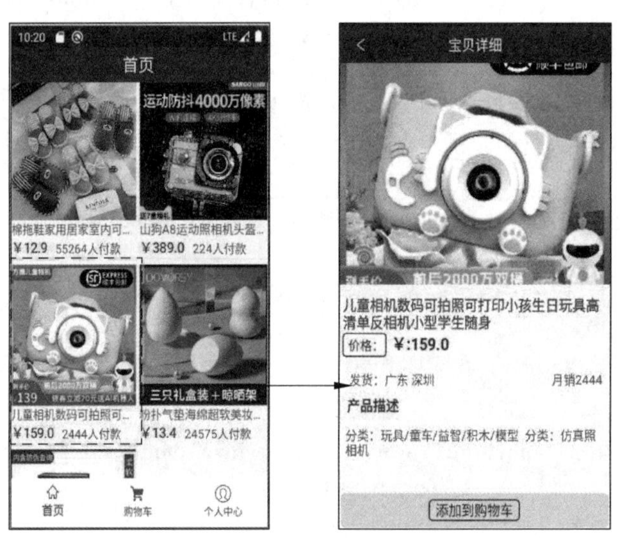

图 6-6 列表监听页面跳转

视频详解

6.2.1 Intent 概述

“Intent”中文翻译为“意图”，是对一次即将运行的操作的抽象描述，包括操作的动作、动作涉及数据、附加数据等，Android 系统则根据 Intent 的描述，负责找到对应的组件，并将 Intent 传递给调用的组件，完成组件的调用。因此，Intent 在这里起着媒体中介的作用，专门提供组件互相调用的相关信息，实现调用者与被调用者之间的解耦。

例如，我们想通过联系人列表查看某个联系人的详细信息，单击某个联系人后，希望能够弹出此联系人的详细信息。为了实现这个目的，联系人 Activity 需要构造一个 Intent，这个 Intent 用于告诉系统，我们要做“查看”动作，此动作对应的查看对象是“具体的某个联系人”，然后调用 startActivity (Intent intent)，将构造的 Intent 传入，系统会根据此 Intent 中的描述，到 AndroidManifest. xml 中找到满足此 Intent 要求的 Activity，最终传入 Intent，对应的 Activity 则会根据此 Intent 中的描述，执行相应的操作。

Intent 实际上就是一系列信息的集合，既包含对接收该 Intent 的组件有用的信息，如即将执行的动作和数据，也包括对 Android 系统有用的信息，如处理该 Intent 的组件的类型以及如何启动一个目标 Activity。

6.2.2　Intent 构成

视频详解

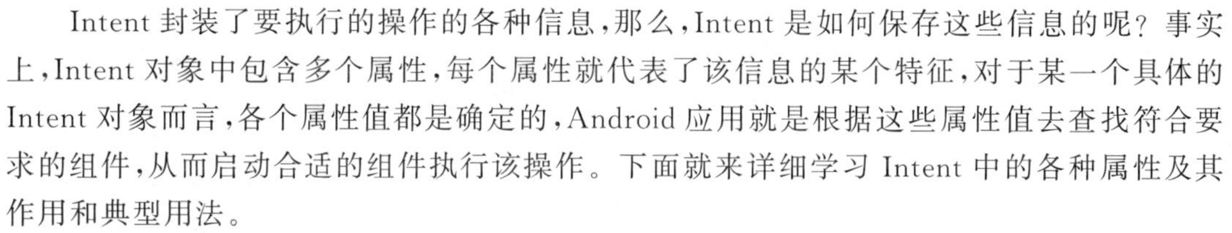
Intent 封装了要执行的操作的各种信息，那么，Intent 是如何保存这些信息的呢？事实上，Intent 对象中包含多个属性，每个属性就代表了该信息的某个特征，对于某一个具体的 Intent 对象而言，各个属性值都是确定的，Android 应用就是根据这些属性值去查找符合要求的组件，从而启动合适的组件执行该操作。下面就来详细学习 Intent 中的各种属性及其作用和典型用法。

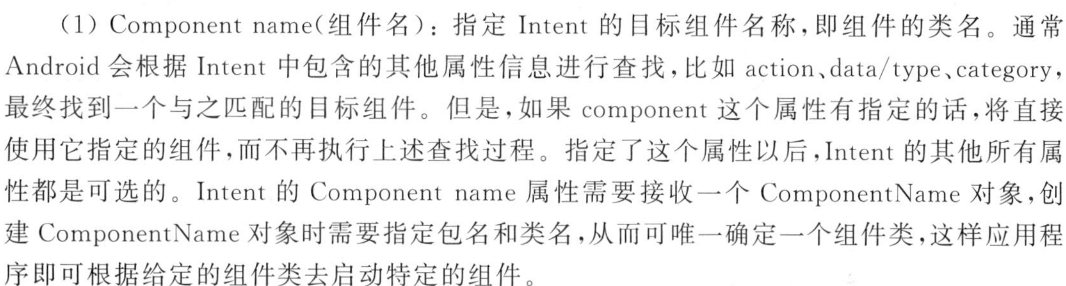
(1) Component name(组件名)：指定 Intent 的目标组件名称，即组件的类名。通常 Android 会根据 Intent 中包含的其他属性信息进行查找，比如 action、data/type、category，最终找到一个与之匹配的目标组件。但是，如果 component 这个属性有指定的话，将直接使用它指定的组件，而不再执行上述查找过程。指定了这个属性以后，Intent 的其他所有属性都是可选的。Intent 的 Component name 属性需要接收一个 ComponentName 对象，创建 ComponentName 对象时需要指定包名和类名，从而可唯一确定一个组件类，这样应用程序即可根据给定的组件类去启动特定的组件。

```
1    ComponentName comp = new ComponentName(Context con,Class class);
                                            →创建一个 ComponentName 对象
2    Intent intent = new Intent();
3    intent.setComponent(comp);             →为 Intent 设置 Component 属性
```

实际上，上面三行代码完全等价于前面所用的创建 Intent 的一行代码：

```
1    Intent intent = new Intent(Context con Class class);
```

在被启动的组件中，通过以下语句即可获取相关 ComponentName 的信息：

```
1    ComponentName comp = getIntent().getComponent();
2    comp.getPackageName();                 →获取组件的包名
3    comp.getClassName();                   →获取组件的类名
```

(2) Action(动作)：Action 代表该 Intent 所要完成的一个抽象“动作”，这个动作具体由哪个组件来完成，Action 这个字符串本身并不管。例如，Android 提供的标准 Action：

Intent. ACTION_VIEW，它只表示一个抽象的查看操作，但具体查看什么，启动哪个 Activity 来查看，它并不知道（这取决于 Activity 的<intent-filter…/>配置，只要某个 Activity 的<intent-filter…/>配置中包含该 ACTION_VIEW，该 Activity 就有可能被启动）。Intent 类中定义了一系列的 Action 常量，具体可查阅 Android SDK→reference 中的 Android. content. intent 类，通过这些常量能调用系统提供的功能。

Intent 类中提供了一些 Action 常量，如表 6-1 所示。

表 6-1　Intent 类中部分 Action 常量

编　号	Action 名称	AndroidManifest. xml 配置名称	描　述
1	ACTION_MAIN	android. intent. action. MAIN	作为应用程序的入口，不需要接收数据
2	ACTION_VIEW	android. intent. action. VIEW	用于数据的显示
3	ACTION_DIAL	android. intent. action. DIAL	调用电话拨号程序
4	ACTION_EDIT	android. intent. action. EDIT	用于编辑给定的数据
5	ACTION_PICK	android. intent. action. PICK	从特定的一组数据中进行数据的选择操作
6	ACTION_RUN	android. intent. action. RUN	运行数据
7	ACTION_SEND	android. intent. action. SEND	调用发送短信程序
8	ACTION_CHOOSER	android. intent. action. CHOOSER	创建文件操作选择器

（3）Category（类别）：执行动作的组件的附加信息。例如，LAUNCHER_CATEGORY 表示 Intent 的接收者应该在 Launcher 中作为顶级应用出现；而 ALTERNATIVE_CATE GORY 表示当前的 Intent 是一系列可选动作中的一个，这些动作可以在同一块数据上执行。同样地，在 Intent 类中定义了一些 Category 常量。

一个 Intent 对象最多只能包括一个 Action 属性，程序可调用 setAction(String str)方法来设置 Action 属性值；但一个 Intent 对象可以包含多个 Category 属性，程序可调用 Intent 的 addCategory(String str)方法来为 Intent 添加 Category 属性。当程序创建 Intent 时，该 Intent 默认启动 Category 属性值为 Intent. CATEGORY_DEFAULT 常量的组件。

Intent 中部分 Category 常量及对应的字符串和作用如表 6-2 所示。

表 6-2　Intent 类中部分 Category 常量

编号	Category 常量	对应字符串	简单描述
1	CATEGORY_DEFAULT	android. intent. category. DEFAULT	默认的 Category
2	CATEGORY_BROWSABLE	android. intent. category. BROWSABLE	指定该 Activity 能被浏览器安全调用
3	CATEGORY_TAB	android. intent. category. TAB	指定 Activity 作为 TabActivity 的 Tab 页
4	CATEGORY_LAUNCHER	android. intent. category. LAUNCHER	Activity 显示在顶级程序列表中
5	CATEGORY_HOME	android. intent. category. HOME	设置该 Activity 随系统启动而运行

(4) Data(数据)：Data 属性通常用于向 Action 属性提供操作的数据。不同的 Action 通常需要携带不同的数据，例如，如果 Action 是 ACTION_CALL，那么数据部分将会是 tel：需要拨打的电话号码。Data 属性接收一个 URI 对象，一个 URI 对象通常通过如下形式的字符串来表示。

```
content://com.android.contacts/contacts/1
tel:13876523467
```

上面所示的两个字符串的冒号前面大致指定了数据的类型(MIME 类型)，冒号后面的是数据部分。因此一个合法的 URI 对象既可决定操作哪种数据类型的数据，又可指定具体的数据值。常见的数据类型及其数据 URI 如表 6-3 所示。

表 6-3　Android 中部分数据

编　　号	操作类型	数据格式	简单示例
1	浏览网页	http://网页地址	http://www.mldn.cn
2	拨打电话	tel：电话号码	tel：01051283346
3	发送短信	smsto：短信接收人号码	smsto：13621384455
4	查找 SD 卡	file:///sdcard/文件或目录	file:///sdcard/mypic.jpg
5	显示地图	geo：坐标，坐标	geo：31.899533,-27.036173

(5) Type(数据类型)：显式指定 Intent 的数据类型(MIME)。一般 Intent 的数据类型能够根据数据本身进行判定，但是通过设置这个属性，可以强制采用显式指定的类型而不再进行推导。通常来说，当 Intent 不指定 Data 属性时 Type 属性才会起作用，否则 Android 系统将会根据 Data 属性来分析数据的类型，因此无须指定 Type 属性。

(6) Extras(附加信息)：其他所有附加信息的集合，以键值对形式保存所有的附加信息。使用 extras 可以为组件提供扩展信息。例如，如果要执行"发送电子邮件"这个动作，可以将电子邮件的标题、正文等保存在 extras 里，传给电子邮件发送组件。Intent 类中包含一系列的 putXxx()方法用于插入各种类型的附加信息，相应地也提供了一系列的 getXxx()方法，用于获取附加信息。这些方法与 Bundle 中的方法相似。事实上，可以把所有的附加信息都放在一个 Bundle 对象中，然后把 Bundle 对象再添加到 Intent 中。

上面详细介绍了 Intent 对象的各个属性及其作用，那么系统又是如何根据 Intent 的属性来找到符合条件的组件的呢？首先，需要为组件配置相应的条件即指定该组件能被哪些 Intent 所启动，这主要是通过<intent-filter…/>元素的配置来实现的。

<intent-filter…/>元素是 AndroidManifest.xml 文件中某一组件的子元素，例如，<activity…/>元素的子元素，该子元素用于配置该 Activity 所能"响应"的 Intent。对于后面所学的 Service、BroadcastReceiver 组件也是类似的。

<intent-filter…/>元素里通常可包含如下子元素。

0～N 个<action…/>子元素。

0～N 个<category…/>子元素。

0～1 个<data…/>子元素。

当<activity…/>元素的<intent-filter…/>子元素里包含多个<action…/>子元素时，表

明该Activity能响应Action属性值为其中任意一个字符串的Intent,能被多个Intent启动。

视频详解

6.2.3 Intent实现Activity数据传递

Activity间数据传递的方法——采用Intent对象。

前面学习了Activity的生命周期、Activity间的跳转,实际应用中,仅有跳转还是不够的,往往还需要进行通信,即数据的传递。在Android中,主要是通过Intent对象来完成这一功能的,Intent对象就是它们之间的信使。

数据传递方向有两个:一个是从当前Activity传递到新启动的Activity,另一个是从新启动的Activity返回结果到当前Activity。下面详细讲解这两种情景下数据的传递。

在介绍Activity启动方式时,我们知道Activity提供了一个startActivityForResult(Intent intent,int requestCode)方法来启动其他Activity,该方法可以将新启动的Activity中的结果返回给当前Activity。如果要使用该方法,还必须做以下操作。

(1) 在当前Activity中重写onActivityResult(int requestCode,int resultCode,Intent intent)方法,其中,requestCode代表请求码,resultCode代表返回的结果码。

(2) 在启动的Activity执行结束前,调用该Activity的setResult(int resultCode,Intent intent)方法,将需要返回的结果写到Intent中。

整个执行过程为:当前Activity调用startActivityForResult(Intent intent,int requestCode)方法启动一个符合Intent要求的Activity之后,执行它相应的方法,并将执行结果通过setResult(int resultCode,Intent intent)方法写入Intent,当该Activity执行结束后,会调用原来Activity的onActivityResult(int requestCode,int resultCode,Intent intent),判断请求码和结果码是否符合要求,从而获取Intent里的数据。

请求码和结果码的作用:因为在一个Activity中可能存在多个控件,每个控件都有可能添加相应的事件处理,调用startActivityForResult()方法,从而就有可能打开多个不同的Activity处理不同的业务。但这些Activity关闭后,都会调用原来Activity的onActivityResult(int requestCode, int resultCode, Intent intent)方法。通过请求码,就知道该方法是由哪个控件所触发的,通过结果码,就知道返回的数据来自于哪个Activity。

Intent保存数据的方法:从当前Activity传递数据到新启动的Activity相对来说比较简单,只需要将需要传递的数据存到Intent即可。上面两种传值方式,都需要将数据存入Intent,那么Intent是如何保存数据的呢?Intent提供了多个重载的方法来存放额外的数据,主要格式如下。

putExtras(String name, Xxx data):其中,Xxx表示数据类型,向Intent中放入Xxx类型的数据,例如int、long、String等。

此外还提供了一个putExtras(Bundle data)方法,该方法可用于存放一个数据包,Bundle类似于Java中的Map对象,存放的是键值对的集合,可把多个相关数据放入同一个Bundle中,Bundle提供了一系列存入数据的方法,方法格式为putXxx(String key, Xxx data),向Bundle中放入int、long、String等各种类型的数据。为了取出Bundle数据携带包中的数据,Bundle还提供了相应的getXxx(String key)方法,从Bundle中取出各种类型的数据。

6.2.4 实战演练——跳转到商品详情页面

视频详解

根据任务分析借助 Intent 实现 Activity 的页面跳转和数据传递，在“欢乐购商城”项目中采取的案例是监听列表子项然后跳转进入商品详情页面。在整体项目中，商品详情数据从服务器中获取，因从服务器获取数据是后面章节内容，在本节中商品数据采用放在本地第一个 Activity 中，封装成集合，然后借助 Intent 跳转传递商品详情页面进行展示，效果如图 6-7 所示。

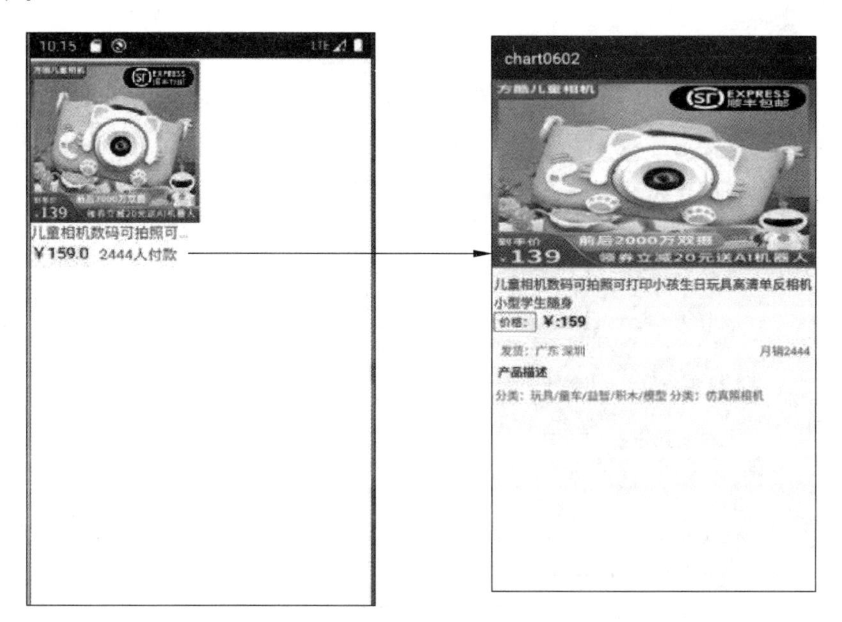

图 6-7　项目案例页面跳转

本节案例项目结构如图 6-8 所示。

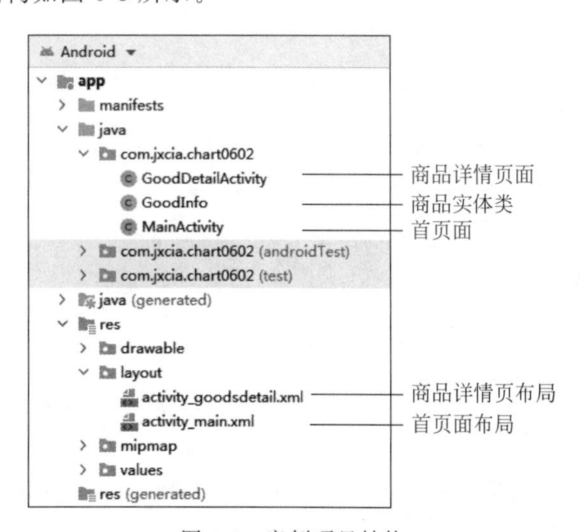

图 6-8　案例项目结构

首页面布局代码如下，其中需要对整体布局监听跳转。

程序清单 6-6：chart0602\app\src\main\res\layout\activity_main.xml

```
1   <RelativeLayout
2       android:id="@+id/itemclick"        →用于整体监听
3       xmlns:android="http://schemas.android.com/apk/res/android"
4       android:orientation="vertical"
5       android:layout_width="match_parent"
6       android:background="@color/white"
7       android:layout_height="match_parent">
8       <ImageView
9           android:id="@+id/iv_iamge"
10           android:src="@drawable/goodimg704"
11           android:layout_width="160dp"
12           android:layout_height="160dp"
13           android:layout_marginTop="4dp" />
14       <TextView
15           android:id="@+id/tv_title"
16           android:layout_width="160dp"
17           android:layout_height="wrap_content"
18           android:layout_below="@+id/iv_iamge"
19           android:ellipsize="end"
20           android:maxLines="1"
21           android:text="儿童相机数码可拍照可打印小孩生日玩具高清单反相机小型学生随身"
22           android:textSize="14dp" />
23       <TextView
24           android:id="@+id/tv_prices"
25           android:layout_width="wrap_content"
26           android:layout_height="wrap_content"
27           android:layout_below="@id/tv_title"
28           android:text="¥159.0"
29           android:textColor="#f00"
30           android:textSize="16sp" />
31       <TextView
32           android:id="@+id/tv_numbers"
33           android:layout_width="wrap_content"
34           android:layout_height="wrap_content"
35           android:layout_alignBottom="@id/tv_prices"
36           android:layout_marginLeft="10dp"
37           android:layout_toRightOf="@id/tv_prices"
38           android:text="2444 人付款" />
39   </RelativeLayout>
```

商品详情页面布局代码如下。

程序清单 6-7：chart0602\app\src\main\res\layout\activity_goodsdetail.xml

```
1   <RelativeLayout xmlns:android="http://schemas.android.com/apk/res/android"
2       android:layout_width="match_parent"
3       android:layout_height="match_parent">
```

```
4        <ImageView
5            android:id="@+id/iv_image"
6            android:layout_width="match_parent"
7            android:layout_height="240dp"
8            android:scaleType="fitXY"
9            android:src="@drawable/timg" />
10        <RelativeLayout
11            android:layout_width="match_parent"
12            android:layout_height="wrap_content"
13            android:layout_below="@+id/iv_image"
14            android:layout_margin="3dp">
15            <LinearLayout
16                android:id="@+id/ll_title"
17                android:layout_width="wrap_content"
18                android:layout_height="wrap_content"
19                android:layout_marginTop="8dp"
20                android:orientation="horizontal">
21                <TextView
22                    android:id="@+id/tv_title_detail"
23                    android:layout_width="wrap_content"
24                    android:layout_height="wrap_content"
25                  android:text="今日秒杀 云南天然干花花束真花勿忘我家居客厅摆设超
                                  大花束"
26                    android:textSize="16dp"
27                    android:textStyle="bold" />
28            </LinearLayout>
29            <LinearLayout
30                android:id="@+id/ll_price"
31                android:layout_width="match_parent"
32                android:layout_height="wrap_content"
33                android:layout_below="@+id/ll_title"
34                android:orientation="horizontal">
35                <TextView
36                    android:id="@+id/tv_s"
37                    android:layout_width="wrap_content"
38                    android:layout_height="wrap_content"
39                    android:background="@drawable/textview_borders"
40                    android:gravity="center"
41                    android:padding="3dp"
42                    android:text=" 价格:"
43                    android:layout_marginBottom="5dp"
44                    android:textColor="@color/colorPrimary" />
45                <TextView
46                    android:id="@+id/tv_price_detail"
47                    android:layout_width="wrap_content"
48                    android:layout_height="wrap_content"
49                    android:layout_marginLeft="5dp"
50                    android:text=" ¥ 1123.00"
51                    android:textColor="@color/colorPrimary"
```

```
52                          android:textSize = "18sp"
53                          android:textStyle = "bold" />
54                  </LinearLayout>
55                  <RelativeLayout
56                      android:id = "@ + id/rl_count"
57                      android:layout_width = "match_parent"
58                      android:layout_height = "wrap_content"
59                      android:layout_below = "@ + id/ll_price"
60                      android:layout_marginTop = "10dp">
61                      <TextView
62                          android:id = "@ + id/tv_address"
63                          android:layout_width = "wrap_content"
64                          android:layout_height = "wrap_content"
65                          android:layout_marginLeft = "8dp"
66                          android:text = "发货:" />
67                      <TextView
68                          android:id = "@ + id/tv_addresss"
69                          android:layout_width = "wrap_content"
70                          android:layout_height = "wrap_content"
71                          android:layout_toRightOf = "@ + id/tv_address"
72                          android:text = "江西" />
73                      <TextView
74                          android:layout_width = "wrap_content"
75                          android:layout_height = "wrap_content"
76                          android:layout_alignBottom = "@ + id/tv_addresss"
77                          android:layout_toLeftOf = "@ + id/tv_counts"
78                          android:text = "月销" />
79                      <TextView
80                          android:id = "@ + id/tv_counts"
81                          android:layout_width = "wrap_content"
82                          android:layout_height = "wrap_content"
83                          android:layout_alignBottom = "@ + id/tv_addresss"
84                          android:layout_alignParentRight = "true"
85                          android:layout_marginRight = "5dp"
86                          android:text = "0" />
87                  </RelativeLayout>
88                  <RelativeLayout
89                      android:id = "@ + id/rv_juan"
90                      android:layout_width = "wrap_content"
91                      android:layout_height = "wrap_content"
92                      android:layout_below = "@ + id/rl_count"
93                      android:layout_marginTop = "8dp">
94                      <TextView
95                          android:layout_width = "wrap_content"
96                          android:layout_height = "wrap_content"
97                          android:layout_marginLeft = "5dp"
98                          android:text = "产品描述"
99                          android:textColor = "@color/colorPrimary"
```

```
100                      android:textSize = "15sp"
101                      android:textStyle = "bold" />
102              </RelativeLayout >
103              < LinearLayout
104                  android:id = "@ + id/ll_copy_coupon"
105                  android:layout_width = "match_parent"
106                  android:layout_height = "wrap_content"
107                  android:layout_below = "@ + id/rv_juan"
108                  android:orientation = "horizontal">
109                  android:orientation = "horizontal">
110                  < TextView
111                      android:id = "@ + id/tv_content"
112                      android:layout_width = "match_parent"
113                      android:layout_height = "wrap_content"
114                      android:layout_margin = "5px"
115                      android:text = "今日秒杀云南天然" />
116              </LinearLayout >
117          </RelativeLayout >
118      </RelativeLayout >
```

由于从首页面跳转到商品详情页面需要携带多项数据，于是创建 GoodInfo 实体类用于封装商品对应信息，可能很多数据属性在本案例中没有用到，为了和"欢乐购商城"项目相吻合，把所有属性全部列出，GoodInfo 实体类代码如下。

程序清单 6-8：chart0602\app\src\main\java\com\jxcia\chart0602\GoodInfo.java

```
1    public class GoodInfo implements Serializable {
2        private int image;                          →图片地址
3        private String productLocation;             →商品生产地
4        private String name;                        →商品名称
5        private String price;                       →商品价格
6        private String info;                        →商品详情
7        private String volume;                      →商品销量
8        private String shopName;                    →商店名称
9        private String totalValue;                  →总额
10       private String count;                       →数量
11       private String dateTime;                    →日期
12       public int getImage() {
13           return image;
14       }
15       public void setImage(int image) {
16           this.image = image;
17       }
18       ...                                         →后面代码类似
19   }
```

在 MainActivity 中主要把数据封装成 goodInfo 实体对象，其中包含商品详细信息，把实体数据添加到 Intent 对象中。Intent 对象实现 Activity 页面跳转时，实现数据传递，具体代码如下。

程序清单 6-9：chart0602\app\src\main\java\com\jxcia\chart0602\MainActivity .java

```
1    public class MainActivity extends Activity{
2        private RelativeLayout itemclick;
3        private GoodInfo goodInfo;
4        @Override
5        protected void onCreate(Bundle savedInstanceState) {
6            super.onCreate(savedInstanceState);
7            setContentView(R.layout.activity_main);
8            goodInfo = new GoodInfo();                         →创建实体类
9            goodInfo.setImage(R.drawable.goodimg704);          →封装数据
10           goodInfo.setProductLocation("广东 深圳");
11           goodInfo.setName("儿童相机数码可拍照可打印小孩生日玩具高清单反相机小型
                        学生随身");
12           goodInfo.setInfo("分类:玩具/童车/益智/积木/模型\t 分类:仿真照相机");
13           goodInfo.setPrice("159");
14           goodInfo.setVolume("2444");
15           goodInfo.setShopName("方拓玩具专营店");
16           itemclick = findViewById(R.id.itemclick);
17           itemclick.setOnClickListener(new View.OnClickListener() {
18               @Override
19               public void onClick(View v) {
20                   Intent intent = new Intent(MainActivity.this,GoodDetailActivity.
  class);                                                       →页面跳转
21                   Bundle bundle = new Bundle();
22                   bundle.putSerializable("goodInfo",goodInfo);
                                                                →把数据封装的 bundle 兑现中
23                   intent.putExtras(bundle);
24                   startActivity(intent);                     →页面跳转
25               }
26           });
27       }
28   }
```

在商品详情页面通过 getIntent()方法接收跳转过来的数据，并把接收的数据分别对应地显示在布局文件上，用于给用户展示，具体代码如下。

程序清单 6-10：chart0602\app\src\main\java\com\jxcia\chart0602\GoodDetailActivity. java

```
1    public class GoodDetailActivity extends AppCompatActivity {
2        private TextView tv_title;
3        private TextView tv_price;
4        private TextView tv_address;
5        private TextView tv_counts;
6        private TextView tv_content;
7        private GoodInfo goodInfo;
8        private ImageView im_image;
9        @Override
10       protected void onCreate(Bundle savedInstanceState) {
11           super.onCreate(savedInstanceState);
```

```
12            setContentView(R.layout.activity_goodsdetail);
13            goodInfo = (GoodInfo) getIntent().getSerializableExtra("goodInfo");
              //接收 Activity 跳转传递过来的数据
14            initView();                    →获取控件 id
15            initData();                    →初始化数据
16        }
17        public void initView() {
18            tv_title = (TextView) findViewById(R.id.tv_title_detail);
19            tv_price = (TextView) findViewById(R.id.tv_price_detail);
20            tv_address = (TextView) findViewById(R.id.tv_addresss);
21            tv_counts = (TextView) findViewById(R.id.tv_counts);
22            tv_content = (TextView) findViewById(R.id.tv_content);
23            im_image = (ImageView) findViewById(R.id.iv_image);
24        }
25        public void initData() {
26            if (goodInfo != null) {
27                im_image.setImageResource(goodInfo.getImage());
                  //在控件上显示传递过来的图片地址
28                tv_title.setText(goodInfo.getName());
29                tv_price.setText("¥:" + goodInfo.getPrice());
30                tv_address.setText(goodInfo.getProductLocation());
31                tv_counts.setText(goodInfo.getVolume());
32                tv_content.setText(goodInfo.getInfo());
33            }
34        }
35    }
```

最后完成 AndroidManifest 清单文件商品详情 Activity 页面注册。

程序清单 6-11：chart0602\app\src\main\AndroidManifest.xml

```
1   <manifest xmlns:android="http://schemas.android.com/apk/res/android"
2       package="com.example.chart0501">
3       <application
4           android:allowBackup="true"
5           android:icon="@mipmap/ic_launcher"
6           android:label="@string/app_name"
7           android:roundIcon="@mipmap/ic_launcher_round"
8           android:supportsRtl="true"
9           android:theme="@style/AppTheme">
10           <activity android:name=".MainActivity">
11               <intent-filter>
12                   <action android:name="android.intent.action.MAIN" />
13                   <category android:name="android.intent.category.LAUNCHER" />
14               </intent-filter>
15           </activity>
16           <activity android:name=".GoodDetailActivity"/>     →注册商品详情注册
17       </application>
18   </manifest>
```

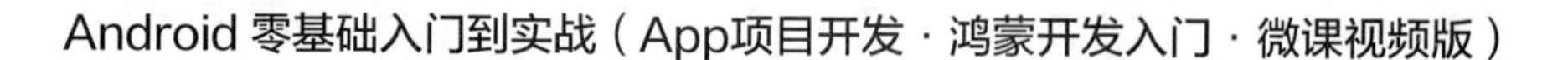

视频详解

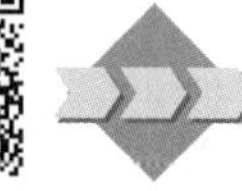

6.3 Fragment 详解

任务陈述

在日常使用的 App 软件中，为了在手机有限的空间中呈现更多的内容，经常用到底部导航栏。在“欢乐购商城”中包含多个页面，其中就使用到了底部导航栏。实现底部导航栏功能方法有多种，但是每种实现方式都要使用到 Fragment，在“欢乐购商城”中采用 Fragment＋Bottombar 进行实现。

分析：在“欢乐购商城”中包含 3 个 Fragment 页面，其中包括首页、购物车、个人中心，这 3 个 Fragment 在 MainActivity 中进行实现，其实现方式为 Fragment＋Bottombar，具体效果如图 6-9 所示。

图 6-9　欢乐购商城底部导航栏

视频详解

相关知识

6.3.1 Fragment 概述

Fragment 中文直译是“碎片”，是 Android 3.0 开始引入的组件，顾名思义，它就是浮在 Activity 上的一个碎块，主要是为了便于大屏 UI 的设计和实现。它有自己的生命周期，但是它的生命周期会受到加载 Fragment 的 Activity 的生命周期的约束。在手机 App 开发中，Fragment 多是用来制作主页的一个重要模块。就如市面上下载使用的大多数 App，首页一般是有 3～5 个大的模块，其中微信的主页就是由消息、通信录、发现、我 4 个大模块构成的。为什么开发者要这样设计 App 呢？因为每个 App 都是一个公司业务的融合，很多

App 的业务是非常庞大的，如果每个业务都设计成 Activity，那必定是非常多的 Activity 跳转，然后回到某个界面又要层层地单击返回。其次就是主页是一个 App 的第一展示界面，这里面应当包含 App 中所有业务的入口，还有一些广告公示之类的模块，那如果设计为一个 Activity 肯定是不够的，所以综合考虑现在的 App 基本都是主页里用多个 Fragment 来展示不同大类的业务板块。

Fragment 的优点如下。

(1) 代码复用。Activity 用来管理 Fragment。因为一个 Fragment 可以被多个 Activity 嵌套，有个共同的业务模块就可以复用了。

(2) 模块化。Fragment 具有自己的生命周期，是模块化 UI 的良好组件。

(3) 可控性。Fragment 的生命周期是寄托到 Activity 中的，Fragment 可以被 Attach 添加和 Detach 释放。

(4) 切换灵活。Fragments 是 View Controllers，它们包含可测试的、解耦的业务逻辑块，由于 Fragments 是构建在 Views 之上的，而 Views 很容易实现动画效果，因此 Fragments 在屏幕切换时具有更好的控制。

(5) 可控性。Fragment 可以像普通对象那样自由地创建和控制，传递参数更加容易和方便，也不用处理系统相关的事情，如显示方式、替换，不管是整体还是部分，都可以做到相应的更改。

6.3.2 Fragment 的管理和处理

视频详解

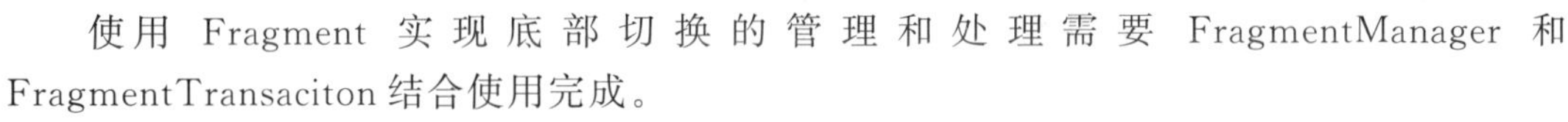

使用 Fragment 实现底部切换的管理和处理需要 FragmentManager 和 FragmentTransaciton 结合使用完成。

FragmentManager 是用来管理 Fragment 的容器，通常在 Activity 中通过接口 getSupportFragmentManager 或 getFragmentManager 获得 FragmentManager 对象，可以认为 Activity 是 FragmentManager 的宿主环境类。之前介绍过 Fragment 是寄生在 Activity 之上的一个块，Activity 也提供给了这样一个对象用来管理寄生在自己里面的 Fragment。

FragmentTransaciton 是 FragmentManager 通过 beginTransaciton()方法得到的一个对象，这个对象是直接用来给 FragmentManager 添加和移除 Fragment 的工具，也就是说，FragmentManager 通过它来控制哪个 Fragment 展示，哪个 Fragment 要退到幕后隐藏起来，其主要的 API 如下。

add(int containerViewId, Fragment fragment, String tag)：向 Activity state 中添加一个 Fragment。参数 containerViewId 一般会传 Activity 中某个视图容器的 id。如果 containerViewId 传 0，则这个 Fragment 不会被放置在一个容器中。(不要认为 Fragment 没添加进来，只是我们添加了一个没有视图的 Fragment，这个 Fragment 可以用来做一些类似于 Service 的后台工作。)

remove(Fragment fragment)：移除一个已经存在的 Fragment。Fragment 被 remove 后，Fragment 的生命周期会一直执行完，之后 Fragment 的实例也会从 FragmentManager 中移除。

hide(Fragment fragment)：隐藏一个已经存在的 Fragment，前提是这个 Fragment 已经被添加到容器里面了。

show(Fragment fragment)：显示一个已经存在的 Fragment，前提是这个 Fragment 已经被添加到容器里面了。如果隐藏和显示控制不得当会出现视图套层的情况。

视频详解

6.3.3 实战演练——底部导航栏

根据任务分析，本节主要实现底部导航栏，其中 3 个 Fragment 中的内容不做过多现实。在本项目中需要完成一个 MainActivity. java，在其中加载布局文件，通过使用 Java 语言逻辑处理，根据用户选择加载 3 个不同的 Fragment，并实现其布局文件，其基本框架结构如图 6-10 所示，具体实现步骤如下。

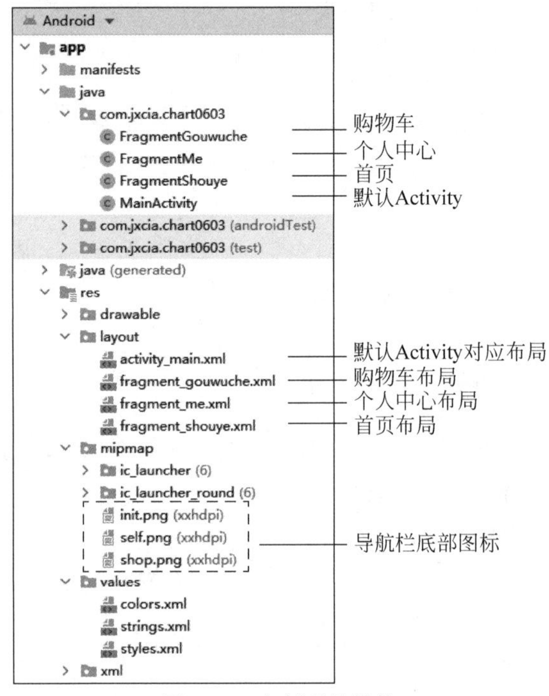

图 6-10　案例项目结构

第一步，导入 bottombar 包。在实现底部导航栏时需要加载第三方包，加载包只需要在 build. gradle 中添加 bottombar 相关依赖信息，并进行同步即可下载 bottombar 包，如图 6-11 所示。

```
dependencies {
    implementation fileTree(dir: 'libs', include: ['*.jar'])
    implementation 'androidx.appcompat:appcompat:1.1.0'
    implementation 'androidx.constraintlayout:constraintlayout:1.1.3'
    testImplementation 'junit:junit:4.12'
    androidTestImplementation 'androidx.test.ext:junit:1.1.1'
    androidTestImplementation 'androidx.test.espresso:espresso-core:3.2.0'
    implementation 'com.roughike:bottom-bar:2.3.1'
}
```

图 6-11　导入 bottombar 包

第二步，定义导航栏布局。在使用 bottombar 实现底部导航栏前，需要定义底部导航栏布局文件，首先在 res 文件夹下新建 xml 文件夹，在其中新建 bottombar_menu. xml 文件定义导航栏布局，代码如下。

程序清单 6-12：chart0603\app\src\main\res\xml\bottombar_menu. xml

```
1    <tabs
2      xmlns:android = "http://schemas.android.com/apk/res/android">
3        <tab
4            id = "@ + id/tab_yi"
5            icon = "@mipmap/init"
6            title = "首页"
7            />
8        <tab
9            id = "@ + id/tab_er"
10            title = "购物车"
11            icon = "@mipmap/shop"
12            />
13         <tab
14            id = "@ + id/tab_san"
15            icon = "@mipmap/self"
16            title = "个人中心"
17            />
18     </tabs>
```

第三步，定义 MainActivity 布局文件。需要定一个 MainActivity 用于放置其他的几个 Fragment，这里需要注意的是，需要继承 AppCompatActivity 或者 FragmentActivity，因为这样才可以使用 getSupportFragmentManager()来初始化 FragmentManager 对象。这里 Android 的编译器会提示，不建议使用老的 getFragmentManager()方法。主页的布局文件由两大部分构成，一部分是处于最下方的底部导航栏，剩余其他位置都属于 Fragment。

MainActivity 中的对应布局如下。

程序清单 6-13：chart0603\app\src\main\res\layout\main_activity. xml

```
1    <LinearLayout xmlns:android = "http://schemas.android.com/apk/res/android"
2        xmlns:app = "http://schemas.android.com/apk/res - auto"
3        xmlns:tools = "http://schemas.android.com/tools"
4        android:layout_width = "match_parent"
5        android:layout_height = "match_parent"
6        android:orientation = "vertical"
7        tools:context = ".MainActivity">
8        <FrameLayout        →用于显示底部导航栏剩余部分
9             android:id = "@ + id/contentContainer"
10             android:layout_width = "match_parent"
11             android:layout_height = "0dp"
12             android:layout_weight = "1"
13             />
14         <com.roughike.bottombar.BottomBar
15             android:id = "@ + id/bottomBar"
16             android:layout_width = "match_parent"
17             android:layout_height = "56dp"
18             app:bb_tabXmlResource = "@xml/bottombar_menu"/>
19     </LinearLayout>
```

第四步，定义 Fragment 类布局文件。三个不同的 Fragment 页面需要对应的布局，需要创建三个布局页面，这里的布局文件可自拟不做具体要求，例子中仅有一个 TextView 进行效果的展示。因为代码相似，在此只列出“首页”Fragment 的布局文件，其他只是内部 TextView 显示内容不一样，具体代码如下。

程序清单 6-14：chart0603\app\src\main\res\layout\fragment_shouye. xml

```
1    <LinearLayout
2        xmlns:android = "http://schemas.android.com/apk/res/android"
3        android:orientation = "vertical"
4        android:gravity = "center"
5        android:layout_width = "match_parent"
6        android:layout_height = "match_parent">
7        <TextView
8            android:text = "首页"
9            android:textSize = "30sp"
10           android:layout_width = "wrap_content"
11           android:layout_height = "wrap_content"/>
12   </LinearLayout>
```

第五步，定义 Fragment 类。接下来要定义 3 个不同的 Fragment，这里需要注意的是，继承的 Fragment，继承之后需要重写 onCreateView()方法，用布局关联器 LayoutInflater 去关联一个布局文件，作为当前 Fragment 展示的界面，并通过 View 对象放置在函数返回值中。因为 3 个 Fragment 类比较类似，在此只展示“首页”Fragment 的代码。

程序清单 6-15：chart0603\app\src\main\java\com\jxcia\chart0603\FragmentShouye. java

```
1    public class FragmentShouye extends Fragment {
2        @Nullable
3        @Override
4        public View onCreateView(@NonNull LayoutInflater inflater,
5            @Nullable ViewGroup container, @Nullable Bundle savedInstanceState) {
6                View view = inflater.inflate(R.layout.fragment_shouye,
                         container, false);
7            return view;
8        }
9    }
```

第六步，完成 MainActivity 逻辑调用。完成了布局文件的编辑以及每个 Fragment 类的编辑以后，开始最繁重的工作——MainActivity 类的代码编辑，具体实现逻辑如下。

(1) 对 bottombar 控件进行监听。

(2) 根据用户单击 Tab 项，进行 switch 判断。

(3) 根据判断结果获取 Fragment 对象。

(4) 通过 FragmentManager 得到一个 FragmentTransaciton 对象进行 Fragment 操作。

(5) 最后提交，呈现关联页面。

程序清单 6-16：chart0603\app\src\main\java\com\jxcia\chart0603\MainActivity. java

```
1    public class MainActivity extends AppCompatActivity {
2        public BottomBar bottomBar;
3        @Override
```

```
4        protected void onCreate(Bundle savedInstanceState) {
5            super.onCreate(savedInstanceState);
6            setContentView(R.layout.activity_main);
7            bottomBar = findViewById(R.id.bottomBar);
8            bottomBar.setOnTabSelectListener(new OnTabSelectListener() {
9                @Override
10               public void onTabSelected(int tabId) {
11                   Object ob = null;
12                   switch (tabId){
13                       case R.id.tab_yi:
14                           ob = new FragmentShouye();
15                           break;
16                       case R.id.tab_er:
17                           ob = new FragmentGouwuche();
18                           break;
19                       case R.id.tab_san:
20                           ob = new FragmentMe();
21                           break;
22                       default:
23                           break;
24                   }
25                   getSupportFragmentManager().beginTransaction().
26                           replace(R.id.contentContainer,(Fragment) ob).commit();
27               }
28           });
29       }
30   }
```

效果图如图 6-12 和图 6-13 所示。

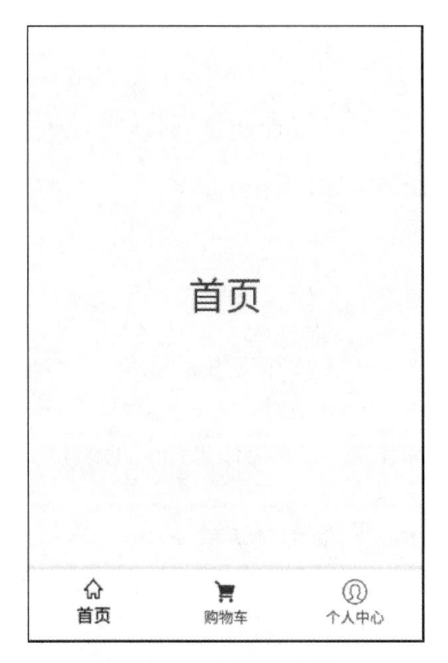

图 6-12 “首页”页面

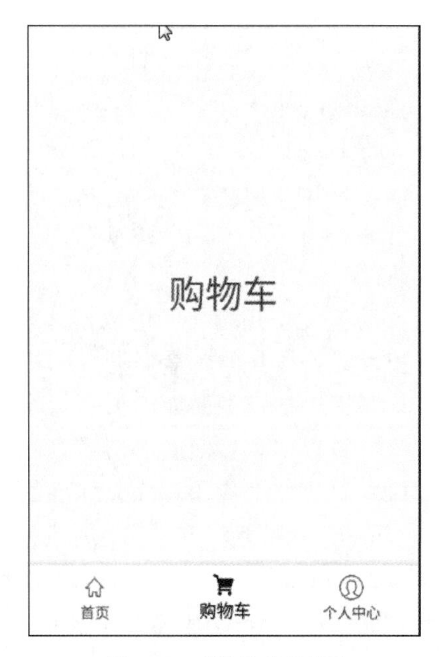

图 6-13 “购物车”页面

本章小结

本章围绕“欢乐购商城”项目主页、商品页面引入 Android 中负责界面和交互的两个重要模块 Activity 和 Fragment。包括 Activity 的生命周期和它的运行机制，完善 Android 时如何管理 Activity 这部分的知识。详细介绍 Activity 的跳转和 Activity 之间的数据传递，包括 Activity 的顺向数据传递、逆向数据传递以及大型数据传递等，让每个 Activity 之间通过这种方式更加紧密地连接起来，从而实现各种复杂的业务设计。

随后介绍了 Fragment 这个使用广泛的模块，以市场上很多 App 进行举例，设计并实现了经典的 App 主页，同时也学习了 Fragment 的生命周期以及 Activity 是如何对 Fragment 进行管理的。学习以后可以自己设计一款 App 的主页进行实现，尽量参考市面上的主流 App（如微信、淘宝等），用来巩固自己的学习内容。

自测习题

1. 以下方法不属于 Activity 生命周期的回调方法的是（　　）。
   A. onStart()　　B. onCreate()　　C. onPause()　　D. onFinish()
2. 以下方法中，在 Activity 的生命周期中不一定被调用的是（　　）。
   A. onCreate()　　B. onStart()　　C. onPause()　　D. onStop()
3. 对于 Activity 中一些重要资源与状态的保存最好在生命周期的（　　）函数中进行。
   A. onPause()　　B. onCreate()　　C. onResume()　　D. onStart()
4. 配置 Activity 时，下列（　　）是必不可少的。
   A. android：name 属性　　B. android：icon 属性
   C. android：label 属性　　D. <intent-filter…/>元素
5. 下列选项（　　）不能启动 Activity 的方法。
   A. startActivity　　B. goToActivity
   C. startActivityForResult　　D. startActivityFromChild
6. Android 中下列属于 Intent 的作用的是（　　）。
   A. 实现应用程序间的数据共享
   B. 是一段长的生命周期，没有用户界面的程序，可以保持应用在后台运行，而不会因为切换页面而消失
   C. 可以实现界面间的切换，可以包含动作和动作数据，连接四大组件的纽带
   D. 处理一个应用程序整体性的工作
7. Intent 的以下（　　）属性通常用于在多个 Action 之间进行数据交换。
   A. Category　　B. Component　　C. Data　　D. Extra
8. 简要描述 Activity 的生命周期。

第7章 Android中的数据存取

一个比较好的应用程序，应该能够为用户提供一些个性化的设置，能够保存用户的使用记录，而这些都离不开数据的存储。Android 系统提供了多种数据存储方式，开发者可根据具体情景选择合适的存储方式，比如数据是仅限于本应用程序访问还是允许其他应用程序访问，数据是结构化的还是非结构化的等。针对不同的使用场景，Android 系统中数据存储的方式大致分为以下五种。

（1）文件存储：Android 提供了 openFileOutput()和 openFileInput()方法读取设备上的文件，本质上是以 Java 的 I/O 流方式读取数据，文件存储的关键是获取相应的输入流和输出流。

（2）SharedPreferences：是 Android 提供的用来存储简单配置信息的一种方式，它采用键值对的形式以 XML 格式将数据存储到设备中。通常使用 ShardPeferences 存储一些应用程序的配置信息，如用户名、密码等。

（3）SQLite 数据库：SQLite 是 Android 自带的一个轻量级的数据库，它没有后台进程，整个数据库对应于一个文件。Android 为访问 SQLite 数据库提供了大量便捷的 API。并且支持基本 SQL 语法，一般使用它作为复杂数据的存储引擎，把相对复杂的结构化数据存储在本地。

（4）ContentProvider（内容提供者）：Android 四大组件之一，用于在应用程序间共享数据，它可以将自己的数据共享给其他应用程序使用，是不同应用程序之间进行数据交换的标准 API。

（5）网络数据读取：通过网络向服务器发送请求，获取响应数据或者将数据存储到服务器上。

本章将重点针对文件存储、SharedPreferences、SQLite 数据库进行讲解。需要注意的是，上述数据存储方式各有优缺点，因此需要根据开发需求选择合适的数据存储方式。

学习目标

本章要点

（1）了解五种不同的存储方式，并掌握不同存储方式的特点。

（2）掌握如何使用文件存储数据。

（3）掌握 SharedPreferences 的使用，实现数据存储功能。

（4）掌握 SQLite 数据库的使用。

视频详解

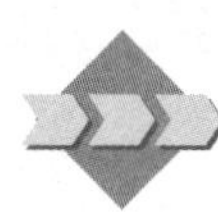

7.1 文件存储

任务陈述

“快乐购商城”项目中在登录页面当用户输入用户名和密码之后，再次登录需要实现自动登录，此时可以使用文件存储的方式实现此功能。

分析：登录页面当用户第一次登录的时候需要输入用户名和密码，当用户输入正确的用户名和密码后将用户名和密码保存在本地设备上，当用户下一次登录时，自动读取保存的用户名和密码，实现自动登录，如图 7-1 所示。

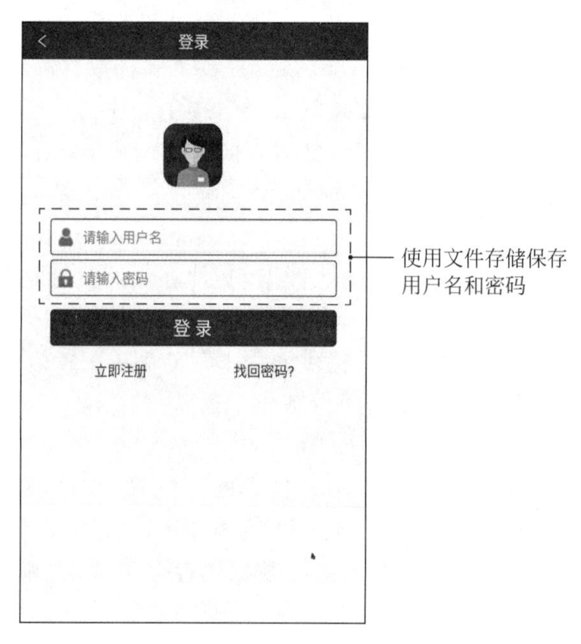

图 7-1　登录页面

相关知识

文件存储是 Android 中最基本的一种数据存储方式。Android 可以使用 Java 语言开发，Java 提供了一套完整的输入/输出流操作体系，与文件相关的有 FileInputStream、FileOutputStream 等，通过这些类可以非常方便地访问磁盘上的文件内容。同样，Android 也支持这种方式来访问手机上的文件。Android 手机中的文件有两个存储位置：内置存储空间和外部 SD 卡。接下来对这两种存储方式进行详细的讲解。

视频详解

7.1.1 内部存储空间文件的存取

内部存储是指将数据以文件的形式存储到应用程序中。对于 Android 应用来说，内部存储的数据属于应用程序的私有数据，如果其他应用程序想要操作本应用程序的文件，需要设置权限。当应用卸载以后,内部存储的数据也就清空了。为了保证内存数据的安全,不让用户直接定位访问,Android 中对内部存储空间中文件的读取进行了封装,用户不需要知道具体的存储路径就可以打开相应的文件输入/输出流。在 Context 类中提供了两个方法来打开文件 I/O 流。

FileInputStream openFileInput(String name):获取应用程序中名为 name 的文件对应的文件输入流。

FileOutputStream openFileOutput(String name, int mode):获取应用程序名为 name 的文件对应的文件输出流。

name 参数表示读取或者存入指定文件的文件名,不能包含路径分隔符“\”。如果文件不存在 Android 会自动创建该文件。mode 参数用于指定操作模式,Context 类中定义了四种操作模式常量,分别如下。

(1) Context.MODE_PRIVATE:为默认操作模式,代表该文件是私有数据,只能被应用本身访问,在该模式下写入的内容会覆盖原文件的内容。

(2) Context.MODE_APPEND:附加模式,会检查文件是否存在,存在就往文件后追加内容,否则创建新文件再写入内容。

(3) Context.MODE_WORLD_READABLE:表示当前文件可以被其他应用读取。

(4) Context.MODE_WORLD_WRITEABLE:表示当前文件可以被其他应用写入。

提示:如果希望文件既能被其他应用读也能写,可以传入:Context.MODE_WORLD_READABLE+Contex.MODE_WORLD_WRITEABLE 或者直接传入数值 3,四种模式中除了 Contex.MODE_ APPEND 会将内容追加到文件末尾,其他模式都会覆盖掉原文件的内容。

注意:在 Android 高版本中已经废弃了 MODE_WORLD_READABLE 和 MODE_WORLD_WRITEABLE 两种模式,因为让其他应用访问具体的文件是一件很危险的事情,容易导致安全漏洞。建议采用更好的机制,例如,通过 ContentProvider 暴露访问接口,或者通过服务或广播。

在手机上创建文件和向文件中追加内容的步骤如下。

(1) 调用 openFileOutput()方法传入文件的名称和操作的模式,该方法将会返回一个文件输出流。

(2) 调用文件输出流的 write()方法,向文件中写入内容。

(3) 调用文件输出流的 close()方法,关闭文件输出流。

读取手机上文件的一般步骤如下。

(1) 调用 openFileInput()方法传入读取数据的文件名,该方法返回一个文件输入流对象。

(2) 调用文件输入流的 read()方法读取文件的内容。

(3) 调用文件输入流的 close()方法关闭文件输入流。

下面以一个简单的例子来演示文件读取的操作,界面包括一个 EditText、一个 TextView 和两个 Button 控件。EditText 用于获取用户的输入,TextView 用于展示从文件中读取的数据。当用户单击第一个 Button 按钮时,将数据写入到文件中。当用户单击第二个 Button 按钮时,获取文件中的内容。

用于控件布局文件的核心代码如下。

程序清单 7-1：chart0701\app\src\main\res\layout\activity_main. xml

```
1    <?xml version = "1.0" encoding = "utf - 8"?>
2    < LinearLayout xmlns:android = "http://schemas.android.com/apk/res/android"
3    xmlns:tools = "http://schemas.android.com/tools"
4        android:layout_width = "match_parent"
5        android:layout_height = "match_parent"
6        android:orientation = "vertical"
7        tools:context = ".MainActivity">
8        < EditText
9             android:id = "@ + id/edit_content"
10            android:layout_width = "match_parent"
11            android:layout_height = "wrap_content"
12            android:hint = "请输入内容" />
13        < Button
14            android:id = "@ + id/btn_save"
15            android:layout_width = "wrap_content"
16            android:layout_height = "wrap_content"
17            android:text = "保存到手机内部存储空间" />
18        < TextView
19            android:id = "@ + id/tv_content"
20            android:layout_width = "match_parent"
21            android:layout_height = "wrap_content"
22            android:lines = "3"
23            android:text = "读取的内容" />
24        < Button
25            android:id = "@ + id/btn_read"
26            android:layout_width = "wrap_content"
27            android:layout_height = "wrap_content"
28            android:text = "读取文件" />
29     </LinearLayout >
```

编写界面交互代码，在 MainActivity 中为按钮绑定单击事件，实现对数据的保存和读取，代码如下。

程序清单 7-2：chart0701\java\com\example\administrator\chart0701\MainActivity. java

```
1    public class MainActivity extends AppCompatActivity {
2        private EditText editText;
3        private TextView textView;
4        private Button buttonSave;
5        private Button buttonRead;
6        private String fileName  =  "content.txt";
7        @Override
8        protected void onCreate(Bundle savedInstanceState) {
9             super.onCreate(savedInstanceState);
10            setContentView(R.layout.activity_main);
11            initView();
```

```
12      }
13
14
15      private void initView() {
16          editText = (EditText) findViewById(R.id.edit_content);
17          textView = (TextView) findViewById(R.id.tv_content);
18          buttonSave = (Button) findViewById(R.id.btn_save);
19          buttonRead = (Button) findViewById(R.id.btn_read);
20            buttonSave.setOnClickListener(new View.OnClickListener() {
21              @Override
22              public void onClick(View v) {
23                  String content = editText.getText().toString();
24                  try {
25                      FileOutputStream fos = openFileOutput(fileName, MODE_PRIVATE);
26                      fos.write(content.getBytes());
27                      fos.close();
28                  } catch (Exception e) {
29                      e.printStackTrace();
32                  }
33              }
34          });
35
36          buttonRead.setOnClickListener(new View.OnClickListener() {
37              @Override
38              public void onClick(View v) {
39                  try {
40                      FileInputStream fis = openFileInput(fileName);
41                      byte[] bytes = new byte[fis.available()];
42                      fis.read(bytes);
43                      String content = new String(bytes);
44                      textView.setText(content);
45                      fis.close();
46                  } catch (Exception e) {
47                      e.printStackTrace();
48                  }
49              }
50          });
51      }
52  }
```

上述代码中，第20～34行代码为“保存到手机内部存储空间”按钮绑定单击事件，在这段代码中首先获取用户的输入，再通过openFileOutput()方法获取到文件输入/输出流对象，并传入两个参数，第一个是文件名称，第二个是操作模式，之后再调用write()方法向文件中写入内容，最后通过close()方法关闭文件。

第36～50行代码为“读取文件”按钮绑定单击事件。在上述代码中首先通过openFileInput()方法获取文件输入流对象。然后通过available()方法获取文件的长度并创建相应大小的byte数组用于存取读入的数据，再通过read()方法将文件内容读取到byte

数组中，最后将读取到的内容转换成指定字符串。

当程序运行之后，首先在文本编辑框中输入"I love China"。单击"保存到手机内部存储空间"按钮。系统会首先查找手机上是否存在该文件，如果不存在就创建该文件，应用程序的数据文件默认保存在\data\data\< package name >\files 目录下。其中，package name 为当前应用程序的包名。

为了验证生成文件是否成功，可以通过 Device File Explorer 视图找到 data/data/com. example. administrator. chart0701/files 目录下的 content. txt 文件，如图 7-2 所示。双击 Device File Explorer 视图中的 content. txt 即可在 Android Studio 编辑框中查看 content. txt 文件中存储的数据，此时说明存储成功。

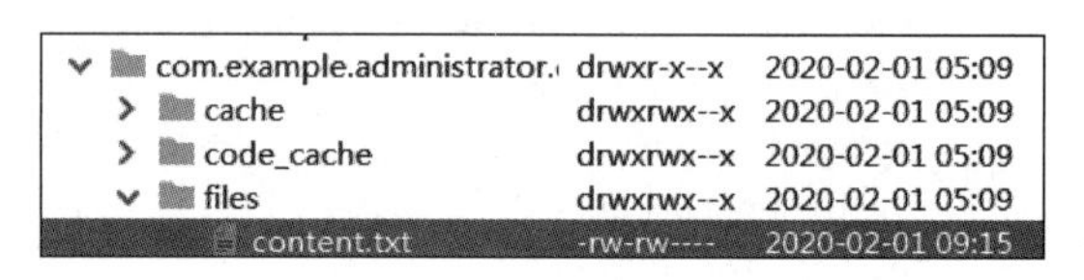

图 7-2　文件的存储位置

之后再单击"读取文件"按钮，将 content. txt 文件的内容读取到应用程序中，并将数据展示在 TextView 控件中。

该代码的显示效果如图 7-3 所示。

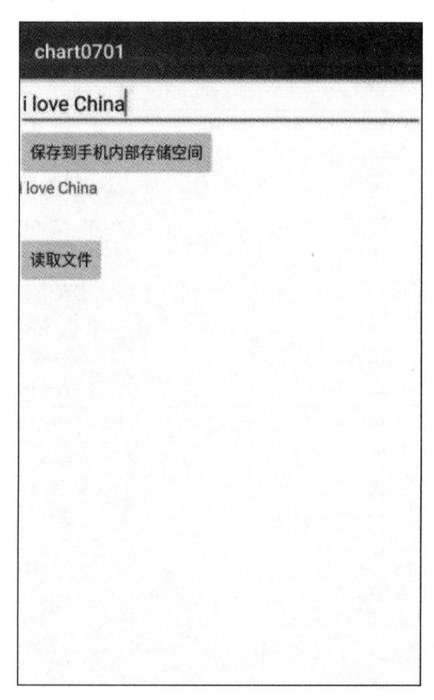

图 7-3　使用手机存储空间读写数据

视频详解

7.1.2　读取 SD 卡上的文件

前面学习了如何读取手机内存中的文件，内存的空间直接会影响到手机的运行速度，通常不建议将数据保存到手机内存中，特别是一些比较大的资源，如图片、音频、视频等。而是将这些数据保存在外部存储，如 SD 卡或者设备内嵌的存储卡中，这种存储属于永久性存储，其中比较常见的就是 SD 卡。

读取 SD 卡上的文件和读取手机上的文件类似，都是通过文件操作流的方式进行读取的，Android 中没有提供单独的 SD 卡文件操作类，直接使用 Java 中的文件操作即可。因为 SD 卡的可移动性，可能被移除或者丢失。并且不是所有手机都有 SD 卡，还有可能 SD 卡损坏或安装不正确等。因此，在访问之前需要验证手机的 SD 卡的状态，Android 提供了 Environment 类来完成这一操作，当外部设备可用并且具有读写权限时，就可以通过 FileInputStream 和 FileOutputStream 对象来读写外部设备中的文件。

SD 卡中的数据涉及用户的隐私，访问时需要申请相关的权限，需要使用到运行权限，即在程序运行时时提示用户进行授权。因此，读、写 SD 卡上文件的主要步骤如下。

(1) 调用 Environment 的 getExternalStorageState()方法判断手机上是否插入了 SD 卡，并且 SD 卡是否正常读写。Environment. getExternalStorageState()方法用于获取 SD

卡的状态，如果手机装有 SD 卡，并且可以进行读写，那么方法返回的状态等于 Environment. MEDIA _MOUNTED。

（2）判断用户是否授权，如果没有授权，请求授权，如果已授权则执行下一步。

（3）调用 Environment 的 getExternalStorageDirectory()方法来获取外部存储器的目录，也就是 SD 卡的目录（如果知道 SD 卡目录，可以使用绝对路径表示，但不提倡，因为不同版本可能路径不同。）

（4）使用 FileInputStream、FileOutputStream 等读写 SD 卡的文件。

提示：为了保证应用程序的安全性，Android 系统规定程序访问系统的一些关键信息时，必须申请权限，否则程序运行时会因为没有访问系统信息的权限而直接崩溃。根据程序适配的 Android SDK 版本的不同，申请权限分为静态申请权限和动态申请权限两种。

1. 静态申请权限

静态申请权限的方式适用于 Android SDK 6.0 以下的版本。该方式是在清单文件（AndroidManifest. xml）的< manifest >节点中声明需要申请的权限。以申请 SD 卡的读权限为例，代码如下。

```
<uses-permission android:name="android.permission.READ_EXTERNAL_STORAGE"/>
```

2. 动态申请权限

当程序适配的 Android SDK 版本为 6.0 及以上时，Android 改变了权限的管理模式，权限被分为正常权限和危险权限，具体如下。

（1）正常权限：表示不会直接给用户隐私权带来风险的权限，如请求网络的权限。

（2）危险权限：表示涉及用户隐私的权限，申请了该权限的应用可能涉及用户隐私信息的数据或资源，也可能对用户存储的数据或其他应用的操作产生影响。危险权限一共有九组，分别为位置（LOCATION）、日历（CALENDAR）、照相机（CAMERA）、联系人（CONTACTS）、存储卡（STORAGE）、传感器（SENSORS）、麦克风（MICROPHONE）、电话（PHONE）和短信（SMS）的相关权限。

申请正常权限时使用静态申请权限的方式即可，但是对于一些涉及用户隐私的危险权限 需要用户的授权才可以使用，因此危险权限不仅需要在清单文件（AndroidManifest. xml）的< manifest >节点中添加权限，还需要在代码中动态申请权限，以动态申请 SD 卡的读权限为例说明：

```
ActivityCompat.requestPermissions(MainActivity.this,new String[]{Manifest.permission.READ_
    EXTERNAL_STORAGE},1);
```

requestPermissions()方法中包含 3 个参数，第 1 个参数为 Context 上下文，第 2 个参数为需要申请的权限，第 3 个参数为请求码。添加完动态申请权限后，运行程序，界面上会弹出是否允许申请权限的对话框，由用户进行授权，如图 7-4 所示。

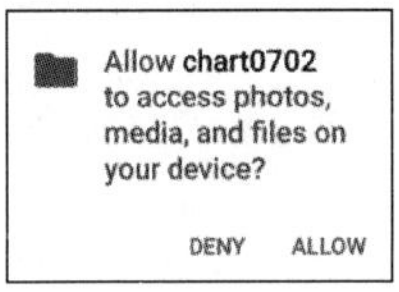

Allow chart0702 to access photos, media, and files on your device?
DENY ALLOW

图 7-4 弹出对话框

当用户单击 ALLOW 按钮时，表示允许授权。此时程序会执行动态申请权限的回调方法 onRequestPermissionsResult()，在该方法中可以获取用户授权申请权限的结果。实例代码如下。

```
public void onRequestPermissionsResult(int requestCode, @NonNull String[] permissions,
  @NonNull int[] grantResults) {
    super.onRequestPermissionsResult(requestCode, permissions, grantResults);
    if (requestCode == 1) {
        for (int i = 0; i < permissions.length; i++) {
            if (permissions[i].equals
                    ("android.permission.READ_EXTERNAL_STORAGE")
                   && grantResults[i] == PackageManager.PERMISSION_GRANTED) {
                Toast.makeText(this, "授权成功", Toast.LENGTH_SHORT).show();
            } else {
                Toast.makeText(this, "拒绝授权", Toast.LENGTH_SHORT).show();
            }
        }
    }
}
```

在上述代码中，onRequstPermissionsResult()方法中包含 3 个参数：requestCode、permissions、grantResults，分别表示请求码、请求的权限和用户授予权限的结果。当用户授予 SD 卡读权限时，对应该权限的 grantResults 数组中的值为 PackageManager.PERMISSION_GRANTED。

仍然以上一小节的程序为例，展示把数据写入到 SD 卡，对于布局页面保持一致，在此就不再列出。

读写 SD 卡上的数据的第一步，必须在清单文件(AndroidManifest.xml)的< manifest >节点中声明需要申请的读写 SD 卡的权限，如下。

```
<uses-permission android:name="android.permission.READ_EXTERNAL_STORAGE" />
<uses-permission android:name="android.permission.WRITE_EXTERNAL_STORAGE" />
```

在 MainActivity 代码页面中首先要判断是否存在 SD 卡，如果存在 SD 卡判断是否有相关的授权，如果没有授权请求授权，授权成功之后利用 IO 输入/输出流对文件进行读写操作。关键代码如下。

程序清单 7-3：char0702\app\src\main\java\com\example\administrator\chart0702\MainActivity.java

```
1    public class MainActivity extends AppCompatActivity {
2        private EditText editText;
3        private TextView textView;
4        private Button buttonSave;
5        private Button buttonRead;
6        private String fileName = "sdtext.txt";
7        @Override
```

```
8        protected void onCreate(Bundle savedInstanceState) {
9            super.onCreate(savedInstanceState);
10           setContentView(R.layout.activity_main);
11           initView();
12       }
13
14
15
16       private void initView() {
17           editText = (EditText) findViewById(R.id.edit_content);
18           textView = (TextView) findViewById(R.id.tv_content);
19           buttonSave = (Button) findViewById(R.id.btn_save);
20           buttonRead = (Button) findViewById(R.id.btn_read);
21           buttonSave.setOnClickListener(new View.OnClickListener() {
22               @Override
23               public void onClick(View v) {
24                   if(Environment.getExternalStorageState().equals(Environment.MEDIA_
  MOUNTED)) {
25                       if ( ContextCompat. checkSelfPermission ( MainActivity. this,
  Manifest.permission
26                       .WRITE_EXTERNAL_STORAGE) != PackageManager.PERMISSION_GRANTED) {
27                               ActivityCompat. requestPermissions ( MainActivity.
  this, new
28                               String[]{Manifest.permission.WRITE_EXTERNAL_STORAGE},
2);
29                       } else {
30                       writeToSD();
31                       }
32                   }
33               }
34           });
35           buttonRead.setOnClickListener(new View.OnClickListener() {
36               @Override
37               public void onClick(View v) {
38                   if(Environment.getExternalStorageState().equals(Environment.MEDIA_
  MOUNTED)) {
39                   if(ContextCompat. checkSelfPermission ( MainActivity. this,
  Manifest.permission
40                               .READ_EXTERNAL_STORAGE) != PackageManager.PERMISSION_
  GRANTED) {
41                           ActivityCompat.requestPermissions(MainActivity.this, new
42                               String[] { Manifest. permission. READ _ EXTERNAL _
  STORAGE}, 1);
43                       } else {
44                           readFromSD();
45                       }
46                   }
47               }
48           });
49       }
```

```
50        @Override
51        public void onRequestPermissionsResult(int requestCode, @NonNull String[]
permissions,
52                @NonNull int[] grantResults) {
53            super. onRequestPermissionsResult ( requestCode, permissions,
grantResults);
54            if (requestCode == 1) {
55                for (int i = 0; i < permissions.length; i++) {
56                    if(permissions[i].equals("android.permission.READ_EXTERNAL_
STORAGE")
57                            && grantResults[i] == PackageManager.PERMISSION_
GRANTED) {
58                        readFromSD();
59                    } else {
60                        Toast.makeText(this, "拒绝授权", Toast.LENGTH_SHORT).show();
61                    }
62                }
63            }
64            if (requestCode == 2) {
65                for (int i = 0; i < permissions.length; i++) {
66                    if(permissions[i].equals("android.permission.WRITE_EXTERNAL_
STORAGE")
67                            && grantResults[i] == PackageManager.PERMISSION_
GRANTED) {
68                        writeToSD();
69                    } else {
70                        Toast.makeText(this, "拒绝授权", Toast.LENGTH_SHORT).show();
71                    }
72                }
73            }
74        }
75        private void readFromSD() {
76            File SDPath = Environment.getExternalStorageDirectory();
77            File file = new File(SDPath, fileName);
78            try {
79                FileInputStream fis = new FileInputStream(file);
80                BufferedReader reader = new BufferedReader(new InputStreamReader(fis));
81                String content = reader.readLine();
82                textView.setText(content);
83                fis.close();
84            } catch (Exception e) {
85                e.printStackTrace();
86            }
87        }
88        private void writeToSD() {
89            String content = editText.getText().toString();
90            File SDPath = Environment.getExternalStorageDirectory();
91            try {
92              File file = new File(SDPath, fileName);
93              FileOutputStream fos = new FileOutputStream(file);
```

```
94                  fos.write(content.getBytes());
95                  fos.close();
96              } catch (Exception e) {
97                  e.printStackTrace();
98              }
99          }
100     }
```

在上述代码中第21～34行代码为“保存到SD卡”注册单击事件，在单击事件中首先判断SD卡是否可用，如果可用检查是否动态授权，如果没有授权第27～28行申请授权。如果已经动态授权，第30行代码直接写入到SD卡的文件中。

第35～49行代码为“读取SD卡文件”注册单击事件，同样地，首先判断SD卡是否可用，如果可用检查是否动态授权，如果没有授权第41～42行申请授权。如果已经动态授权，第44行代码直接写入到SD卡的文件中。

第51～74行代码，在用户手动选择是否授权之后执行回调函数，在回调函数中判断用户是否授权，如果授权则执行相应的方法执行读写文件的操作。如果没有授权则提示用户“拒绝授权”。

第75～99行代码封装读取SD卡文件和写入到SD卡文件的方法，在方法中利用Java中的输入/输出流实现对文件的读取和存储。

先在文本输入框中输入“I love Android”，单击“保存到SD卡”按钮将数据保存在SD卡中。为了验证生成文件是否成功，可以通过Device File Explorer视图找到storage\emulated\0\目录下sdtext.txt文件，如图7-5所示。双击Device File Explorer视图中的sdtext.txt即可在Android Studio编辑框中查看sdtext.txt文件中存储的数据，此时说明存储成功。

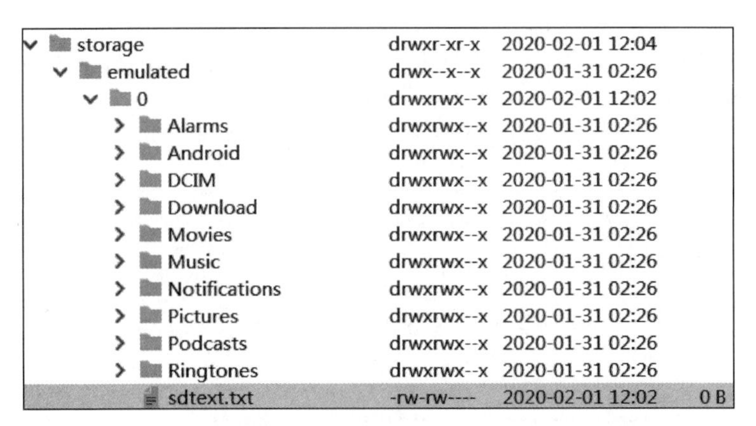

图7-5 文件存储的位置

之后再单击“读取SD卡文件”按钮，将sdtext.txt文件的内容读取到应用程序中，并将数据展示在TextView控件中。

代码的运行效果如图7-6所示。

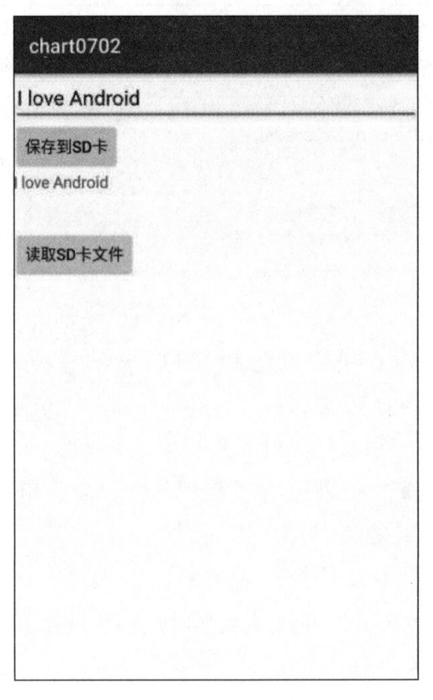

图 7-6 使用 SD 卡读写文件

视频详解

7.1.3 实战演练——文件存储实现自动登录

通过对文件存储的学习，实现登录页面用户名和密码的自动保存。布局页面的代码如下。

程序清单 7-4：chart0703\app\src\main\res\layout\activity_main.xml

```
1   <?xml version = "1.0" encoding = "utf - 8"?>
2   <LinearLayout xmlns:android = "http://schemas.android.com/apk/res/android"
3       xmlns:tools = "http://schemas.android.com/tools"
4       android:layout_width = "match_parent"
5       android:layout_height = "match_parent"
6       android:orientation = "vertical"
7       tools:context = ".MainActivity">
8       <ImageView
9           android:id = "@ + id/iv_head"
10          android:layout_width = "70dp"
11          android:layout_height = "70dp"
12          android:layout_gravity = "center_horizontal"
13          android:layout_marginTop = "70dp"
14          android:background = "@drawable/defaultuser_icon" />
15
16          android:id = "@ + id/et_user_name"
17          android:layout_width = "fill_parent"
18          android:layout_height = "40dp"
19          android:layout_gravity = "center_horizontal"
```

```
20              android:layout_marginLeft = "35dp"
21              android:layout_marginRight = "35dp"
22              android:layout_marginTop = "35dp"
23              android:background = "@drawable/textview_borders"
24              android:drawableLeft = "@drawable/user_name_icon"
25              android:drawablePadding = "10dp"
26              android:gravity = "center_vertical"
27              android:hint = "请输入用户名"
28              android:paddingLeft = "8dp"
29              android:singleLine = "true"
30              android:textColor = "#000000"
31              android:textColorHint = "#a3a3a3"
32              android:textSize = "14sp" />
33          <EditText
34              android:id = "@+id/et_psw"
35              android:layout_width = "fill_parent"
36              android:layout_height = "40dp"
37              android:layout_gravity = "center_horizontal"
38              android:layout_marginLeft = "35dp"
39              android:layout_marginRight = "35dp"
40              android:background = "@drawable/textview_borders"
41              android:layout_marginTop = "5dp"
42              android:drawableLeft = "@drawable/psw_icon"
43              android:drawablePadding = "10dp"
44              android:hint = "请输入密码"
45              android:inputType = "textPassword"
46              android:paddingLeft = "8dp"
47              android:singleLine = "true"
48              android:textColor = "#000000"
49              android:textColorHint = "#a3a3a3"
50              android:textSize = "14sp" />
51          <Button
52              android:id = "@+id/btn_login"
53              android:layout_width = "fill_parent"
54              android:layout_height = "40dp"
55              android:layout_gravity = "center_horizontal"
56              android:layout_marginLeft = "35dp"
57              android:layout_marginRight = "35dp"
58              android:layout_marginTop = "15dp"
59              android:background = "@color/colorPrimary"
60              android:text = "登 录"
61              android:textColor = "@android:color/white"
62              android:textSize = "18sp" />
63          <LinearLayout
64              android:layout_width = "fill_parent"
65              android:layout_height = "fill_parent"
66              android:layout_marginLeft = "35dp"
67              android:layout_marginRight = "35dp"
68              android:layout_marginTop = "8dp"
69              android:gravity = "center_horizontal"
```

```
70              android:orientation = "horizontal">
71              < TextView
72                  android:id = "@ + id/tv_register"
73                  android:layout_width = "0dp"
74                  android:layout_height = "wrap_content"
75                  android:layout_weight = "1"
76                  android:gravity = "center_horizontal"
77                  android:padding = "8dp"
78                  android:text = "立即注册"
79                  android:textColor = "@color/colorPrimary"
80                  android:textSize = "14sp" />
81              < TextView
82                  android:id = "@ + id/tv_find_psw"
83                  android:layout_width = "0dp"
84                  android:layout_height = "wrap_content"
85                  android:layout_weight = "1"
86                  android:gravity = "center_horizontal"
87                  android:padding = "8dp"
88                  android:text = "找回密码?"
89                  android:textColor = "@color/colorPrimary"
90                  android:textSize = "14sp" />
91          </LinearLayout >
92      </LinearLayout >
```

新建工具类实现对用户名和密码的保存,类名为 saveFile,在该类中实现对用户名和密码的保存。

程序清单 7-5:char0703\app\src\main\java\com\example\administrator\chart0703\saveFile.java

```
1    public class saveFile {
2        public static boolean saveInfoToFile ( Context context, String userName, String
 userPwd) {
3            FileOutputStream fos = null;
4            try {
5                fos = context.openFileOutput("user.txt", Context.MODE_PRIVATE);
6                 fos.write((userName + "&" + userPwd).getBytes());
7                 fos.close();
8                 return true;
9            } catch (Exception e) {
10                e.printStackTrace();
11                return false;
12           }
13       }
14    public static Map < String, String > getInfoFromFile(Context context){
15            FileInputStream fis = null;
16            try {
17                fis = context.openFileInput("user.txt");
18                byte[] bytes = new byte[fis.available()];
19                fis.read(bytes);
```

```
20                  String content = new String(bytes);
21                  String[] splitStr = content.split("&");
22                  HashMap<String, String> userMap = new HashMap<>();
23                  userMap.put("userName", splitStr[0]);
24                  userMap.put("userPwd", splitStr[1]);
25                  fis.close();
26                  return userMap;
27              } catch (Exception e) {
28                  e.printStackTrace();
29                  return null;
32              }
33          }
34      }
```

在上述代码中，第2～13行代码创建一个saveInfoToFile()方法，用于将用户名和密码保存在user.txt文件中，第5行通过调用openFileOutput()方法获取fos对象，之后通过该对象的write()方法将用户名和密码以字节的形式写入到user.txt文件中。

第14～33行代码创建一个getInfoFromFile()方法用于读取保存在user.txt文件中的用户名和密码。第17行代码通过调用openFileInput()方法获取fis对象，之后通过该对象的read()方法将user.txt文件中的内容读取到字节数组中。在将字节数组转换成字符串并对字符串进行分隔，最终获取到用户名和密码。

逻辑代码页面MainActivity中的代码如下。

程序清单7-6：chart0703\app\src\main\java\com\example\administrator\chart0703\MainActivity.java

```
1       public class MainActivity extends AppCompatActivity {
2           private TextView tvRegister;
3           private TextView tvFindPwd;
4           private EditText editName;
5           private EditText editPwd;
6           private Button btnLogin;
7           private String userName;
8           private String psw;
9           @Override
10          protected void onCreate(Bundle savedInstanceState) {
11              super.onCreate(savedInstanceState);
12              setContentView(R.layout.activity_main);
13              initView();
14              initListener();
15              Map<String, String> uerInfo = saveFile.getInfoFromFile(this);
16              if (uerInfo != null) {
17                  editName.setText(uerInfo.get("userName"));
18                  editPwd.setText(uerInfo.get("userPwd"));
19              }
20          }
21          private void initView() {
```

```
22            tvRegister = (TextView) findViewById(R.id.tv_register);
23            editName = (EditText) findViewById(R.id.et_user_name);
24            editPwd = (EditText) findViewById(R.id.et_psw);
25            tvRegister = (TextView) findViewById(R.id.tv_register);
26            tvFindPwd = (TextView) findViewById(R.id.tv_find_psw);
27            btnLogin = (Button) findViewById(R.id.btn_login);
28        }
29
30
31
32        private void initListener() {
33            btnLogin.setOnClickListener(new View.OnClickListener() {
34                @Override
35                public void onClick(View v) {
36           userName = editName.getText().toString().trim();
37                          psw = editPwd.getText().toString();
38                    if (userName.isEmpty() || psw.isEmpty()) {
39                        Toast.makeText(MainActivity.this, "用户名或密码不能为空",
                          Toast.LENGTH_SHORT).show();
40                        return;
41                    } else {
42                         Toast.makeText(MainActivity.this, "登录成功", Toast.LENGTH_
  SHORT).show();
43                         boolean isSave = saveFile.saveInfoToFile(MainActivity.this,
  userName, psw);
44                        if (isSave) {
45                           Toast.makeText(MainActivity.this, "保存成功", Toast.LENGTH
  _SHORT).show();
46                        }
47                    }
48                }
49            });
50        }
51    }
```

在上述代码中，第 15～19 行首先通过工具类 saveFile 中的 getInfoFromFile()方法获取保存在文件中的用户名和密码，如果之前保存了用户名和密码，则将读取的用户名和密码展示在登录框和密码框中。

第 33～49 行代码为“登录”按钮注册单击事件，在单击事件内部，首先获取用户输入的用户名和密码，如果获取的用户名或者密码为空，则提示用户“用户名或密码不能为空”，否则调用 saveFile 类中的 saveInfoToFile()方法将用户名和密码保存在本地文件中，并提示用户“登录成功”和“保存成功”。

运行以上程序，在用户名和密码框中分别输入“123”，单击“登录”按钮，弹出“登录成功”和“保存成功”表示用户名和密码已经保存在本地文件中，也可以通过 Device File Explorer 浏览设备文件，从该应用程序目录中查找该文件。运行结果如图 7-7 所示。

当关闭程序再次打开程序之后，界面上会自动加载出用户名和密码，如图 7-8 所示。

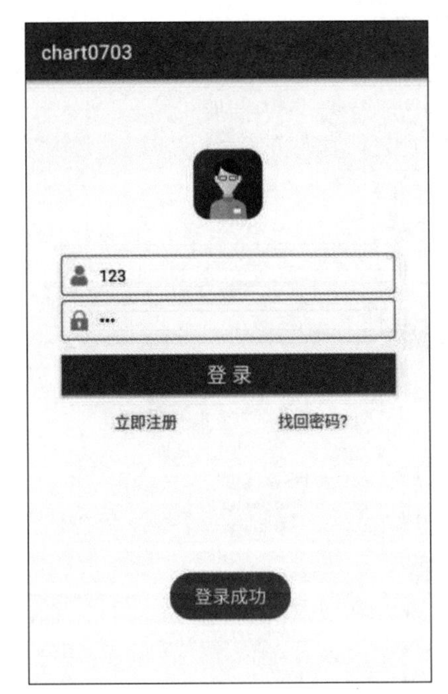

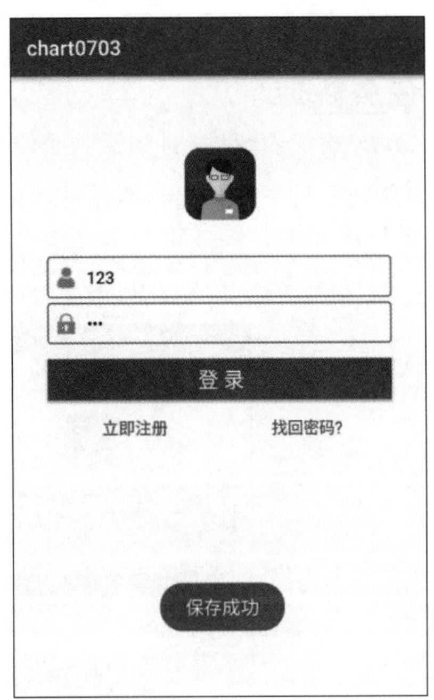

图 7-7　登录成功并保存登录信息

图 7-8　再次登录自动加载登录信息

视频详解

7.2 SharedPreferences 存储

任务陈述

在 7.1 节中“欢乐购商城”的登录页面使用了文件存储的方式实现了，但是在实际应用中，通常采用 SharedPreferences 类进行存储。

分析：在登录页面上实现保存用户名和密码的功能，可以采用 SharedPreferences 类，SharedPreferences 类很适合保存用户名、密码之类的简单配置信息，如图 7-9 所示。

图 7-9　登录页面

相关知识

通常用户在使用 Android 应用时，都会根据自己的爱好进行简单的设置，例如，设置背景颜色、记录用户名和密码、登录状态等，为了使用户下次打开应用时不需要重复设置，应用程序需要保存这些设置信息。Android 提供了一个 SharedPreferences 接口，来保存配置参数。应用程序使用 SharedPreferences 接口可以快速而高效地以键值对的形式保存数据，信息以 XML 文件的形式存储在 Android 设备上。SharedPreferences 本身是一个接口，不能直接实例化，但 Android 的 Context 类提供了方法可以获取 SharedPreferences 实例。

视频详解

7.2.1 SharedPreferences 数据操作

1. 使用 SharedPreferences 存储数据

使用 SharedPreferences 类存储数据时，首先需要调用 getSharedPreferences(String name,int mode)方法获取实例对象。在该方法中需要传递两个参数，第一个参数表示保存信息的文件名，不需要后缀，在同一个应用中可以使用多个文件保存不同的信息；第二个参数表示 SharedPreferenes 的访问权限，和 7.1 节中读取应用程序的文件类似，包括只能被本应用程序读、写和能被其他应用程序读写。由于该对象本身只能获取数据，不能对数据进行

存储和修改,因此需要调用 SharedPreferences 类的 edit()方法获取可编辑的 Editor 对象,最后通过该对象的 putXXX()方法存储数据,示例代码如下。

```
1    SharedPreferences preferences = getSharedPreferences("data", MODE_PRIVATE);
2    SharedPreferences.Editor editor = preferences.edit();
3    editor.putString("name", "James");
4    editor.putInt("age", 25);
5    editor.commit();
```

上述代码中,第 1 行代码通过 getSharedPreferences()获取 SharedPreferences 对象,之后通过该对象的 edit()方法获取 Editor 对象并通过相应的方法存储数据,最后通过 commit()方法提交。通过上述代码,Editor 对象是以 key/value 的形式保存数据的,其中 value 值只能是 float、int、long、boolean、String、Set <String>类型数据。并且根据数据类型的不同,会调用不同的方法。需要注意的是,最后一定要调用 commit()方法进行数据提交,否则所有操作不生效。

2. 使用 SharedPreferences 读取数据

读取 SharedPreferences 中的数据只需要获取 SharedPreferences 对象,然后通过该对象的 getXXX()方法根据相应的 key 值获取 value 值。该方法需要传递两个参数:第一个参数为关键字,即保存的时候使用的 key 值;第二个参数为默认值,即如果根据 key 值没有找到 value 值时,返回该默认值。程序所读取的 SharedPreferences 文件不存在时,程序也会返回默认值,并不会抛出异常。SharedPreferences 数据总是保存在 data/data/<packagename>/shared_prefs 目录下,并且 SharedPreferences 数据总是以 XML 格式保存,实例代码如下。

```
1    SharedPreferences preferences = getSharedPreferences("data", MODE_PRIVATE);
2    preferences.getString("name", "");
3    preferences.getInt("age", 0);
```

上述代码中,第 1 行代码通过 getSharedPreferences()获取 SharedPreferences 对象,之后通过该对象的 getString()方法和 getInt()方法获取用户名和年龄。

3. 使用 SharedPreferences 删除数据

使用 SharedPreferences 删除数据的时候,需要先获取 SharedPreferences 对象,并通过该对象的 edit()方法获取 Editor 对象。通过调用 Editor 对象的 remove(String key)方法删除数据。该方法中的参数为关键字,即键-值对的 key 值。该方法会删除 key 相对应的数据。如果要删除所有数据可采用 clear()方法。实例代码如下。

```
1    SharedPreferences preferences = getSharedPreferences("data", MODE_PRIVATE);
2    SharedPreferences.Editor editor = preferences.edit();
3    editor.remove("name");
4    editor.clear();
```

上述代码中，第1行代码通过 getSharedPreferences()获取 SharedPreferences 对象，之后通过该对象的 edit()方法获取 Editor 对象，通过 Editor 对象中的 remove()和 clear()方法删除数据。

视频详解

7.2.2 实战演练——SharedPreferences 保存用户名和密码

通过对 SharedPreferences 的学习，可以对"欢乐购商城"中的登录页面进行实现。界面布局与7.1节布局相同，在此不再重复演示。在程序中创建一个工具类 SPSaveInfo 保存用户名和密码。

程序清单 7-7：chart0704\app\src\main\java\com\example\administrator\chart0704\SPSaveInfo.java

```
1    public class SPSaveInfo {
2       public static boolean saveInfoToSP(Context context, String userName, String userPwd) {
3            SharedPreferences preferences = context.getSharedPreferences("data",
           Context.MODE_PRIVATE);
4            SharedPreferences.Editor editor = preferences.edit();
5            editor.putString("userName", userName);
6            editor.putString("userPwd", userPwd);
7            editor.commit();
8            return true;
9        }
10      public static Map<String, String> getInfoFromSP(Context context) {
11           HashMap<String, String> hashMap = new HashMap<>();
12           SharedPreferences preferences = context.getSharedPreferences("data",
   Context.MODE_PRIVATE);
13           String userName = preferences.getString("userName", "");
14           String userPwd = preferences.getString("userPwd", "");
15           hashMap.put("userName", userName);
16           hashMap.put("userPwd", userPwd);
17           return hashMap;
18       }
19   }
```

在上述代码中，第2～9行创建 saveInfoToSP()方法，用于保存用户名和密码到 data.xml 文件中。在该方法中首先通过 getSharedPreferences()方法获取到 SharedPreferences 对象，并通过该对象的 edit()方法获取到 Editor 对象，并通过该对象将数据保存在文件中。

第10～18行创建一个 getInfoFromSP()方法用来获取 data.xml 文件中保存的用户名和密码。在该方法中首先通过 getSharedPreferences()方法获取到 SharedPreferences 对象，并通过该对象的 getString()方法获取到用户名和密码，并将获取的数据存放在 Map 集合中。

MainActivity 页面的代码如下。

程序清单 7-8：chart0704\app\src\main\java\com\example\administrator\chart0704\MainActivity. java

```
1    public class MainActivity extends AppCompatActivity {
2        private TextView tvRegister;
3        private TextView tvFindPwd;
4        private EditText editName;
5        private EditText editPwd;
6        private Button btnLogin;
7        private String userName;
8        private String psw;
9        @Override
10       protected void onCreate(Bundle savedInstanceState) {
11           super.onCreate(savedInstanceState);
12           setContentView(R.layout.activity_main);
13           initView();
14           initListener();
15            Map<String, String> uerInfo = SPSaveInfo.getInfoFromSP(this);
16           if (uerInfo != null) {
17               editName.setText(uerInfo.get("userName"));
18               editPwd.setText(uerInfo.get("userPwd"));
19           }
20        }
21        private void initView() {
22           tvRegister = (TextView) findViewById(R.id.tv_register);
23           editName = (EditText) findViewById(R.id.et_user_name);
24           editPwd = (EditText) findViewById(R.id.et_psw);
25           tvRegister = (TextView) findViewById(R.id.tv_register);
26           tvFindPwd = (TextView) findViewById(R.id.tv_find_psw);
27           btnLogin = (Button) findViewById(R.id.btn_login);
28        }
29        private void initListener() {
30           btnLogin.setOnClickListener(new View.OnClickListener() {
31               @Override
32               public void onClick(View v) {
33          userName = editName.getText().toString().trim();
34                         psw = editPwd.getText().toString();
35                   if (userName.isEmpty() || psw.isEmpty()) {
36                        Toast.makeText(MainActivity.this, "用户名或密码不能为空",
   Toast.LENGTH_SHORT).show();
37                       return;
38                   } else {
39                       Toast.makeText(MainActivity.this, "登录成功", Toast.LENGTH_
   SHORT).show();
40                        boolean isSave = SPSaveInfo.saveInfoToSP(MainActivity.this,
   userName, psw);
41                       if (isSave) {
42                           Toast.makeText(MainActivity.this, "保存成功", Toast.LENGTH
   _SHORT).show();
43                       }
44                   }
45               }
46           });
47        }
48   }
```

运行以上程序,在用户名和密码框中分别输入"234",单击"登录"按钮,弹出"登录成功"和"保存成功"表示用户名和密码已经保存在本地文件中,运行结果如图 7-10 所示。

图 7-10 使用 sp 登录成功并保存登录信息

当关闭程序再次打开程序之后,界面上会自动加载出用户名和密码,如图 7-11 所示。

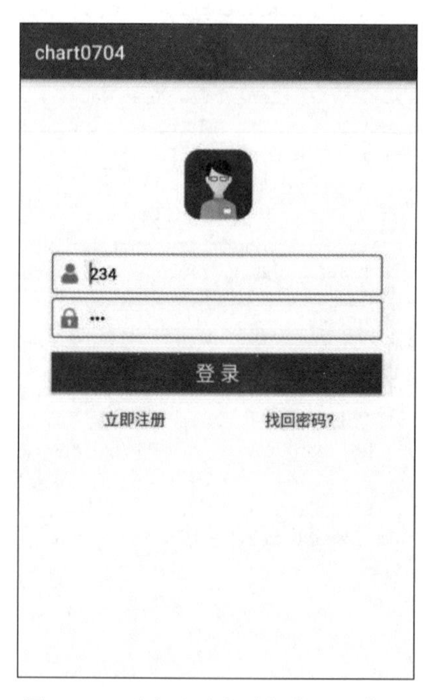

图 7-11 再次登录自动加载登录信息

为了验证用户信息是否成功保存到了 SharedPreferences 中，可以在 Device File Explorer 视图中找到 data\data\com. example. administrator. chart0704\shared_prefs 目录，然后找到 data. xml 文件，data. xml 文件目录如图 7-12 所示。

com.example.administrator.chart0704	drwxr-x--x	2020-02-02 10:34
cache	drwxrwx--x	2020-02-02 10:34
code_cache	drwxrwx--x	2020-02-02 10:34
shared_prefs	drwxrwx--x	2020-02-02 10:34
data.xml	-rw-rw----	2020-02-02 10:34

图 7-12　文件保存的位置

打开 data. xml 文件。

```
1    <?xml version = '1.0' encoding = 'utf - 8' standalone = 'yes' ?>
2    < map >
3        < string name = "userPwd"> 234 </string >
4        < string name = "userName"> 234 </string >
5    </map >
```

使用 saveInfoToSP 类成功地把用户名和密码保存在文件中。从 data. xml 文件可知，该 XML 文件是以 map 为根节点，map 元素的每个子元素代表一个 key-value 对，子元素的 name 属性值为 value 对应的类型名。

视频详解

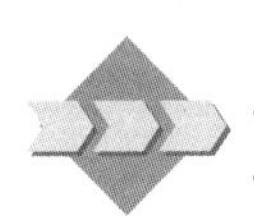

7.3　SQLite 数据库

任务陈述

“欢乐购商城”项目中订单页面、购物车页面的订单数据是在用户将商品加入购物车的时候将商品数据保存在数据库中，当打开订单页面和购物车页面的时候从数据库读取商品数据，如图 7-13 所示。

通过读取SQLite数据库获取加入购物车的商品

读取SQLite数据库购物车内商品的信息数据并加载出来

图 7-13　使用 SQLite 加载数据

前面介绍了如何使用文件存储和 SharedPreferences 存储数据，但这两种方式只适合存储一些简单的数据。如果想存储结构复杂的数据，需要用到数据库。在 Android 平台上，嵌入了一个轻量级的关系型数据库 SQLite，它可以存储应用程序中的大量数据，包含操作本地数据的所有功能，简单易用、反应快。

视频详解

7.3.1 SQLite 数据库简介

SQLite 内部只支持 NULL、INTEGER、REAL（浮点数）、TEXT（字符文本）和 BLOB（二进制对象）这五种数据类型，但实际上，SQLite 也接受 varchar(n)、char(n)、decimal(p,s) 等数据类型，只不过在运算或保存时会转成上面对应的数据类型。

SQLite 最大的特点是可以把各种类型的数据保存到任何字段中，而不用关心字段声明的数据类型是什么。例如，可以把字符串类型的值存入 INTEGER 类型字段中，或者在布尔型字段中存放数值类型等。但有一种情况例外：定义为 INTEGER PRIMARYKEY 的字段只能存储 64 位整数，当向这种字段保存除整数以外的数据时，SQLite 会产生错误。由于 SQLite 允许存入数据时忽略底层数据列实际的数据类型，因此 SQLite 在解析建表语句时，会忽略建表语句中跟在字段名后面的数据类型信息。因此在编写建表语句时可以省略数据列后面的类型声明。SQLite 允许开发者使用 SQL 语句操作数据库中的数据，并且 SQLite 数据库不需要安装、启动服务进程，其底层只是一个数据库文件。本质上，SQLite 的操作方式只是一种更为便捷的文件操作。

视频详解

7.3.2 SQLite 数据库的创建

在 Android 系统中提供了相关的类帮助我们创建数据库，其中，SQLiteOpenHelper 是 Android 提供的管理数据的工具类，主要用于数据库的创建、打开和版本更新。一般用法是创建 SQLiteOpenHelper 类的子类，并重写父类的 onCreate() 和 onUpgrade() 方法（这两个方法是抽象的，必须重写）。

SQLiteOpenHelper 包含如下常用方法。

(1) abstract void onCreate(SQLiteDatabase db)：当数据库第一次被创建的时候调用该方法。

(2) abstract void onUpgrade(SQL iDatabre db, int oldVersion, int newVersion)：当数据库需要更新的时候调用该方法。

(3) void onOpen(SQLiteDatabase db)：当数据库打开时调用该方法。

(4) SQLiteDatabase getWritableDatabase()：以写的方式打开数据库对应的 SQLiteDatabase 对象，一旦打开成功，将会缓存该数据库对象。

(5) SQLiteDatabase getReadableDatabase()：以读写的方式打开数据库对应的 SQLiteDatabase 对象，该方法内部调用 getWritableDatabase() 方法，返回对象与 getWritableDatabase() 返回对象一致，除非数据库的磁盘空间满了，此时 getWritableDatabase() 打开数据库就会出错，当打开失败后 getReadableDatabase() 方法会继续尝试以只读方式打开数据库。

（6）当调用 SQLiteOpenHelper 的 getWritableDatabase()或者 getReadableDatabase()方法获取 SQLiteDatabase 实例的时候，如果数据库不存在，Android 系统会自动生成一个数据库，然后调用 onCreate()方法，在 onCreate()方法中可以初始化表结构。onUpgrade()方法在数据库版本号发生变化时会被调用，一般在软件升级需要修改表结构的时候需要升级数据库版本号，假设数据库原有版本是 1，当要修改表结构时就需要升级版本号，此时可以设置数据库版本号 2，并在 onUpgrade()方法中实现表结构的更新。onUpgrade()方法在数据库版本号增加时会被调用，并做出相应表结构和数据的更新，如果版本号不增加，则该方法不会被调用。

在项目中新建类 SQLiteHelper，用于生成数据库，示例代码如下。

程序清单 7-9：chart0705\app\src\main\java\com\example\administrator\chart0705\sqlite\SQLiteHelper. java

```
1     public class SQLiteHelper extends SQLiteOpenHelper {
2         public SQLiteHelper(Context context) {
3
4             super(context, "userInfo.db", null, 1);
5         }
6
7         @Override
8         public void onCreate(SQLiteDatabase db) {
9             db.execSQL("create table if not exists" + " userinfo" + "(" + "_id integer
   primary key " +
10                    "autoincrement," + "user_name varchar," + "user_pwd varchar," +
   "user_nick varchar"
11                    + ")");
12          }
13     public void onUpgrade(SQLiteDatabase db, int oldVersion, int newVersion) {
15     }
```

由上述代码中，首先创建了一个 SQLiteHelper 类继承自 SQLiteOpenHelper，并重写该类的构造方法 SQLiteHelper()，在该方法中通过 super()调用父类 SQLiteOpenHelper 的构造方法，并传入 4 个参数，分别表示上下文对象、数据库名称、游标工厂（通常是 null）、数据库版本。然后重写了 onCreate() 和 onUpgrade()方法，其中，onCreate()方法是在数据库第一次创建时调用，该方法通常用于初始化表结构。MainActivity. java 中的代码如下。

程序清单 7-10：chart0705\app\src\main\java\com\example\administrator\chart0705\MainActivity. java

```
1     public class MainActivity extends AppCompatActivity {
2         @Override
3         protected void onCreate(Bundle savedInstanceState) {
4             super.onCreate(savedInstanceState);
5             setContentView(R.layout.activity_main);
6             SQLiteHelper sqLiteHelper = new SQLiteHelper(this);
7             SQLiteDatabase database = sqLiteHelper.getWritableDatabase();
8         }
9     }
```

在上述代码中，首先生成 SQLiteHelper 对象，调用该对象的 getWritableDatabase()生成数据库。程序运行后，在 Device File Explorer 视图中找到数据库文件所在的路径 data/data/com.example.administrator.chart0705/databases/userInfo.db，如图 7-14 所示。

com.example.administrator.chart0705	drwxr-x--x	2020-02-03 10:28
cache	drwxrwx--x	2020-02-03 10:28
code_cache	drwxrwx--x	2020-02-03 10:28
databases	drwxrwx--x	2020-02-03 10:28
userInfo.db	-rw-rw----	2020-02-03 10:28
userInfo.db-journal	-rw-------	2020-02-03 10:28

图 7-14　数据库文件存储的位置

右键单击数据库文件 userInfo.db，选择 Save as 可以将数据库文件保存在本地磁盘上。如果要查看数据库中的数据，可以使用 SQLite Expert Personal 可视化工具。可在官网 http://www.sqliteexpert.com/download.html 下载 SQLite Expert Personal 工具并进行安装，安装完成运行程序，结果如图 7-15 所示。

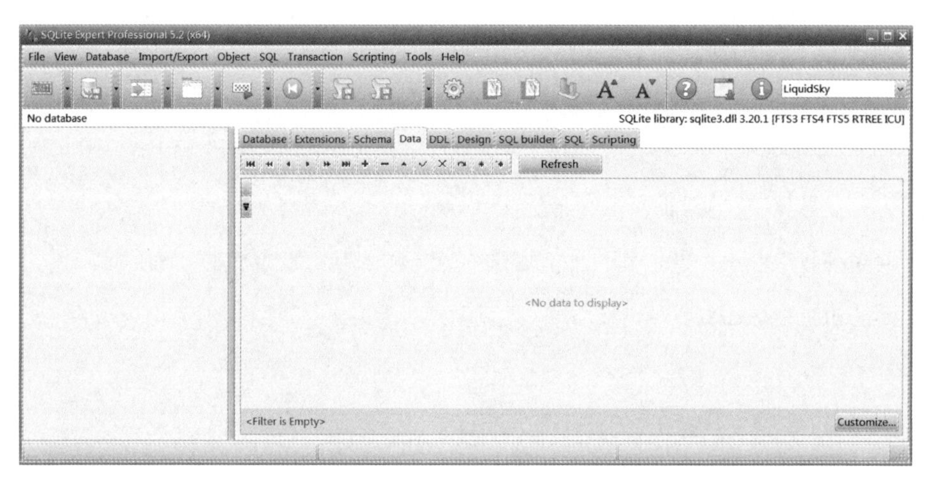

图 7-15　SQLite Expert Personal 界面

在 SQLite Expert Personal 工具中单击 File→Open Database 选项，选择需要查看的数据库文件，结果如图 7-16 所示。

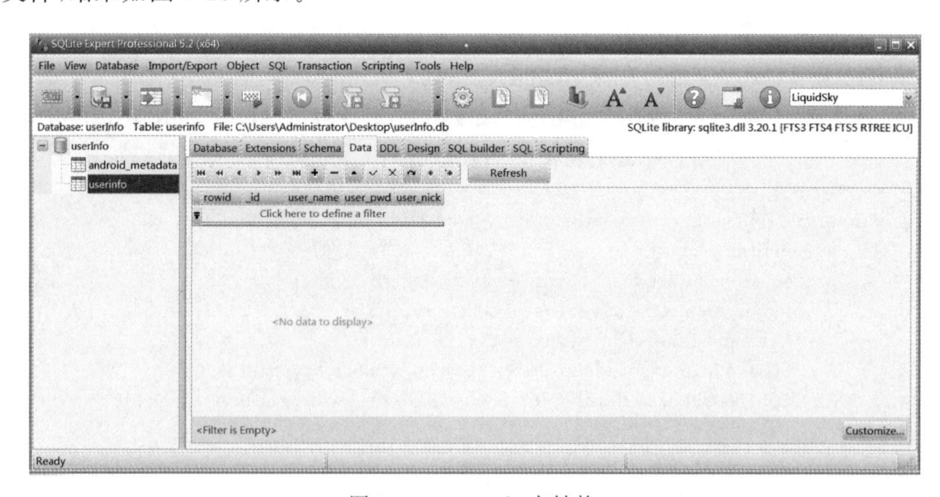

图 7-16　userinfo 表结构

由图 7-16 可知，创建的数据库 userInfo.db 中的各个字段已经展示出来，当数据库中有新添加的数据时，也可通过 SQLite Expert Personal 可视化工具可以进行查看。

7.3.3 SQLite 数据库的升级

视频详解

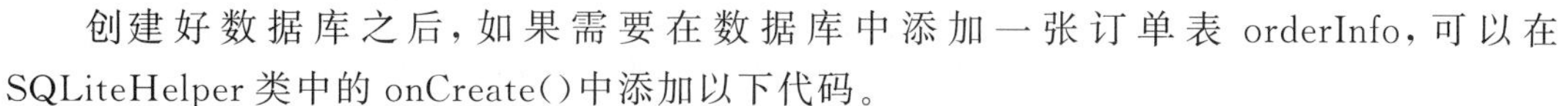

创建好数据库之后，如果需要在数据库中添加一张订单表 orderInfo，可以在 SQLiteHelper 类中的 onCreate()中添加以下代码。

```
public void onCreate(SQLiteDatabase db) {
        db.execSQL("create table if not exists" + " orderInfo" + "(" + "_id integer
primary key " +"autoincrement," + "goodsName varchar(50)," + "goodsPrice varchar(10)," +
"goodsLocation varchar(50)," + "shopName varchar(50)," + "dateTime varchar(50)" + ")");
}
```

重新运行程序的时候，发现 orderInfo 表并没有创建成功，原因在于再次运行程序的时候 userInfo 数据库已经存在。再次运行程序的时候 onCreate()并不会再次执行，因此新添加的表 orderInfo 就不会创建成功。

如果在数据库已经存在的情况下需要更新表时，就可以用 onUpgrade()方法，实例代码如下。

程序清单 7-11：chart0705\app\src\main\java\com\example\administrator\chart0705\sqlite\SQLiteHelper.java

```
1    public class SQLiteHelper extends SQLiteOpenHelper {
2        public SQLiteHelper(Context context) {
3            super(context, "userInfo.db", null, 2);
4        }
5
6        @Override
7        public void onCreate(SQLiteDatabase db) {
8            db.execSQL("create table if not exists" + " userinfo" + "(" + "_id integer
9                    primary key " +"autoincrement," + "user_name varchar(20)," + "user_
10                   pwd varchar(20)," +"user_nick varchar(20) " + ")");
11           db.execSQL("create table if not exists" + " orderInfo" + "(" + "_id integer
12                   primary key " + "autoincrement," + "goodsName varchar(50)," +
13                   "goodsPrice varchar(10)," +"goodsLocation varchar(50)," +
14                   "shopName varchar(50)," + "dateTime varchar(50)" +")");
15       }
16       @Override
17       public void onUpgrade(SQLiteDatabase db, int oldVersion, int newVersion) {
18           db.execSQL("drop table if exists userinfo");
19           db.execSQL("drop table if exists orderInfo");
20           onCreate(db);
21       }
22   }
```

在上述代码中,第 17～21 行代码 onUpgrade()方法中,如果发现数据库中已经存在 userinfo 表和 orderInfo 表就将这两张表删除,之后再调用 onCreate()方法创建这两张表。此外,如果想要 onUpgrade()能够执行,在 SQLiteHelper 构造函数中要增加版本号,原来的版本号为 1,所以修改 SQLiteHelper 构造函数中版本号为 2,表示对数据库的升级。

再次运行程序,将数据库文件导出到本地磁盘,使用 SQLite Expert Personal 可视化工具查看数据库中的表,发现 orderInfo 表创建成功,如图 7-17 所示。

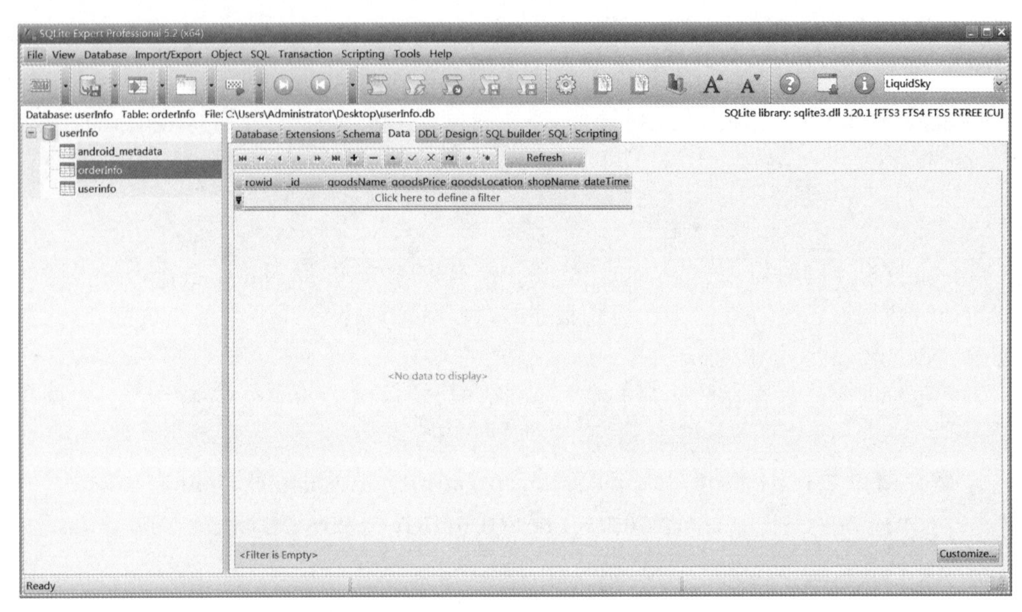

图 7-17 orderInfo 表结构

视频详解

7.3.4 SQLite 数据库的基本操作

前面介绍了如何创建 SQLite 数据库,接下来针对 SQLite 数据的增、删、改、查操作进行详细的讲解。实现对数据的增删改查主要使用 SQLiteDatabase 类。该类封装了一些操作数据库的 API,使用该类可以完成对数据进行添加(Create)、查询(Retrieve)、更新(Update)和删除(Delete)操作。在该类中重点掌握 execSQL()和 rawQuery()方法。execSQL()方法可以执行 insert、delete、update 和 create table 之类有更改行为的 SQL 语句,而 rawQuery()方法用于执行查询语句。

(1) execSQL(String sql,Object[] bindArgs):执行带占位符的 SQL 语句,如果 SQL 中没有占位符,则第二个参数可传 null。

(2) execSQL(String sql):执行 SQL 语句。

(3) rawQuery(String sql,String[] selectionArgs):执行带占位符的 SQL 查询。

除了 execSQL()和 rawQuery()方法,SQLiteDatabase 还专门提供了对应于添加、删除、更新、查询的操作方法 insert()、delete()、update()和 query()。实际上,这些方法内部也是执行 SQL 语句,由底层用 StringBuilder 根据这些参数拼接成完整的 SQL 语句。

例如:Cursor query(String table, String[] columns, String selection, String[] slectionArgs, String groupBy, String having, String orderBy, String limit)方法各参数的

含义如下。

table：表名，相当于 select 语句 from 关键字后面的部分。

columns：要查询的列名，可以是多列，相当于 select 语句 select 关键字后面的部分。

selection：查询条件子句，相当于 select 语句 where 关键字后面的部分，在条件子句中允许使用占位符“?”。

selectionArgs：对应于 selection 语句中占位符的值，值在数组中的位置与占位符在语句中的位置必须一致，否则就会有异常。

groupBy：相当于 select 语句 group by 关键字后面的部分。

having：相当于 select 语句 having 关键字后面的部分。

orderBy：相当于 select 语句 order by 关键字后面的部分。

limit：指定偏移量和获取的记录数，相当于 select 语句 limit 关键字后面的部分。

Cursor 接口主要用于存放查询记录的接口，Cursor 是结果集游标，用于对结果集进行访问。该接口提供了如下方法来移动查询结果的记录指针。

move(int offset)：将记录指针向上或向下移动指定的行数。offset 为正数就向下移动，为负数就向上移动。

moveToNext()：可以将游标从当前记录移动到下一记录，如果已经移过了结果集的最后一条记录，返回结果为 false，否则为 true。

moveToPrevious()：用于将游标从当前记录移动到上一记录，如果已经移过了结果集的第一条记录，返回值为 false，否则为 true。

moveToFirst()：用于将游标移动到结果集的第一条记录，如果结果集为空，返回值为 false，否则为 true。

moveToLast()：用于将游标移动到结果集的最后一条记录，如果结果集为空，返回值为 false，否则为 true。

使用 SQLiteDatabase 进行数据库操作的步骤如下。

(1) 定义一个数据库操作辅助类，从 SQLiteOpenHelper 继承，重写 onCreate() 和 onUpdate()方法，在 onCreate()方法中执行建表语句和初始化数据。

(2) 创建 SQLiteOpenHelper 类对象，指定数据库的名称和版本后，调用该类的 getReadableDatabase()或者 getWritableDatabase()方法，获取 SQLiteDatabase 对象，该对象代表了与数据库的连接。

(3) 调用 SQLiteDatabase 对象的相关方法来执行增、删、查、改操作。

(4) 对数据库操作的结果进行处理，例如，判断是否插入、删除或者更新成功，将查询结果记录转换成列表显示等。

(5) 关闭数据库连接，回收资源。

1. 添加数据

接下来以 userInfo 表为例，介绍如何使用 SQLiteDatabase 对象的 insert()方法向表中插入一条数据，实例代码如下。

程序清单 7-12：chart0705\app\src\main\java\com\example\administrator\chart0705\MainActivity. java

```
1    public void insert(String name, String pwd, String nick) {
2        SQLiteHelper sqLiteHelper = new SQLiteHelper(this);
3        SQLiteDatabase database = sqLiteHelper.getWritableDatabase();
4        ContentValues values = new ContentValues();
5        values.put("user_name", name);
6        values.put("user_pwd", pwd);
7        values.put("user_nick", nick);
8        long id = database.insert("userinfo", null, values);
9        database.close();
10   }
```

在上述代码中，首先通过 sqLiteHelper 的 getWritableDatabase()获取 SQLiteDatabase 对象，之后获取 ContentValues 对象并将数据添加到 ContentValues 对象中。最后调用 insert()方法将数据添加到 userInfo 表中，返回值代表新行的 ID。其中，insert()方法接收 3 个参数，第 1 个参数是表名，第 2 个参数表示如果发现将要插入的行为空行时，会将这个列名的值设为 null，第 3 个参数为 ContentValues 对象。其中，ContentValues 类类似 Map 类，通过键-值对的形式存入数据，这里的 key 表示插入数据的列名，value 表示要插入的数据。

需要注意的是，使用完 SQLiteDatabase 对象后一定要调用 close()方法关闭数据库连接，否则数据库连接会一直存在，不断消耗内存，当系统内存不足时将获取不到 SQLiteDatabase 对象，并且会报出数据库未关闭异常。

2. 修改数据

SQLitDatabase 类中存在一个 update()方法，用于修改数据库表中的数据，以 userInfo 表 为例，如果想要修改该表中的某一条数据时，可直接调用 SQLiteDatabase 对象的 update()方法，示例代码如下。

程序清单 7-13：chart0705\app\src\main\java\com\example\administrator\chart0705\MainActivity. java

```
1    public int update(String name, String pwd, String nick) {
2        SQLiteHelper sqLiteHelper = new SQLiteHelper(this);
3        SQLiteDatabase database = sqLiteHelper.getWritableDatabase();
4        ContentValues values = new ContentValues();
5        values.put("user_pwd", pwd);
6        values.put("user_nick", nick);
7        int num = database.update("userinfo", values, "user_name = ?", new String[]
{name});
8        database.close();
9        return num;
10   }
```

上述代码中，首先通过 sqLiteHelper 的 getWritableDatabase()获取 SQLiteDatabase 对象。之后调用该对象的 update()方法，其中，update()方法接收 3 个参数，第 1 个参数是表

名，第 2 个参数表示查询条件的子句，第 3 个参数为查询条件子句中对应的值。该方法的返回值表示受影响的行数。

3. 删除数据

SQLiteDatabase 类中提供了一个 delete() 方法，用于删除数据库表中的数据。以 userInfo 表为例说明，示例代码如下。

程序清单 7-14：chart0705\app\src\main\java\com\example\administrator\chart0705\MainActivity.java

```
1    public int delete(String name) {
2        SQLiteHelper sqLiteHelper = new SQLiteHelper(this);
3        SQLiteDatabase database = sqLiteHelper.getWritableDatabase();
4        int num = database.delete("userinfo", "user_name = ?", new String[]{name});
5        database.close();
6        return num;
7    }
```

上述代码中，首先通过 sqLiteHelper 的 getWritableDatabase()获取 SQLiteDatabase 对象。之后调用该对象的 delete()方法，其中，delete()方法接收 3 个参数，第 1 个参数是表名，第 2 个参数表示查询条件的子句，第 3 个参数为查询条件子句中对应的值。该方法的返回值表示受影响的行数。

4. 查询数据

在进行数据查询时使用的是 query()方法，该方法返回的是一个行数集合 Cursor，Cursor 是一个游标接口，提供了遍历查询结果的方法。需要注意的是，在使用完 Cursor 对象后，一定要及时关闭，否则会造成内存泄漏。以 userInfo 表为例说明，示例代码如下。

程序清单 7-15：chart0705\app\src\main\java\com\example\administrator\chart0705\MainActivity.java

```
1    public List<UserInfo> query(String name) {
2        List<UserInfo> infoList = new ArrayList<UserInfo>();
3        SQLiteHelper sqLiteHelper = new SQLiteHelper(this);
4        SQLiteDatabase database = sqLiteHelper.getWritableDatabase();
5        Cursor cursor = database.query("userinfo", null, "user_name = ?", new String[]
6                {name}, null,null, null, null);
7        if (cursor != null && cursor.getCount() > 0) {
8            while (cursor.moveToNext()) {
9                UserInfo info = new UserInfo();
10                info.setUser_name(cursor.getString(1));
11                info.setUser_pwd(cursor.getString(2));
12                info.setUser_nick(cursor.getString(3));
13                infoList.add(info);
14            }
15        }
16        cursor.close();
17        database.close();
18        return infoList;
19    }
```

上述代码中，封装一个查询表 userInfo 的方法，其中返回为一个集合，集合中每个对象的类型为 UserInfo。UserInfo 是根据 userInfo 表封装的实体类。

其中，第 1～6 行代码首先通过 sqLiteHelper 的 getWritableDatabase( ) 获取 SQLiteDatabase 对象，再通过 SQLiteDatabase 对象的 query()方法查询 userInfo 表中的数据，并返回 Cursor 对象。query()方法包含 7 个参数，第 1 个参数表示表名称，第 2 个参数指明查询哪几列，如果不指明，用 null 表示则表示查询所有列，第 3 个参数表示的是接收查询条件的子句，第 4 个参数接收查询子句对应的条件值，如果第 3,4 个参数不指明，用 null 表示则表示查询所有行，第 5 个参数表示分组方式，第 6 个参数接收 having 条件，即对 group By 之后的数据再添加过滤器，第 7 个参数表示排序方式，如果不指明则按照默认的排序方式。

第 7～13 行代码首先通过 getCount()方法获取到查询结果的总数，然后循环取出每一列的值，通过 moveToNext()方法不断移动游标到下一行数据，接着通过为 getString()方法传入列索引获取对应的数据，并将数据存入到 UserInfo 对象中，最后把 UserInfo 对象存入到集合中，最后关闭 Cursor 和 SQLiteDatabase 对象，并将集合数据返回。

视频详解

7.3.5 使用 SQL 操作数据库

虽然 Android 已经提供了非常方便的 API 用于操作数据库，但是对于多表连接查询的时候使用 Android 封装的 API 不是很方便，因此在操作数据库的时候还可以使用 execSQL()方法通过 SQL 语句对数据库进行操作，示例代码如下。

1. 添加数据

```
database.execSQL("insert into userinfo (user_name,user_pwd,user_nick) values(?,?,?)", new
                String[]{"小明", "123", "Jack"});
```

2. 修改数据

```
database.execSQL("update userinfo set user_nick = ? where user_name = ?", new
                String[]{"James", "小明"});
```

3. 删除数据

```
database.execSQL("delete from userinfo where user_name = ?", new String[]{"小明"});
```

4. 查询数据

```
Cursor cursor  =  database.rawQuery("select  *  from userinfo where user_name = ?", new
                String[]{"小明"});
```

从上述代码中可知，增、删、改都是调用 execSQL()方法执行 SQL 语句，而查询使用

rawQuery()方法,因为 execSQL()方法没有返回值,而查询需要返回一个结果集 Cursor。采用上述方式的执行结果和使用 Android 提供的 API 执行结果相同,可以根据实际需求选择使用合适的方式。

7.3.6 实战演练——订单处理

视频详解

通过对 SQLite 数据库的学习,我们可以对“购物商城”中订单页面进行实现,如图 7-1 所示。由于在“欢乐购商城中”订单数据是在别的页面添加到数据库中的,涉及页面跳转等内容。这里在页面中添加两个按钮,一个是添加数据按钮,模拟在其他页面向数据库里面添加数据,一个是加载数据按钮,模拟在当前页面启动的时候加载订单表中的所有数据。其中布局页面如下。

程序清单 7-16:chart0706\app\src\main\res\layout\activity_main.xml

```
1    <?xml version = "1.0" encoding = "utf - 8"?>
2    <RelativeLayout xmlns:android = "http://schemas.android.com/apk/res/android"
3    xmlns:tools = "http://schemas.android.com/tools"
4        android:layout_width = "match_parent"
5        android:layout_height = "match_parent"
6        tools:context = ".MainActivity">
7        <Button
8            android:id = "@ + id/btn_add"
9            android:layout_width = "match_parent"
10           android:layout_height = "wrap_content"
11           android:onClick = "addData"
12           android:text = "添加数据" />
13       <Button
14           android:id = "@ + id/btn_load"
15           android:layout_width = "match_parent"
16           android:layout_height = "wrap_content"
17           android:layout_below = "@id/btn_add"
18           android:onClick = "loadData"
19           android:text = "加载数据" />
20       <ListView
21           android:id = "@ + id/order_ListView"
22           android:layout_width = "match_parent"
23           android:layout_height = "match_parent"
24           android:layout_below = "@ + id/btn_load"></ListView>
25   </RelativeLayout>
```

order_list_items 的布局文件,实例代码如下。

程序清单 7-17:chart0706\app\src\main\res\layout\order_list_items.xml

```
1    <?xml version = "1.0" encoding = "utf - 8"?>
2    <RelativeLayout xmlns:android = "http://schemas.android.com/apk/res/android"
3        android:layout_width = "match_parent"
4        android:layout_height = "match_parent">
5        <TextView
```

```
6                android:id = "@ + id/goods_title"
7                android:layout_width = "match_parent"
8                android:layout_height = "wrap_content"
9                android:layout_margin = "8dp"
10               android:text = "云南天然干花花束真花"
11               android:textSize = "15sp"
12               android:textStyle = "bold" />
13           < TextView
14               android:id = "@ + id/order_status"
15               android:layout_width = "wrap_content"
16               android:layout_height = "wrap_content"
17             android:layout_below = "@ id/goods_title"
18               android:layout_marginLeft = "5dp"
19               android:text = "订单状态:已结算" />
20           < TextView
21               android:id = "@ + id/order_value"
22               android:layout_width = "wrap_content"
23               android:layout_height = "wrap_content"
24             android:layout_alignParentRight = "true"
25             android:layout_below = "@ id/goods_title"
26               android:layout_marginRight = "8dp"
27               android:text = "总价:80 元" />
28           < TextView
29               android:id = "@ + id/order_time"
32               android:layout_width = "wrap_content"
33               android:layout_height = "wrap_content"
34           android:layout_below = "@ id/order_status"
35               android:layout_margin = "8dp"
36               android:text = "创建时间:" />
37     </RelativeLayout >
```

定义一个 GoodInfo 类，实例代码如下。

程序清单 7-18：chart0706\app\src\main\java\com\example\administrator\chart0706\bean\GoodInfo.java

```
1     public class GoodInfo implements Serializable {
2         private String id;
3         private String goodsName;
4         private String totalValue;
5         private String dateTime;
6         private int isPlay;
7         public GoodInfo() {
8         }
9         public GoodInfo(String goodsName, String totalValue, int isPlay) {
10            this.goodsName  =  goodsName;
11            this.totalValue  =  totalValue;
12            this.isPlay  =  isPlay;
13        }
```

```
14        public String getId() { return id; }
15        public void setId(String id) { this.id = id; }
16        public int getIsPlay() {
17            return isPlay;
18        }
19        public void setIsPlay(int isPlay) {
20            this.isPlay = isPlay;
21        }
22        public String getGoodsName() {
23            return goodsName;
24        }
25        public void setGoodsName(String goodsName) {
26            this.goodsName = goodsName;
27        }
28        public String getDateTime() {
29            return dateTime;
32        }
33        public void setDateTime(String dateTime) {
34            this.dateTime = dateTime;
35        }
36        public String getTotalValue() {
37            return totalValue;
38        }
39        public void setTotalValue(String totalValue) {
40            this.totalValue = totalValue;
41        }
42    }
```

定义一个SQLiteHelper类，用于生成数据库，实例代码如下。

程序清单7-19：chart0706\app\src\main\java\com\example\administrator\chart0706\sqlite\SQLiteHelper.java

```
1     public class SQLiteHelper extends SQLiteOpenHelper {
2         private static final int DB_VERSION = 1;
3         private static final String DB_NAME = "goodInfo.db";
4         public SQLiteHelper(Context context) {
5             super(context, DB_NAME, null, DB_VERSION);
6         }
7         @Override
8         public void onCreate(SQLiteDatabase db) {
9             db.execSQL("create table if not exists" + " orderInfo" + "(" + "_id integer
10                    primary key " + "autoincrement," + "goodsName varchar(100)," +
11                    "isPlay int," + "totalValue " + "varchar(50)," + "dateTime varchar
                      (50)" + ")");
12        }
13        @Override
14        public void onUpgrade(SQLiteDatabase db, int oldVersion, int newVersion) {
15            onCreate(db);
16        }
17    }
```

在上述代码中，创建了一个 SQLiteHelper 类继承自 SQLiteOpenHelper，并重写该类的构造方法 SQLiteHelper()，在该方法中通过 super()调用父类 SQLiteOpenHelper 的构造方法，并传入 4 个参数，分别表示上下文对象、数据库名称、游标工厂（通常是 null）、数据库版本。然后重写了 onCreate() 和 onUpgrade()方法，其中，onCreate()方法是在数据库第一次创建时调用，用于初始化表结构。

之后定义一个 DBUtils 类，封装对表数据的增、删、查功能，实例代码如下。

程序清单 7-20：chart0706\app\src\main\java\com\example\administrator\chart0706\sqlite\DBUtils. java

```
1    public class DBUtils {
2        private static SQLiteHelper helper;
3        private static SQLiteDatabase db;
4        private static DBUtils instance = null;
5        public DBUtils(Context context) {
6            helper = new SQLiteHelper(context);
7            db = helper.getWritableDatabase();
8        }
9        public static DBUtils getInstance(Context context) {
10           if (instance == null) {
11               instance = new DBUtils(context);
12           }
13           return instance;
14       }
15       public boolean saveGoodInfo(GoodInfo goodInfo) {
16           ContentValues values = new ContentValues();
17           values.put("goodsName", goodInfo.getGoodsName());
18           values.put("totalValue", goodInfo.getTotalValue());
19           values.put("isPlay", goodInfo.getIsPlay());
20           SimpleDateFormat df = new SimpleDateFormat("yyyy-MM-dd HH:mm:ss");
21           values.put("dateTime", df.format(new Date()));
22           long i = db.insert("orderInfo", null, values);
23           if (i > 0) {
24               return true;
25           } else {
26               return false;
27           }
28       }
29       public List<GoodInfo> getGoodsInfoList() {
30           List<GoodInfo> goodInfoList = new ArrayList<GoodInfo>();
31           GoodInfo goodInfo = null;
32           Cursor cursor = db.query("orderInfo", null, null, null, null, null, null);
33           if (cursor.getCount() > 0 && cursor != null) {
34               while (cursor.moveToNext()) {
35                   goodInfo = new GoodInfo();
36                   goodInfo.setId(cursor.getString(cursor.getColumnIndex("_id")));
37                   goodInfo.setGoodsName(cursor.getString(cursor.getColumnIndex
("goodsName")));
```

```
38                      goodInfo. setIsPlay ( cursor. getInt ( cursor. getColumnIndex
   ("isPlay")));
39                      goodInfo. setTotalValue ( cursor. getString ( cursor. getColumnIndex
   ("totalValue")));
40                      goodInfo. setDateTime ( cursor. getString ( cursor. getColumnIndex
   ("dateTime")));
41                      goodInfoList. add(goodInfo);
42                  }
43                  cursor. close();
44                  return goodInfoList;
45              } else {
46                  return null;
47              }
48          }
49          public void deleteGoodsInfo(String id) {
50              db. delete("orderInfo", "_id = ?", new String[ ]{id + ""});
51          }
52      }
```

在上述代码中定义一个 DBUtils 类，实现对表 orderInfo 的增删改查。其中，第 9～14 行定义 getInstance()方法，实现单例模式。第 15～28 行代码中定义 saveGoodInfo()方法实现向数据库中添加一条记录。第 29～48 行代码定义方法 getGoodsInfoList()实现查询 orderInfo 表中的记录。第 49～51 行定义方法 deleteGoodsInfo()实现根据 id 删除 orderInfo 表中的某条记录。

MainActivity 页面的代码如下。

程序清单 7-21：chart0706\app\src\main\java\com\example\administrator\chart0706\MainActivity. java

```
1    public class MainActivity extends AppCompatActivity {
2        private ListView orderList;
3        private List < GoodInfo > goodInfoList  =  new ArrayList < GoodInfo >();
4        private myListAdapter adapter;
5
6        @Override
7        protected void onCreate(Bundle savedInstanceState) {
8            super. onCreate(savedInstanceState);
9            setContentView(R. layout. activity_main);
10           orderList  =  (ListView) findViewById(R. id. order_ListView);
11       }
12
13       public void addData(View view) {
14           GoodInfo goofInfo1  =  new GoodInfo("鞋袋子装鞋子的收纳袋旅行鞋包收纳包
                 防尘袋家用鞋罩束口鞋袋鞋套", "45", 1);
15           GoodInfo goofInfo2  =  new GoodInfo("2019 初秋新款潮韩版洋气女装
                 宽松秋季短款衬衫女秋装方领长袖上衣", "132", 1);
```

```
16            GoodInfo goofInfo3 = new GoodInfo("儿童数码照相机可拍照拍立得
                  玩具可打印迷你小单反宝生日礼物 wifi", "399", 0);
17            GoodInfo goofInfo4 = new GoodInfo("母婴包轻便双肩包女手提书包多
                  功能妈妈包大容量外出旅行防盗背包", "118", 1);
18            DBUtils.getInstance(this).saveGoodInfo(goofInfo1);
19            DBUtils.getInstance(this).saveGoodInfo(goofInfo2);
20            DBUtils.getInstance(this).saveGoodInfo(goofInfo3);
21            DBUtils.getInstance(this).saveGoodInfo(goofInfo4);
22    }
23
24        public void loadData(View view) {
25            goodInfoList = DBUtils.getInstance(this).getGoodsInfoList();
26            adapter = new myListAdapter();
27            orderList.setAdapter(adapter);
28            orderList.     setOnItemLongClickListener     (   new      AdapterView.
  OnItemLongClickListener() {
29                @Override
30                public boolean onItemLongClick(AdapterView<?> parent, View view, final
  int position,
31                                              long id) {
32                    new AlertDialog.Builder(MainActivity.this).setMessage("确定要删
  除吗?")
33                            .setPositiveButton("确定",     new     DialogInterface.
  OnClickListener() {
34                                @Override
35                                public  void  onClick ( DialogInterface  dialog,  int
  which) {
36                                DBUtils.      getInstance       (    MainActivity.
  this).deleteGoodsInfo
37                                (goodInfoList.get(position).getId());
38                        goodInfoList.remove(goodInfoList.get(position));
39                                    adapter.notifyDataSetChanged();
40                                }
41                            }).setNegativeButton("取消",     new     DialogInterface.
  OnClickListener() {
42                        @Override
43                        public void onClick(DialogInterface dialog, int which) {
44                            dialog.dismiss();
45                        }
46                    }).create().show();
47                    return true;
48                }
49            });
50        }
51        private class myListAdapter extends BaseAdapter {
52            @Override
53            public int getCount() {
54                return goodInfoList != null ? goodInfoList.size() : 0;
55            }
```

```
56            @Override
57            public Object getItem(int position) {
58                return goodInfoList.get(position);
59            }
60            @Override
61            public long getItemId(int position) {
62                return position;
63            }
64            @Override
65            public View getView(int position, View convertView, ViewGroup parent) {
66                 View inflate = View.inflate(MainActivity.this, R.layout.order_list_
  items, null);
67                TextView orderTitle = (TextView) inflate.findViewById(R.id.goods_
  title);
68                TextView totalValue = (TextView) inflate.findViewById(R.id.order_
  value);
69                TextView orderTime = (TextView) inflate.findViewById(R.id.order_time);
70                TextView orderStatus = (TextView) inflate.findViewById(R.id.order_
  status);
71                orderTitle.setText(goodInfoList.get(position).getGoodsName());
72                totalValue.setText("总价:" + goodInfoList.get(position)
73                        .getTotalValue() + "元");
74                orderTime.setText("创建时间:" + goodInfoList.get(position)
75                        .getDateTime());
76                orderStatus.setText("订单状态:" + (goodInfoList.get(position).
  getIsPlay() == 1 ? "已结算":
77                        "未付款"));
78                return inflate;
79            }
80        }
81    }
```

在上述代码中,第13~22行为"添加数据"按钮的单击事件添加逻辑操作,首先生成4个对象,并将每个对象逐一添加到数据库orderInfo表中。第24~50行为"加载数据"按钮的单击事件添加代码。第25行代码中使用DBUtils类中的getGoodsInfoList()方法查询数据库中orderInfo表中的数据并将其保存到goodInfoList集合中。之后为ListView控件绑定适配器。第28~49行为ListView中每一项绑定长按的单击事件,当用户长按每一项的时候,首先弹出对话框提示用户"确定要删除吗?",当用户选择"确定"时,调用DBUtils类中的deleteGoodsInfo()方法将该条记录从数据库中删除,并同时在goodInfoList集合中删除该对象,并实时更新ListView的适配器对象。

第51~80行代码定义一个类myListAdapter类继承自BaseAdapter类,并重写了BaseAdapter类中的getCount()、getItem()、getItemId()和getView()方法。在getView()方法中通过inflate()方法将layout.order_list_items布局转换成视图对象,并通过findViewById()获取到layout.list_item布局中的各个控件,最后通过setText()方式将各类信息展示出来。

运行上述程序,先单击"添加数据"按钮,将数据添加到数据库,然后单击"加载数据"将数据库表中的数据读取出来并展示在 ListView 上，当长按某一项的时候弹出对话框，提示用户“确定要删除吗?”，当用户单击“确定”按钮时该项从列表项里面删除，如图 7-18 所示。

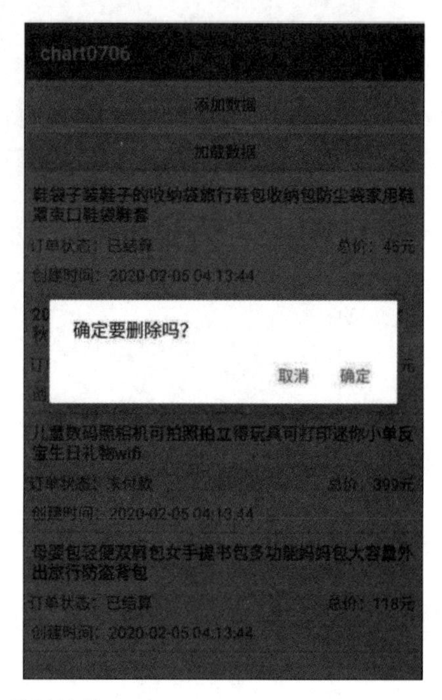

图 7-18 运行结果

本章小结

本章通过“欢乐购商城”项目中的登录页面和订单页面引入了 Android 中的数据存储。详细介绍了 Android 中数据存储的几种方式，首先简单介绍了 Android 中的五种数据存储方式，然后详细介绍了文件存储，使用简单的文件流的形式读取手机内存和 SD 卡。之后介绍了用于保存用户个性化设置、程序参数等之类的 SharedPreferences 工具类，再详细介绍了用于保存比较复杂有一定结构关系的 SQLite 数据库。数据存储是 Android 开发中非常重要的基础知识，一般在应用程序中都会经常涉及数据存储的知识，因此要求必须熟练掌握本章的内容。

自测习题

1. 向手机内置存储空间内文件中写入新的内容时首先调用的方法是(　　)。

   A. openFileOutput()　　　　B. read()

   C. write()　　　　D. openFileInput()

2. 下列选项中，不属于 getSharedPreferences()方法的文件操作模式参数的是(　　)。

   A. Context. MODE_PRIVATE

   B. Context. MODE_PUBLIC

C. Context. MODE_WORLD READABLE

D. Context. MODE_WORLD WRITEABLE

3. SharedPreferences 保存文件的路径和扩展名是()。

A. /data/data/shared prefs/ *. txt

B. /data/data/package name/shared prefs/ *. xml

C. /mntsdcard/指定文件夹下指定扩展名

D. 任意路径/任意扩展名

4. 下列方法中,(　　)方法是 sharedPreferences 获取其编辑器的方法。

A. getEdit()　　B. edit()　　C. setEdit()　　D. edits()

5. 下列关于 SQLiteOpenHelper 的描述不正确的是(　　)。

A. SQLiteOpenHelper 是 Android 中提供的管理数据库的工具类,主要用于数据库的创建、打开、版本更新等,它是一个抽象类

B. 继承 SQLiteOpenHelper 的类,必须重写它的 onCreate()方法

C. 继承 SQLiteOpenHelper 的类,必须重写它的 onUpgrade()方法

D. 继承 SQLiteOpenHelper 的类,可以提供构造方法也可以不提供构造方法

6. Android 对数据库的表进行查询操作时,会使用 SQLiteDatabase 类中的(　　)方法。

A. insert()　　B. execSQL()　　C. update()　　D. query( )

7. 编写一个简单备忘录程序,实现在界面中以列表的形式显示备忘录的信息,备忘录信息包括编号、标题、内容,并实现对备忘录中的信息进行增加和查询,长按 ListView 中某一个条目可以删除,如图 7-19 所示。

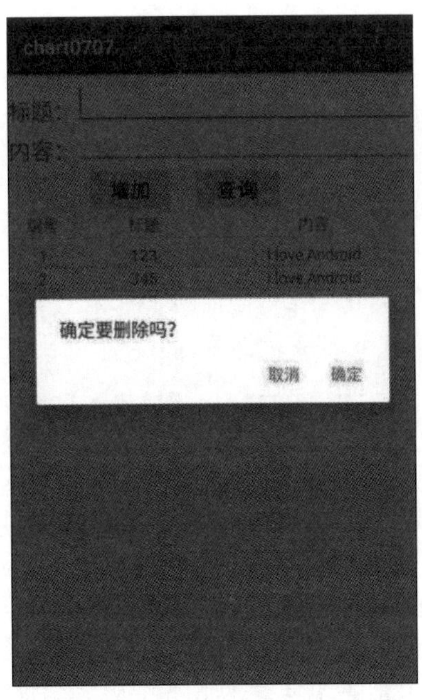

图 7-19　程序截图

第8章 Android客户端与服务器端交互

随着智能手机的普及，社会步入了互联网高速发展的时代，而移动应用作为互联网的主力军，渗透在人们生活中的方方面面，成为人们连接网络的入口。App 作为移动应用中的主力军，更是发挥着巨大的作用。学习 Android 开发的读者，大部分都会从事互联网的相关开发工作，既然是互联网，网络连接和交互就是必然的。严格意义上来说，App 是 C/S 平台架构中的 C 端，其主要负责交互并向后台服务器获取和提交相关数据。这也成为一个 App 开发者开发工作中占比最重的内容之一，即本章的主题为 Android 客户端与服务器端交互。

学习目标

本章要点

(1) 了解网络请求协议。

(2) 了解服务器的运行机制。

(3) 了解网络请求过程。

(4) Get 请求和 Post 请求。

(5) HttpURLConnectiont 的使用方法。

(6) Android 多线程在网络请求中的使用。

(7) JSON 格式数据的解析和生成。

(8) OkHttp 网络请求架构。

视频详解

8.1 HttpURLConnection 框架

任务陈述

在"欢乐购商城"App 中，首页展示的是商品的列表，其中，商品的数据是请求网络得到的。为了方便读者进行任务实现，事先写好了一个可以部署在本机上的客户端，当然在互联网的环境下也部署了相同的服务。服务器提供了一个简单的接口，接口使用的是 Get 请求方式，可以使用本章的知识请求服务器获取数据，并将数据包装成一个对象数组放到我们熟悉的列表控件中，展示在 App 的界面中。使用 JDK 中内置的 HttpURLConnection 类进行

请求,本节使用 HttpURLConnection 对服务器进行请求访问数据。

分析:在"欢乐购商城"中存在多个地方请求服务器数据,在项目中为了方便读者学习,服务器端采用 JTM 封装的数据。本节围绕客户端和服务器端之间的交互,以实现欢乐购产品列表为例进行实现,如图 8-1 所示。

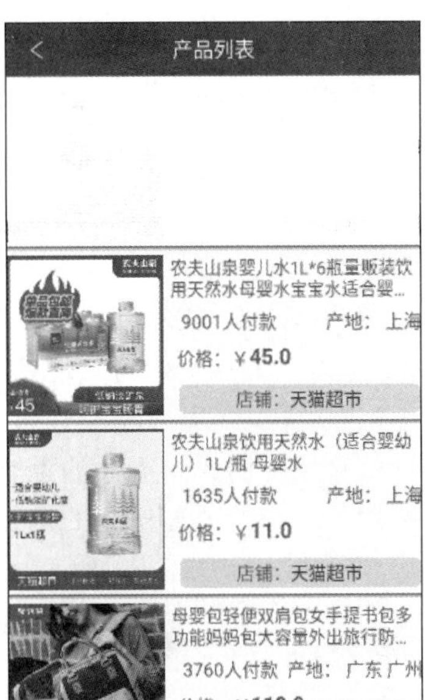

图 8-1 欢乐购产品列表

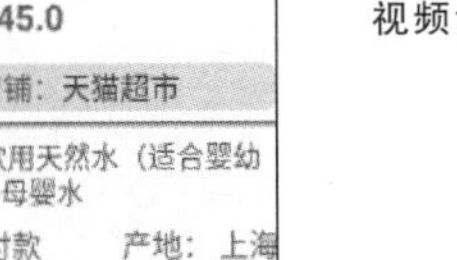

8.1.1 HTTP

HTTP(Hypertext Transfer Protocol,超文本传送协议)是 Web 联网的基础,也是手机联网常用的协议之一。HTTP 是建立在 TCP 之上的一种应用,它是一种基于请求/响应的通信协议,每一次连接只做一次请求/响应,服务器响应完客户端之后,就不会再记得客户端的一切,更不会去维护客户端的状态,因此 HTTP 又称为无状态的通信协议。

由于 HTTP 在每次请求结束后都会主动释放连接,因此 HTTP 连接是一种"短连接""无状态",要保持客户端程序的在线状态,需要不断地向服务器发起连接请求。通常的做法是即使不需要获得任何数据,客户端也保持每隔一段固定的时间向服务器发送一次"保持连接"的请求;服务器在收到该请求后对客户端进行回复,表明知道客户端"在线"。若服务器长时间无法收到客户端的请求,则认为客户端"下线";若客户端长时间无法收到服务器的回复,则认为网络已经断开。

HTTP 的主要特点如下。

支持客户端/服务器模式。

(1) 简单快速:客户端向服务器请求服务时,只需传送请求方法和路径。请求方法常用的有 GET、HEAD、POST。每种方法规定了客户端与服务器联系的类型不同。由于 HTTP 简单,使得 HTTP 服务器的程序规模小,因而通信速度很快。

(2) 灵活:HTTP 允许传输任意类型的数据对象。正在传输的类型由 Content-Type 加以标记。

(3) 无连接:无连接的含义是限制每次连接只处理一个请求。服务器处理完客户端的请求,并收到客户端的应答后,即断开连接。采用这种方式可以节省传输时间。

(4) 无状态:HTTP 是无状态协议。无状态是指协议对于事务处理没有记忆能力。缺少状态意味着如果后续处理需要前面的信息,则它必须重传,这样可能导致每次连接传送的数据量增大。另一方面,在服务器不需要先前信息时,它的应答就较快。

HTTP 中共定义了八种方法(动作)来表明请求指定的资源的不同操作方式:OPTIONS、HEAD、GET、POST、PUT、DELETE、TRACE、CONNECT。其中最常用的就是 GET 和 POST 方法。

GET 和 POST 的区别如下。

(1) GET 是用来从服务器获得数据,而 POST 是用来向服务器传递数据。GET 将需要传递的数据按照“键值”的形式，添加到URL的后面，并且两者使用“?”连接，而多个变量之间使用“&”连接。POST是将传递的数据放在请求的数据体中，不会在URL中显示。

（2）GET是不安全的，因为在传输过程中，数据被放在请求的URL中，而如今现有的很多服务器、代理服务器或者用户代理都会将请求URL记录到日志文件中，然后放在某个地方，这样就可能会有一些隐私的信息被第三方看到。另外，用户也可以在浏览器上直接看到提交的数据，一些系统内部消息将会一同显示在用户面前。POST的所有操作对用户来说都是不可见的。

（3）GET传输的数据量小，这主要是因为受URL长度限制，而POST可以传输大量的数据，所以在上传文件时只能使用POST。

（4）现在GET请求多数用在不需要请求参数或者请求参数固定的地方，一条GET请求可以直接复制为浏览器的地址从而获取到响应。很多网站的地址、下载App的链接、网络图片和文件的地址使用的都是GET请求，而其他大部分请求使用的还是更安全的POST请求。对于App开发者而言，大部分的接口也是POST请求。

视频详解

8.1.2 JSON协议

1. 接口组成示例

一个完整的接口请求一般由4个部分构成，当然有些请求可能不需要返回或不需要验证，需要的东西会少一些。

（1）请求参数：网络请求是传给服务器的数据。

（2）请求地址：接口的地址，通常情况下是一串HTTP地址，由IP（域名）、端口号、服务名、应用名、具体接口标识构成。

（3）请求头：请求头是为了减轻服务器负担而设计的。大部分服务器都是对请求进行加密限制的，这样可以避免服务器被黑客攻击，这时每次通信都需要一个密文（token）进行身份确认。以前把密文放在请求参数中，服务器需要去请求参数中找出这个密文；现在把它放在头部里面，也就像人们到火车站进站一样，直接刷一下验证就可以放行了。

（4）返回参数：返回参数是得到的数据，一般有响应码（用于标识请求的结果）、响应信息（解释请求后的结果，如成功则为success），返回数据（具体得到的响应数据，一般是一串JSON格式数据，或者加密后的JSON数据）。

2. JSON格式数据

JSON（JavaScript Object Notation）是一种数据交互格式。JSON之前，大家都用XML传递数据。XML是一种纯文本格式，所以适合在网络上交换数据，但是XML格式比较复杂，于是道格拉斯·克罗克福特（Douglas Crockford）发明了JSON这种超轻量级的数据交换格式。

普通应用开发大多使用的是面向对象的语言，面向对象也逐渐成为主流的编程思想。可以把JSON格式的数据理解为对象型数据，因为每个JSON类型的数据都可以转换为一个对象或一组对象。一般拿到JSON数据以后，也是把它转换为一个对象再通过对象点属性的方式进行使用。

3. JSON格式的数据举例

JSON格式的数据本质是字符串，其主要是key:value，也就是键和值的方式组合的，有点儿类似字典类型的数据。其中，{}表示里面是一个对象，[]表示里面是一个数组，key：value 可以理解为对象中的属性和值。下面举几个例子。

（1）一个对象的 JSON 格式数据：

```
{ "Name":"John" ,"age":14,"sex":0}
```

（2）多个对象的 JSON 格式数据：

```
{ "Name":"John" , "age":14,"pet":{"name":"xiaobai","kind":"dog",age":3}}
```

（3）包含数组的 JSON 格式数据：

```
{"employees": [{ "firstName":"John" , "lastName":"Doe" },
{ "firstName":"Anna" , "lastName":"Smith" },
{ "firstName":"Peter" , "lastName":"Jones" }]}
```

4. JSON 格式的数据解析

目前主流使用的 JSON 格式的数据解析，大多采用的是 Google 提供的 GSON 包进行解析，虽然 Android 原生也提供了解析的工具类 JSONObject，但由于其过于复杂烦琐，本书中不推荐使用。GSON 的数据解析完全是使用类和对象对数据进行解析，其使用的原理是字符串的识别和 Java 中的映射和反射的原理。

首先新建一个类，并为每个属性添加 get 和 set 方法，其代码如下。

```
1    public class Person implements Serializable{
2        private String name;
3        private int age;
4        private boolean sex;
5        private double weight;
6        public String getName() {
7            return name;
8        }
9        public void setName(String name) {
10            this.name = name;
11        }
12        public int getAge() {
13            return age;
14        }
15        public void setAge(int age) {
16            this.age = age;
17        }
18        public boolean isSex() {
19            return sex;
20        }
21        public void setSex(boolean sex) {
22            this.sex = sex;
```

```
23          }
24          public double getWeight() {
25              return weight;
26          }
27          public void setWeight(double weight) {
28              this.weight = weight;
29          }
30      }
```

接下来需要导入 GSON 的 jar 包，可以下载 GSON 的包放入 libs 文件夹中，当然也可以通过 Gradle 的网络包配置进行导入。为了提高代码的复用性，把 JSON 解析的方法包装成为一个工具类 JsonUtils，这样以后只要使用这个类进行 JSON 数据处理就可以了。

```
1   public class JsonUtil {            //单例引用
2       public static JsonUtil instance;
3       private static Gson gson;
4       public static JsonUtil getInstance(){
5           if(instance == null){
6               synchronized (JsonUtil.class){
7                   if(instance == null){
8                       instance = new JsonUtil();
9                       synchronized (Gson.class){
10                          gson = new Gson();
11                      }
12                  }
13              }
14          }
15          return instance;
16      }
17      public<T> String beanToJsonStr(T t){        //将对象转换为 JSON 的字符串
18          String str = "";
19          if(t != null){
20              str = gson.toJson(t);
21          }
22          return str;
23      }
24      public<T> T jsonToBean(String jsonStr, Class<T> tClass){
        //将 JSON 格式的字符转换为对象
25          T t = null;
26          if(jsonStr != null || jsonStr != ""){
27              try {
28                  t = gson.fromJson(jsonStr,tClass);
29              }catch (JsonIOException e){
30              }
31          }
32          return t;
33      }
34      public static <T> ArrayList<T> jsonToArrayList(String json, Class<T> clazz) {
```

```
35            //将 JSON 的字符转换为对象数组
36            Type type = new TypeToken<ArrayList<JsonObject>>() {
37            }.getType();
38            ArrayList<JsonObject> jsonObjects = new Gson().fromJson(json, type);
39            ArrayList<T> list = new ArrayList<>();
40            for (JsonObject jsonObject: jsonObjects) {
41                list.add(new Gson().fromJson(jsonObject, clazz));
42            }
43            return list;
44        }
45    }
```

使用起来也非常简单，有时候解析出来空的对象，使用的时候需加上非空判断。

```
1    public void test(){
2        Person person = new Person();
3        person.setName("jack");
4        person.setAge(18);
5        person.setSex(false);
6        person.setWeight(60.5);
7        String jsonStr = JsonUtil.getInstance().beanToJsonStr(person);
8        Person personTwo = JsonUtil.getInstance().jsonToBean(jsonStr,Person.class);
9    }
```

8.1.3　HttpURLConnection 介绍

视频详解

早些时候，Android 上发送 HTTP 请求一般有两种方式：HttpURLConnection 和 HttpClient。不过由于 HttpClient 存在 API 数量过多、扩展困难等缺点，Android 团队越来越不建议使用这种方式。在 Android 6.0 系统中，HttpClient 的功能被完全移除了。因此，在这里只简单介绍 HttpURLConnection 的使用。

在某些情况下，会使用 Java 程序来模拟浏览器发送请求。因此，在 JDK 的 java.net 包中已内置了访问 HTTP 的类：java.net.HttpURLConnection。

HttpURLConnection 类继承自 UrlConnection。UrlConnection 是一个抽象类，表示 URL 指向资源的链接，其子类包含诸如 HttpURLConnection、FtpURLConnection、FileURLConnection 等各种的连接类。

java.net.HttpURLConnection 类是一种访问 HTTP 资源的方式。HttpURLConnection 类具有完全的访问能力，可以取代 HttpClient 中的 HttpGet 类和 HttpPost 类。使用 HttpURLConnection 访问 HTTP 资源可以使用如下六步。

（1）使用 java.net.URL 封装 HTTP 资源的 URL，并使用 openConnection()方法获得 HttpURLConnection 对象，代码如下。

```
URL url = new URL("请求网址");
HttpURLConnection httpURLConnection = (HttpURLConnecton) url.openConnection();
```

(2) 设置请求方法，例如 GET、POST 等，代码如下。

```
httpURLConnection.setRequestMethod("POST");
```

注意：setRequestMethod()方法的参数值必须大写，例如 GET、POST 等。

(3) 设置输入/输出及其他权限。如果要下载 HTTP 资源或向服务器端上传数据，需要使用如下代码进行设置。

下载 HTTP 资源，需要将 setDoInput()方法的参数值设为 true。

```
httpURLConnection.setDoInput(true);
httpURLConnection.setDoOutput(true);
```

HttpURLConnection 类还包含更多的选项。例如，使用下面的代码可以禁止 HttpURLConnection 使用缓存。

```
httpURLConnection.setUseCaches(false);
```

(4) 设置 HTTP 请求头。在很多情况下，要根据实际情况设置一些 HTTP 请求头，例如，下面的代码设置了 Charset 请求头的值为 UTF-8。

```
httpURLConnection.setRequestProperty("Charset","UTF - 8");
```

(5) 输入和输出数据。这一步是对 HTTP 资源的读写操作，也是通过 InputStream 和 OutputStream 读取和写入数据。下面的代码获得了 InputStream 对象和 OutputStream 对象。

```
InputStream is = httpURLConnection.getInputStream();
OutputStream os = httpURLConnection.getOutputStream();
```

至于是先读取还是先写入数据，需要根据具体情况而定。

(6) 关闭输入和输出流. 虽然关闭输入/输出流并不是必需的，在应用程序结束后，输入/输出流会自动关闭，但显式关闭输入/输出流是一个好习惯。关闭输入/输出流的代码如下。

```
is.close();
os.close();
```

视频详解

8.1.4 实战演练——采用 HttpURLConnection 访问服务器端

根据任务分析实现客户端和服务器端数据交互，在“欢乐购商城”项目中采取的案例是访问网络数据实现商品列表。在整体项目中，服务器端实现不作为本书讲解内容，只需要了解如何启动即可，重点放在客户端请求服务器端数据和数据展示上，效果如图 8-2 所示。客户端的项目结构如图 8-3 所示。

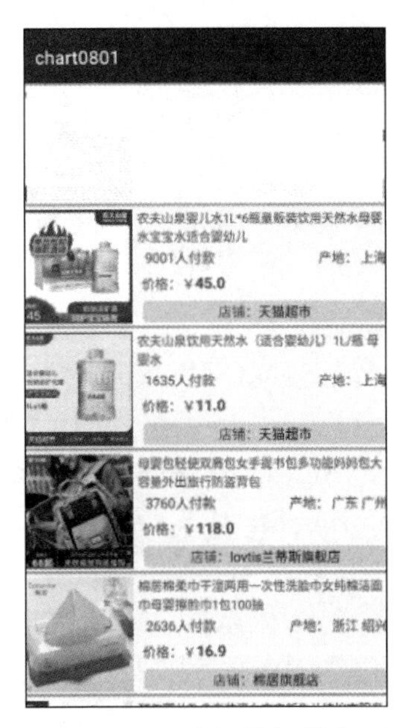

图 8-2　HttpClient 访问服务器端运行效果

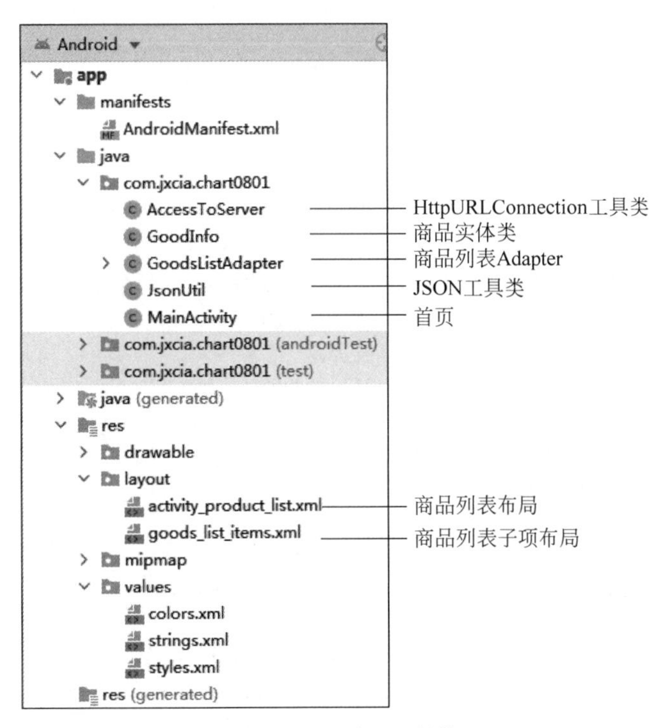

图 8-3　客户端项目结构

网络请求步骤如下。

(1) 配置 Androidmainface. xml 的网络权限。

（2）开启独立的网络请求线程。

（3）配置网络请求 HttpURLConnection 的参数。

（4）发送网络请求，添加回调函数。

（5）解析网络数据，并将数据抛到主线程或全局变量中。

（6）使用 handle 在主线程更新 UI。

第一步，配置 Androidmainface. xml。Android 系统对权限管理非常严格，前面使用存储时接触到了一些权限，现在也要在 AndroidManifest. xml 中增加一些新的网络相关权限，只有配置了这些权限，App 才能正常上网和检查网络状态。主要的权限如下。

android. permission. ACCESS_NETWORK_STATE：允许程序访问有关 GSM 网络信息（Allows applications to access information about networks）。

android. permission. ACCESS_WIFI_STATE：允许程序访问 Wi-Fi 网络状态信息（Allows applications to access information about Wi-Fi networks）。

android. permission. INTERNET：允许程序连接网络的权限（Allows applications to connect Internet）。

这里最主要的是第三个权限，直接决定了我们的应用是否能够联网，当然其他的权限也是非常有用的。

从 Android 9.0(API 28)开始，NetworkSecurityPolicy. getInstance(). isCleartextTrafficPermitted()，将返回 false，这表示 Android 默认将禁止明文访问网络，只允许使用 HTTPS URL 访问。为了避免强制启用 HTTPS，可在 Androidmainface. xml 中添加 android:usesCleartextTraffic="true"，添加代码如下。

程序清单 8-1：chart0801\app\src\main\AndroidManifest. xml

```
1   <manifest xmlns:android = "http://schemas.android.com/apk/res/android"
2       xmlns:tools = "http://schemas.android.com/tools"
3       package = "com.jxcia.chart0801">
4       <application
5           android:allowBackup = "true"
6           android:icon = "@mipmap/ic_launcher"
7           android:label = "@string/app_name"
8           android:roundIcon = "@mipmap/ic_launcher_round"
9           android:supportsRtl = "true"
10          android:usesCleartextTraffic = "true"
11          android:theme = "@style/AppTheme">
12          <activity android:name = ".MainActivity">
13              <intent-filter>
14                  <action android:name = "android.intent.action.MAIN" />
15                  <category android:name = "android.intent.category.LAUNCHER" />
16              </intent-filter>
17          </activity>
18      </application>
19      <uses-permission android:name = "android.permission.ACCESS_NETWORK_STATE"/>
20      <uses-permission android:name = "android.permission.WRITE_EXTERNAL_STORAGE"/>
```

```
21        <uses-permission android:name="android.permission.INTERNET" />
22    </manifest>
```

第二步，导入项目库。导入项目有两种方式，第一种是导入本地依赖包，以导入 GSON 包进行演示；另外一种是导入网络依赖包，以导入 recyclerview 和 glide 进行演示。

第一种方式操作步骤如下。

(1) 下载需要的 jar 包。

(2) 将 jar 包复制到 Project 下的 app→libs 目录下(没有 libs 目录就新建一个)，如图 8-4 所示。

(3) 右击该 jar 包，选择 Add As Library(如图 8-5(a)所示)，添加依赖包，弹出如图 8-5(b)所示对话框，单击 OK 即可。

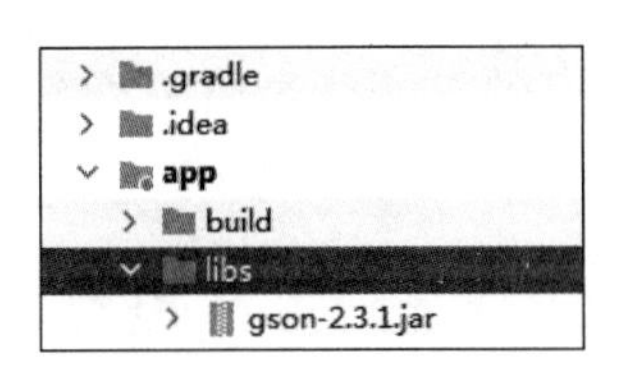

图 8-4 复制 jar 包到 libs 文件夹下

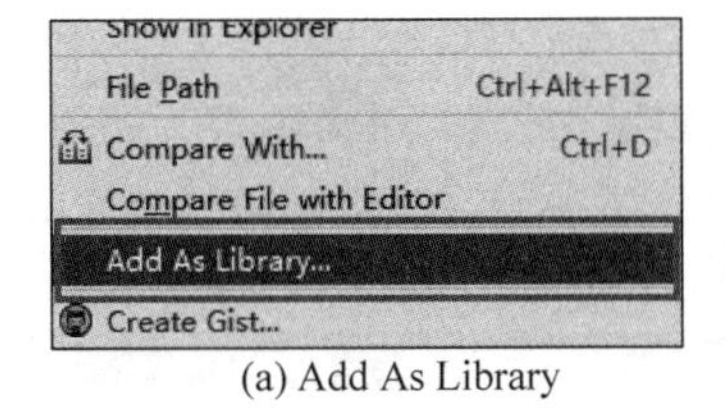

(a) Add As Library

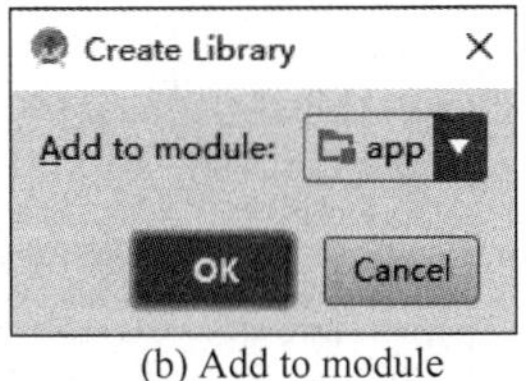

(b) Add to module

图 8-5 添加依赖包

(4) 打开 build.gradle 文件，在 dependencies()方法中出现 jar 包名称，说明导入成功，如图 8-6 所示。

```
dependencies {
    implementation fileTree(include: ['*.jar'], dir: 'libs')
    implementation 'androidx.appcompat:appcompat:1.1.0'
    implementation 'androidx.constraintlayout:constraintlayout:1.1.3'
    testImplementation 'junit:junit:4.13'
    androidTestImplementation 'androidx.test.ext:junit:1.1.1'
    androidTestImplementation 'androidx.test.espresso:espresso-core:3.2.0'
    implementation files('libs/gson-2.3.1.jar')
```

图 8-6 jar 包导入成功

第二种方式操作步骤如下。

(1) 单击工具栏中的 Project Structure 按钮，如图 8-7 所示。

图 8-7 单击 Project Structure 按钮

(2) 选择 Dependencies，单击右边的加号，选择 1 Library Dependency，如图 8-8 所示。

(3) 搜索 com.android.support，选择 recyclerview-v7，选择对应的版本，导入资源，如图 8-9 所示。

(4) 按照上面同样的步骤导入 glide 库，用于 ImagView 直接显示网络图片，如果显示结果如图 8-10 所示，表示导入成功。

第三步，编写该案例布局文件。其中包括首页布局文件和列表子项布局文件。

首页布局文件因为只显示列表，需添加 RecyclerView 列表控件。

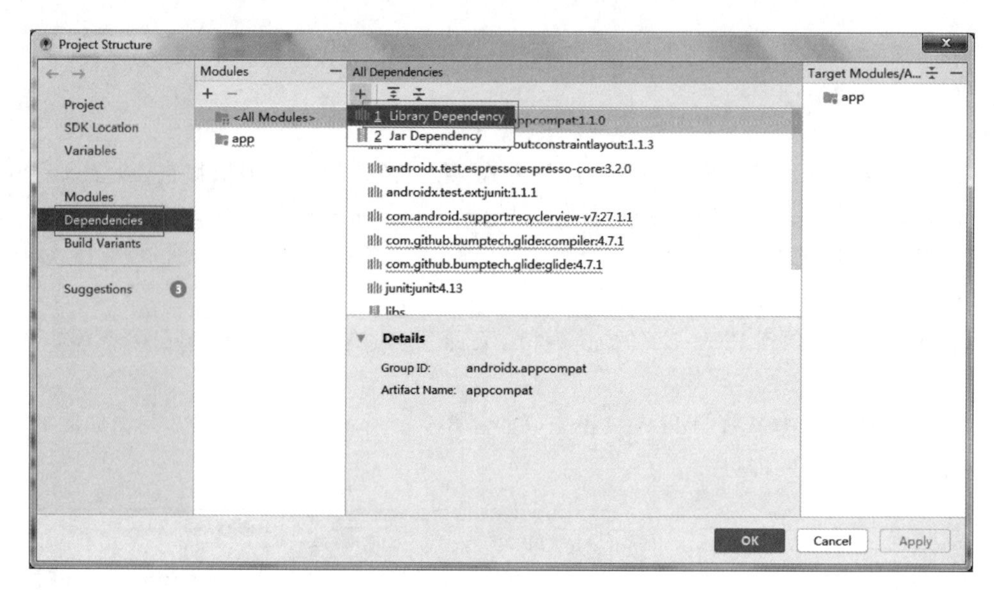

图 8-8 选择 1 Library Dependency

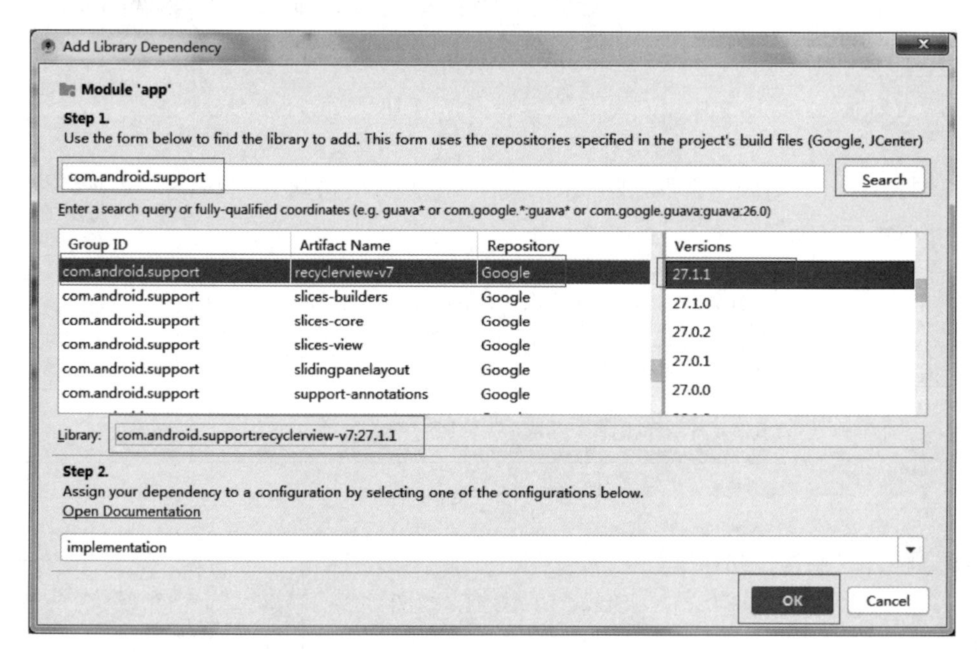

图 8-9 添加 recyclerview-v7 资源

```
dependencies {
    implementation fileTree(include: ['*.jar'], dir: 'libs')
    implementation 'androidx.appcompat:appcompat:1.1.0'
    implementation 'androidx.constraintlayout:constraintlayout:1.1.3'
    testImplementation 'junit:junit:4.13'
    androidTestImplementation 'androidx.test.ext:junit:1.1.1'
    androidTestImplementation 'androidx.test.espresso:espresso-core:3.2.0'
    implementation files('libs/gson-2.3.1.jar')
    implementation 'com.android.support:recyclerview-v7:27.1.1'
    implementation 'com.github.bumptech.glide:glide:4.7.1'
    annotationProcessor 'com.github.bumptech.glide:compiler:4.7.1'
}
```

图 8-10 资源导入后效果

程序清单 8-2：chart0801\app\src\main\res\layout\activity_product_list.xml

```
1    <LinearLayout xmlns:android = "http://schemas.android.com/apk/res/android"
2        xmlns:app = "http://schemas.android.com/apk/res - auto"
3        xmlns:tools = "http://schemas.android.com/tools"
4        android:layout_width = "match_parent"
5        android:layout_height = "match_parent"
6        android:orientation = "vertical"
7        tools:context = ".MainActivity">
8        <android.support.v7.widget.RecyclerView
9            android:id = "@ + id/recycle_Viewes"
10           android:layout_width = "match_parent"
11           android:layout_height = "match_parent">
12       </android.support.v7.widget.RecyclerView >
13   </LinearLayout >
```

列表子项布局文件如下。

程序清单 8-3：chart0801\app\src\main\res\layout\goods_list_items.xml

```
1    < RelativeLayout
2        xmlns:android = "http://schemas.android.com/apk/res/android"
3        android:layout_width = "match_parent"
4        android:layout_height = "match_parent">
5        < ImageView
6            android:id = "@ + id/iv_iamge"
7            android:layout_width = "120dp"
8            android:layout_height = "120dp"
9            android:layout_marginTop = "4dp"
10           android:src = "@drawable/a000" />
11       < TextView
12           android:id = "@ + id/tv_title"
13           android:layout_width = "match_parent"
14           android:layout_height = "wrap_content"
15           android:layout_alignTop = "@ + id/iv_iamge"
16           android:layout_marginLeft = "5dp"
17           android:layout_toRightOf = "@id/iv_iamge"
18           android:ellipsize = "end"
19           android:maxLines = "2"
20           android:text = "测试测试测试测试测试测试测试测试测试测试测试"
21           android:textSize = "13dp" />
22       < TextView
23           android:id = "@ + id/tv_old_price_title"
24           android:layout_width = "wrap_content"
25           android:layout_height = "wrap_content"
26           android:layout_alignLeft = "@ + id/tv_title"
27           android:layout_below = "@id/tv_title"
28           android:layout_marginLeft = "5dp"
29           android:layout_marginTop = "5dp"
30           android:text = " "
```

```
31              android:textSize = "13sp" />
32          <TextView
33              android:id = "@ + id/tv_count"
34              android:layout_width = "wrap_content"
35              android:layout_height = "wrap_content"
36              android:layout_alignBottom = "@ + id/tv_old_price_title"
37              android:layout_toRightOf = "@ + id/tv_old_price_title"
38              android:text = "1100" />
39          <TextView
40              android:layout_width = "wrap_content"
41              android:layout_height = "wrap_content"
42              android:layout_alignBottom = "@ + id/tv_old_price_title"
43              android:layout_marginRight = "3dp"
44              android:layout_toLeftOf = "@ + id/tv_address"
45              android:text = "产地：" />
46          <TextView
47              android:id = "@ + id/tv_address"
48              android:layout_width = "wrap_content"
49              android:layout_height = "wrap_content"
50              android:layout_alignBottom = "@ + id/tv_old_price_title"
51              android:layout_alignParentRight = "true"
52              android:text = "河南 商丘" />
53          <TextView
54              android:id = "@ + id/tv_new_price_title"
55              android:layout_width = "wrap_content"
56              android:layout_height = "wrap_content"
57              android:layout_alignLeft = "@ + id/tv_old_price_title"
58              android:layout_below = "@ + id/tv_old_price_title"
59              android:layout_marginTop = "8dp"
60              android:text = "价格：¥ " />
61          <TextView
62              android:id = "@ + id/tv_new_price"
63              android:layout_width = "wrap_content"
64              android:layout_height = "wrap_content"
65              android:layout_alignBottom = "@ + id/tv_new_price_title"
66              android:layout_toRightOf = "@ + id/tv_new_price_title"
67              android:text = "12.00"
68              android:textSize = "16sp"
69              android:textStyle = "bold" />
70          <LinearLayout
71              android:layout_width = "match_parent"
72              android:layout_height = "25dp"
73              android:layout_alignBottom = "@ + id/iv_iamge"
74              android:layout_alignLeft = "@ + id/tv_new_price_title"
75              android:background = "@drawable/textview_border_style"
76              android:gravity = "center"
77              android:orientation = "horizontal">
78              <TextView
79                  android:layout_width = "wrap_content"
80                  android:layout_height = "wrap_content"
```

```
81                android:text = "店铺:" />
82            < TextView
83                android:id = "@ + id/tv_shopname"
84                android:layout_width = "wrap_content"
85                android:layout_height = "wrap_content"
86                android:text = "童波的店铺"
87                android:textColor = "@color/colorPrimary" />
88        </LinearLayout >
89        < View
90            android:id = "@ + id/view"
91            android:layout_width = "match_parent"
92            android:layout_height = "1dp"
93            android:layout_below = "@id/iv_iamge"
94            android:layout_marginTop = "3dp"
95            android:background = "@color/grey" />
96    </RelativeLayout >
```

第四步，定义必要的工具类、其中包括 JSON 解析类、GoodInfo 商品实体类、HttpClient 联网工具类。在日常开发中，JSON 解析和封装在项目中反复使用，为了提高效率，一般在项目中创建 JSON 解析类。

程序清单 8-4：chart0801\app\src\main\java\com\jxcia\chart0801\JsonUtil. java

```
1     public class JsonUtil {
2         public static JsonUtil instance;
3         private static Gson gson;
4         public static JsonUtil getInstance(){
5             if(instance == null){
6                 synchronized (JsonUtil.class){
7                     if(instance == null){
8                         instance = new JsonUtil();
9                         synchronized (Gson.class){
10                            gson = new Gson();
11                        }
12                    }
13                }
14            }
15            return instance;
16        }
17        public < T > String beanToJsonStr(T t){
18            String str = "";
19            if(t != null){
20                str = gson.toJson(t);
21            }
22            return str;
23        }
24        //把字符变为对象实体
25        public < T > T jsonToBean(String jsonStr, Class < T > tClass){
26            T t = null;
```

```
27          if(jsonStr != null || jsonStr != ""){
28              try {
29                  t = gson.fromJson(jsonStr,tClass);
30              }catch (JsonIOException e){
31              }
32          }
33          return t;
34      }
35      public static <T> ArrayList<T> jsonToArrayList(String json, Class<T> clazz) {
36          Type type = new TypeToken<ArrayList<JsonObject>>() {
37          }.getType();
38          ArrayList<JsonObject> jsonObjects = new Gson().fromJson(json, type);
39          ArrayList<T> list = new ArrayList<>();
40          for (JsonObject jsonObject: jsonObjects) {
41              list.add(new Gson().fromJson(jsonObject, clazz));
42          }
43          return list;
44      }
45  }
```

客户端和服务器端进行数据传递，需要先用实体类进行封装，商品详情实体类如下。

程序清单 8-5：chart0801\app\src\main\java\com\jxcia\chart0801\GoodInfo.java

```
1   public class GoodInfo implements Serializable {
2       private int image;                         →图片地址
3       private String productLocation;            →商品生产地
4       private String name;                       →商品名称
5       private String price;                      →商品价格
6       private String info;                       →商品详情
7       private String volume;                     →商品销量
8       private String shopName;                   →商店名称
9       private String totalValue;                 →总额
10      private String count;                      →数量
11      private String dateTime;                   →日期
12      public int getImage() {
13          return image;
14      }
15      public void setImage(int image) {
16          this.image = image;
17      }
18      ...                                        →后面代码类似
19  }
```

AccessToServer 访问服务器端工具类中含有 doGet()、doPost()两个方法，调用这两个方法时，传入参数向服务端发送请求，就能得到响应结果，代码如下。

程序清单 8-6：chart0801\app\src\main\java\com\jxcia\chart0801\AccessToServer.java

```
1   public class AccessToServer {
2       private static String TAG = "CustomHttpUrlConnection";
3       private static HttpURLConnection conn;
4       public AccessToServer() {
5       }
6       /**
7        * 向服务器发送 Get 请求,获取响应结果
8        * @param strUrl 表示需要访问的资源网址
9        * @param names 表示需传递的多个参数名称集合
10       * @param values 表示传递的每个参数所对应的值
11       * @return 返回字符串(通常是 JSON 格式)
12       */
13      public static String doGet(String strUrl, String[] names, String[] values) {
14          String result = "";
15          if (names != null) {
16              //当有参数时,将参数拼接在地址后面,并用?隔开,参数间用 & 隔开
17              strUrl += "?";      //在网址后面添加?号
18              for (int i = 0; i < names.length; i++) {
19                  //循环遍历参数名和参数值,将其拼接
20                  strUrl += names[i] + "=" + values[i];
21                  if (i != (names.length - 1)) {
22                      //如果不是最后一个参数则添加 & 符号
23                      strUrl += "&";
24                  }
25              }
26          }
27          try {
28              URL url = new URL(strUrl);
29              conn = (HttpURLConnection) url.openConnection();
30              conn.setDoInput(true);
31              conn.setConnectTimeout(3000);
32              conn.setReadTimeout(4000);
33              conn.setRequestProperty("accept", "*/*");
34              conn.connect();
35              InputStream stream = conn.getInputStream();
36              InputStreamReader inReader = new InputStreamReader(stream);
37              BufferedReader buffer = new BufferedReader(inReader);
38              String strLine = null;
39              while((strLine = buffer.readLine())!= null)
40              {
41                  result += strLine;
42              }
43              inReader.close();
44              conn.disconnect();
45              return result;
46          } catch (MalformedURLException e) {
47              Log.e(TAG, "getFromWebByHttpUrlCOnnection:" + e.getMessage());
48              e.printStackTrace();
49              return null;
50          } catch (IOException e) {
51              //TODO Auto-generated catch block
```

```
52                  Log.e(TAG, "getFromWebByHttpUrlCOnnection:" + e.getMessage());
53                  e.printStackTrace();
54                  return null;
55              }
56          }
57          /**
58            * 向服务器发送 Post 请求,获取响应结果
59            * @param strUrl 表示需要访问的资源网址
60            * @param names 表示需传递的多个参数名称集合
61            * @param values 表示传递的每个参数所对应的值
62            * @return 返回字符串(通常是 JSON 格式)
63            */
64          public static String doPost(String strUrl,String[] names, String[] values) {
65              String result = "";
66              if (names != null) {
67                  //当有参数时,将参数拼接在地址后面,并用?隔开,参数间用 & 隔开
68                  strUrl += "?";      //在网址后面添加?号
69                  for (int i = 0; i < names.length; i++) {
70                      //循环遍历参数名和参数值,将其拼接
71                      strUrl += names[i] + "=" + values[i];
72                      if (i != (names.length - 1)) {
73                          //如果不是最后一个参数则添加 & 符号
74                          strUrl += "&";
75                      }
76                  }
77              }
78              try {
79                  URL url = new URL(strUrl);
80                  conn = (HttpURLConnection) url.openConnection();
81                  //设置是否从 httpUrlConnection 读入,默认情况下是 true
82                  conn.setDoInput(true)
83                  //设置是否向 httpUrlConnection 输出,因为这个是 post 请求,参数要放
84                  //在 HTTP 正文内,因此需要设为 true, 默认情况下是 false
85                  conn.setDoOutput(true);
86                  //设定请求的方法为 POST,默认是 GET
87                  conn.setRequestMethod("POST");
88                  //设置超时
89                  conn.setConnectTimeout(3000);
90                  conn.setReadTimeout(4000);
91                  //POST 请求不能使用缓存
92                  conn.setUseCaches(false);
93                  conn.setInstanceFollowRedirects(true);
94                  //设定传送的内容类型是可序列化的 Java 对象
95                  //(如果不设此项,在传送序列化对象时,当 Web 服务默认的不是这种类型时
//可能抛出 java.io.EOFException 异常)
96                  conn.setRequestProperty("Content-Type",
97                          "application/x-www-form-urlencoded");
98                  InputStream in = conn.getInputStream();
99                  InputStreamReader inStream = new InputStreamReader(in);
```

```
100                BufferedReader buffer = new BufferedReader(inStream);
101                String strLine = null;
102                while((strLine = buffer.readLine())!= null)
103                {
104                    result += strLine;
105                }
106                return result;
107            } catch (IOException ex) {
108                Log.e(TAG,"PostFromWebByHttpURLConnection:" + ex.getMessage());
109                ex.printStackTrace();
110                return null;
111            }
112        }
113    }
```

第五步，定义实现商品列表 RecyclerView 需要的适配器 GoodsListAdapter。

程序清单 8-7：chart0801\app\src\main\java\com\jxcia\chart0801\GoodsListAdapter.java

```
1      public class GoodsListAdapter extends RecyclerView.Adapter < GoodsListAdapter.
  ViewHolder > {
2          private List < GoodInfo > mList;
3          private Activity mActivity;
4          private ViewHolder viewHolder;
5          private GoodInfo goodInfo;
6          public GoodsListAdapter(Activity activity, List < GoodInfo > list) {
7              mList = list;
8              mActivity = activity;
9          }
10         @Override
11         public ViewHolder onCreateViewHolder(ViewGroup parent, int viewType) {
12             viewHolder = new ViewHolder(View.inflate(mActivity,
                                            R.layout.goods_list_items, null));
13             return viewHolder;
14         }
15         @Override
16         public void onBindViewHolder(ViewHolder holder, int position) {
17             goodInfo = mList.get(position);
18             Glide.with(mActivity).load(goodInfo.getImage()).into(holder.imageView);
19             holder.title.setText(goodInfo.getName());
20             holder.price.setText("" + goodInfo.getPrice());
21             holder.count.setText(goodInfo.getVolume() + "人付款");
22             holder.address.setText(goodInfo.getProductLocation());
23             holder.shopName.setText(goodInfo.getShopName());
24         }
25         @Override
26         public int getItemCount() {
27             return mList.size();
28         }
```

```
29        class ViewHolder extends RecyclerView.ViewHolder {
30            TextView shopName;
31            TextView title;
32            TextView count;
33            TextView address;
34            TextView price;
35            ImageView imageView;
36            View functionView;
37            public ViewHolder(View itemView) {
38                super(itemView);
39                functionView = itemView;
40                imageView = itemView.findViewById(R.id.iv_iamge);
41                title = (TextView)itemView.findViewById(R.id.tv_title);
42                count = (TextView)itemView.findViewById(R.id.tv_count);
43                address = (TextView)itemView.findViewById(R.id.tv_address);
44                price = (TextView)itemView.findViewById(R.id.tv_new_price);
45                shopName = (TextView) itemView.findViewById(R.id.tv_shopname);
46            }
47        }
48    }
```

第六步，使用本书提供的 JTM 服务器端集成包，进行启动，开启服务器端，如图 8-11 所示。

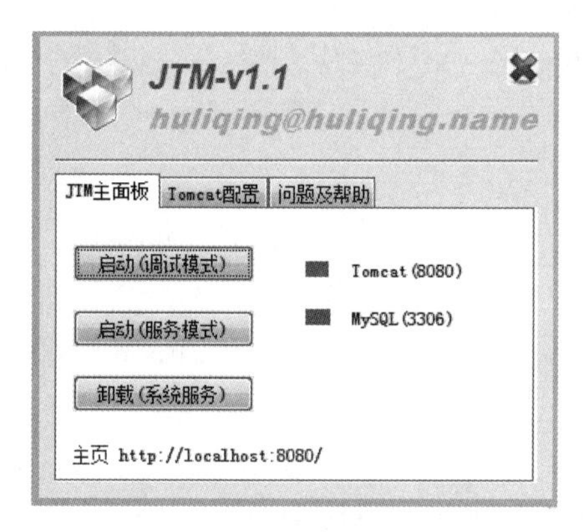

图 8-11　启动 JTM 服务端集成包

第七步，从服务器端获取数据，实现列表。访问网络是耗时的操作，需要开启子线程通过 HttpUrlConnection 框架访问网络。由于子线程不能直接更改主线程 UI，需要借助 Handler 帮助修改界面，呈现列表。

程序清单 8-8：chart0801\app\src\main\java\com\jxcia\chart0801\MainActivity.java

```
1    public class MainActivity extends AppCompatActivity {
2        private LinearLayoutManager linearLayoutManager;
3        private RecyclerView recyclerView;
4        private Handler mHandler;
```

```
5          private String url = "localhost:8080/HappyShop/shopSer";
6          private List < GoodInfo > goodInfos;
7          @Override
8          protected void onCreate(Bundle savedInstanceState) {
9              super.onCreate(savedInstanceState);
10              setContentView(R.layout.activity_product_list);
11              recyclerView = (RecyclerView) findViewById(R.id.recycle_Viewes);
12              getDataFromSever();
13              mHandler = new Handler(){
14                  @Override
15                  public void handleMessage(@NonNull Message msg) {
16                      linearLayoutManager = new LinearLayoutManager(MainActivity.this);
17                      recyclerView.setLayoutManager(linearLayoutManager);
18                      //为 recycleViews 设定动画
19                      recyclerView.setItemAnimator(new DefaultItemAnimator());
20                      //为 recycleViews 添加分隔线
21                      recyclerView.addItemDecoration(new DividerItemDecoration
22                          (MainActivity.this,DividerItemDecoration.VERTICAL));
23                      recyclerView.setAdapter(new GoodsListAdapter(MainActivity.this,
                            goodInfos));
24                  }
25              };
26          }
27          public void getDataFromSever() {
28              new Thread() {
29                      public void run() {//线程执行体
30
31                          String response = AccessToServer .doGet(url, new String[]
                          { "keyword"}, new String[] {"1"});
32                          System.out.println("result = " + response);
33                          goodInfos = JsonUtil.getInstance().jsonToArrayList(response,
                          GoodInfo.class);
34                          Message message = Message.obtain();
35                          message.what = 0x11;
36                          mHandler.sendMessage(message);
37                      }
38              }.start();
39          }
40      }
41              list.add(new Gson().fromJson(jsonObject, clazz));
42          }
43          return list;
44      }
45  }
```

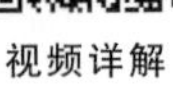
视频详解

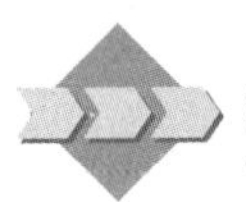

8.2 OkHttp 框架

任务陈述

使用 HttpClient 有点儿复杂，特别是每次进行网络交互时，一些复杂的网络操作及异常处理都非常烦琐。现在 Google 公司的 Android 开发工具 Android Studio 已经对 HttpClient 的使用做了一些限制，本节就来学习一个新的框架 OkHttp。

分析：在“欢乐购商城”中请求服务器网络使用的是 OkHttp 框架，本节将对 8.1 节使用的 HttpClient 的项目案例进行修改，达到同样的访问服务器的目的。

视频详解

8.2.1 OkHttp 框架介绍

OkHttp 是一个处理网络请求的开源项目，是 Android 端最火热的轻量级框架，由移动支付 Square 公司开发，用于替代 HttpURLConnection 和 Apache HttpClient。可以说，当前很多 Android 开发者最喜欢的网络框架就是 OkHttp 了，为此大家还对框架进行了修改和完善，衍生出了包含线程池操作和 RxJava 模式的 OkGo。那么 OkHttp 究竟是有怎么样的魔力让这么多的开发者都爱不释手呢？本节就来好好研究一下。

OkHttp 是非常容易上手的一款框架，很多初学者可能只是因为比较简单才选择它，但我们也要分析它的一些内置性能和容错方面的优势，其主要优点如下。

（1）允许连接到同一个主机地址的所有请求，提高请求效率。

（2）共享 Socket 减少对服务器的请求次数。

（3）通过连接池减少了请求延迟。

（4）缓存响应数据减少重复的网络请求。

（5）减少了对数据流量的消耗。

（6）自动处理 GZip 压缩。

（7）支持各个设计的模式的拓展。

1. 使用流程

（1）当通过 OkHttpClient 创建一个 Call，并发起同步或异步请求时；OkHttp 会通过 Dispatcher 对所有的 RealCall（Call 的具体实现类）进行统一管理，并通过 execute()及 enqueue()方法对同步或异步请求进行处理。

（2）execute() 及 enqueue（ ）这两个方法会最终调用 RealCall 中的 getResponseWithInterceptorChain()方法，从拦截器链中获取返回结果。

（3）在拦截器链中，依次通过 RetryAndFollowUpInterceptor（重定向拦截器）、BridgeInterceptor（桥接拦截器）、CacheInterceptor（缓存拦截器）、ConnectInterceptor（连接拦截器）、CallServerInterceptor（网络拦截器）对请求依次处理。与服务建立连接后，获取返回数据，再经过上述拦截器依次处理后，最后将结果返回给调用方。

2. 相关的类和 API 介绍

Request 用于包装请求参数的类，不能直接通过 new 对象的方式进行初始化，Request

类提供了一个静态类 Builder，通过 Builder. build()方法进行构建，其主要调用的方法如下。

```
url(String url)                          请求的地址
addHerader(String key,String value)      请求的头部，可以添加多个，以键 - 值对的方式进行添加.
add(String key,String value)             请求的请求体，以键 - 值对的方式进行传输.
```

OkHttpClient 是用于进行网络请求的核心类，内部有多个参数，通过方法可以进行初始化，大部分参数都有初始值。如果没有业务需要可以不进行设置，当需要的时候可以调用相关的方法进行设置。主要设置的参数和方法如下。

```
setConnectTimeout()        设置连接等待的超时时长
setReadTimeout()           设置读取文件的超时时长
setWriteTimeout()          设置写入文件的超时时长
setCookieHandler()         设置 Cookie
```

除了设置请求参数，OkHttpClinet 最主要的方法就是发起网络请求，主要方法是 newCall()、execute()方法和 enqueue()方法。

```
newCall(Request request)        建立一个网络请求，需要把包装请求数据的 request 类放在里面.
execute()                       同步接口，如果在主线程里需要新建一个子线程再请求.
enqueue(CallBack callback)      异步请求，有一个 callback 参数专门用于接收请求返回的结果.
```

8.2.2 实战演练——采用 OkHttp 框架访问服务器端

视频详解

根据任务分析，在 8.1 节中使用 HttpClient 通过客户端和服务器端交互实现"欢乐购商城"列表，本节将使用 OkHttp 进行替换达到实现同样的目的。本节项目案例在 chart0801 中进行修改，其中对项目的变动不是太多，在此只展示修改部分和替换部分内容。

要使用 OkHttp 框架需要先导入 OkHttp 库，在 build. gradle 文件中添加以下内容，并进行同步就可导入 OkHttp 库。

```
implementation 'com.squareup.okhttp3:okhttp:3.14.2'
implementation 'com.squareup.okio:okio:1.17.4'
```

剩下的操作只是在 MainActivity 中替换 HttpURLConnection 框架的使用，具体修改如下。

程序清单 8-9：chart0802\app\src\main\java\com\jxcia\chart0802\MainActivity. java

```
1    public class MainActivity extends AppCompatActivity {
2        private LinearLayoutManager linearLayoutManager;
3        private RecyclerView recyclerView;
4        private String url = "localhost:8080/BookShop/shopSer?keyword = 1";
5        private List < GoodInfo > goodInfos;
6        @Override
7        protected void onCreate(Bundle savedInstanceState) {
8            super.onCreate(savedInstanceState);
9            setContentView(R.layout.activity_product_list);
```

```
10              recyclerView = (RecyclerView) findViewById(R.id.recycle_Viewes);
11              getData(url);
12          }
13          public void getData(String url) {
14              OkHttpClient client = new OkHttpClient();
15              Request request = new Request.Builder().url(url).build();
16              Call call = client.newCall(request);
17              call.enqueue(new Callback() {
18                  @Override
19                  public void onFailure(Call call, IOException e) {
20                      Toast.makeText(MainActivity.this,"获取数据失败
                              ",Toast.LENGTH_LONG).show();
21                  }
22                  @Override
23                  public void onResponse(Call call, Response response) throws IOException {
24                      if(response.isSuccessful()){
25                          String result = response.body().string();
26                          System.out.println("result = " + result);
27                          //处理 UI 需要切换到 UI 线程处理
28                          goodInfos = JsonUtil.getInstance().jsonToArrayList(result,
                                GoodInfo.class);
29                          if (goodInfos != null) {
30                              runOnUiThread(new Runnable() {
31                                  @Override
32                                  public void run() {
33                                      linearLayoutManager = new
                                          LinearLayoutManager(MainActivity.this);
34  recyclerView.setLayoutManager(linearLayoutManager);
35                                      //为 recycleViews 设定动画
36                                      recyclerView.setItemAnimator(new
                                            DefaultItemAnimator());
37                                      //为 recycleViews 添加分隔线
38                                      recyclerView.addItemDecoration(new
                                          DividerItemDecoration(MainActivity.this,
39                                              DividerItemDecoration.VERTICAL));
40                                      recyclerView.setAdapter(new
                                      GoodsListAdapter(MainActivity.this, goodInfos));
41
42                                  }
43                              });
44                          }
45                      }
46                  }
47              });
48          }
49  }
```

本章小结

本章主要讲解了 Android 客户端和服务器之间进行网络数据交互的相关知识。访问网络数据是每一个联网应用不可缺失也是极其重要的板块，每个 App 应用都是通过网络数据的交互从而实现一个个业务交互。Android 开发中，请求网络数据是一整套流程，由于网络数据交互过程中存在很多不确定性，故请求网络需要严谨的异常控制也即错误判断。Android 中提供了很多访问网络的相关 API，同时很多开发者也自己开发了相关的网络请求工具，以方便其他开发者使用。

随后介绍了一整套网络请求的流程图，通过流程图大致了解了网络数据交互的每一个过程，为接下来的代码编写提供了思路。同时，网络交互中获取的数据需要进行标准格式处理，也就是通过使用 JSON 格式，将字符串数据直接转换为对象，从而转变为面向对象的编程。

最后使用当前流行的第三方网络请求框架 OkHttp 对其进行了封装和重写，将请求网络编写成了一个工具类，也就是以后都可以使用这个工具类进行网络请求。有了请求网络的模块，最后进行了商城首页的网络编程，请求到了商品数据，同时将商品数据进行了展示。

通过本章的学习，读者应熟悉 Android 客户端的基础，同时可以去百度上搜索一些公开接口，比如查询天气、查询快递等接口，接入自己的 App。

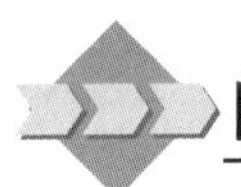

自测习题

1. 在 Android 开发中，通常在主线程中更新 UI，也就是在 Activity 的各个生命周期的函数中，而请求网络等操作仅允许自己新建一个子线程去请求。你觉得 Android 这样设计的原因是什么？

2. 关于网络请求，以下描述错误的是（　　）。
   A. 一般的网络请求都遵循 TCP/IP 网络协议
   B. 网络请求一般分为 GET 和 POST 两种，GET 一般用作数据的获取，POST 一般用作数据的提交
   C. GET 网络请求与 POST 网络请求的区别是：GET 无法包装参数
   D. 网络请求可以采用 HTTP 和 HTTPS 两种，一般来说，后者使用了认证，所以更安全

3. 关于 JSON 格式的数据描述错误的是（　　）。
   A. JSON 格式的数据基本的数据单元是 key：value 组成的键-值对
   B. JSON 格式的数据本质上是字符串
   C. 任何对象数据都可以被序列化成 JSON 字符串，反之，正确格式的 JSON 字符串也可以被反序列化为特定对象
   D. 所有数据的通信都被设计成了 JSON 数据的传递

4. 关于 Android 中的网络请求描述错误的是（　　）。
   A. Android 发送数据请求必须先给 App 加上网络权限

B. Android 网络请求必须要在子线程中进行，获取结果后把结果发给主线程

C. 网络请求中，可以查看网络状态，获取当前连接的是哪种网络，如 Wi-Fi、移动数据等

D. Android 网络内置的 HttpClient 是目前最常用的网络请求工具

5. 开发一个 App 的登录界面，其中用户数据保存在服务器端，简述 Android 端开发的流程。

第9章 综合项目

为巩固前几章所学习的 Android 基础知识，本章开发一款商城购物项目，该项目展示的内容包括商品展示、购物车、个人中心等模块。商品展示模块包括推荐商品展示、商品分类展示和商品详情展示。购物车模块包括购物车列表展示和结算功能。个人中心模块包括订单信息和收货地址等。为了让读者能够熟练掌握商城购物项目中使用到的知识点，接下来从项目分析开始，一步一步开发商城购物项目的各个功能。

学习目标

本章要点

(1) 了解商品购物项目模块结构和功能。

(2) 掌握商品列表模块的和商品分类展示和商品详情显示。

(3) 掌握购物车功能开发，能够独立实现购物车功能并实现结算。

(4) 掌握个人中心中订单展示和收货地址的展示和登录、注册功能。

9.1 项目分析

视频详解

9.1.1 项目概述

"欢乐购商城"项目是一个购物网站的项目，其中包含商品列表、商品分类、商品详情、购物车、订单支付、订单列表、收货地址、用户登录、用户注册和退出登录等模块。项目启动之后，首先展示首页内容，在首页中展示商品分类图标和推荐商品列表，单击商品分类图标可进入不同类别商品列表，单击商品列表中的每一项可进入商品详情页面。在商品详情页面可以将商品添加至购物车。在购物车页面可展示加入购物车的商品并可修改商品数量，确定之后可进行结算。在个人中心页面可进入订单列表，在订单列表页面可展示所有订单。在个人中心页面也可进入收货地址页面，查看并修改收货地址。在个人中心页面还可进行登录和注册。项目模块结构如图 9-1 所示。

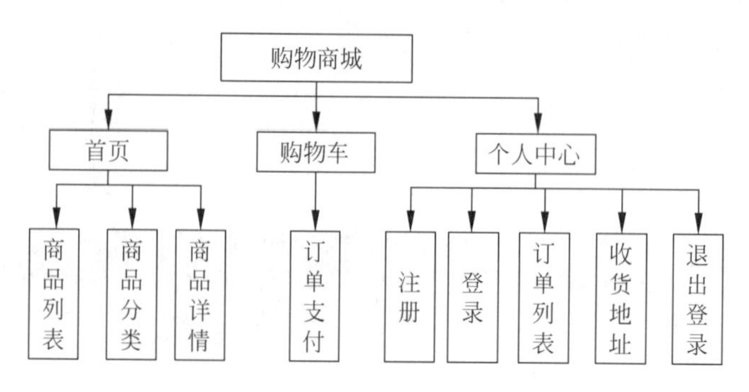

图 9-1　项目模块结构

视频详解

9.1.2　效果图展示

1. 首页模块

当程序启动之后，首先会进入首页。首页展示商品分类图标和推荐商品列表，如图 9-2(a)所示；当单击某一个商品类别图标时，跳转到此类别下的商品列表展示页面，如图 9-2(b)所示；当单击某一商品时，跳转到该商品的详情展示页面，如图 9-2(c)所示。

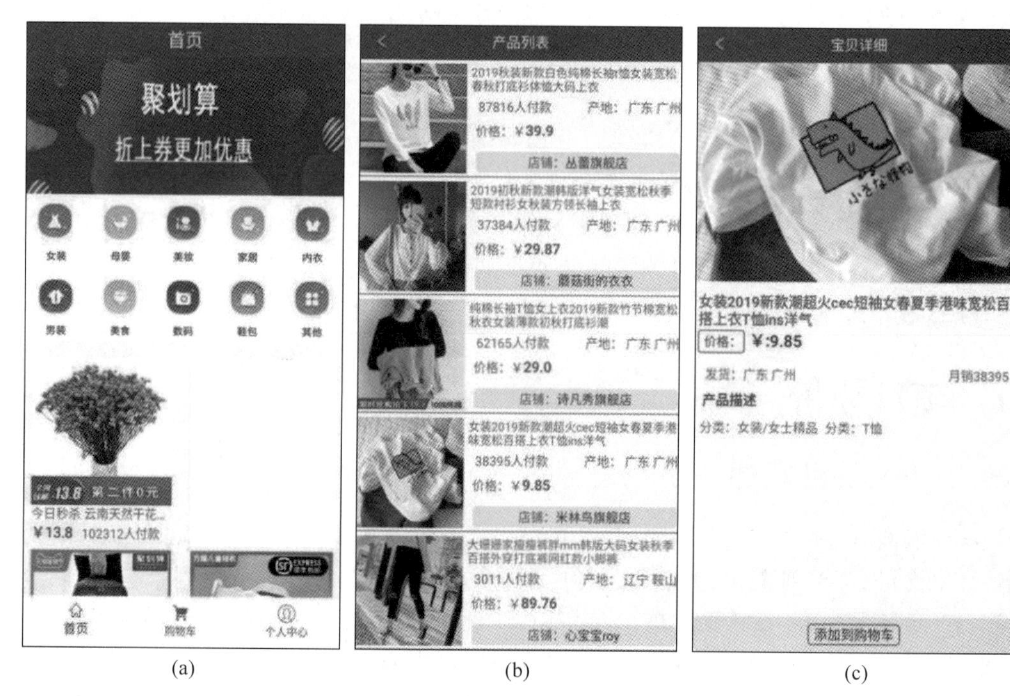

图 9-2　首页模块展示

2. 购物车模块

单击底部导航栏中的“购物车”，可跳转到购物车页面。购物车页面显示已加入购物车的商品信息，并可对商品数量进行修改。当选择好需要结算的商品之后，单击“结算”按钮可进行结算，单击“结算”按钮显示付款方式，并可确认支付，效果如图 9-3 所示。

图 9-3　购物车模块展示

3. 个人中心模块

单击底部导航栏中的"个人中心",可跳转到个人中心页面,如图 9-4(a)所示。个人中心顶部显示用户名头像和用户名。如果没有登录,单击默认头像可跳转到登录页面进行登录,如图 9-4(b)所示。如果还未注册,可单击"立即注册"跳转到注册页面,如图 9-4(c)所示。

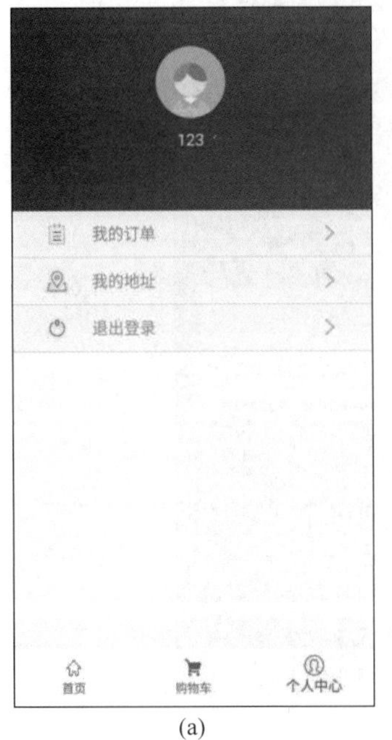

(a)

(b)

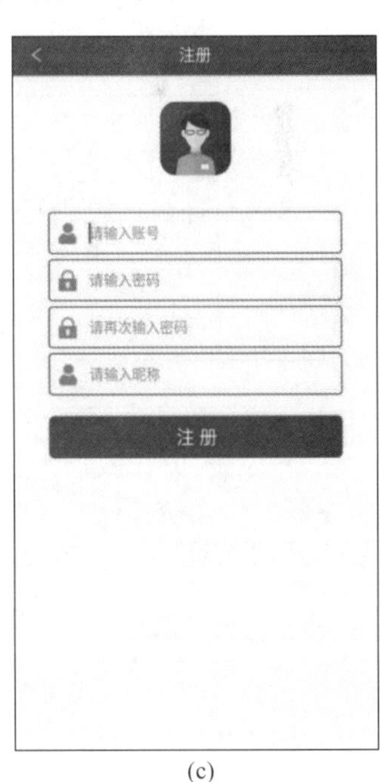

(c)

图 9-4　个人中心模块展示

在“个人中心”中有“我的订单”“我的地址”和“退出登录”三个列表项。单击“我的订单”可跳转到“订单中心”页面，如图 9-5(a)所示。在订单页面展示用户订单信息。单击“我的地址”可跳转到地址列表页面，在该页面展示地址列表，如图 9-5(b)所示。单击地址列表中的项可对地址进行修改，如图 9-5(c)所示。单击“退出登录”，当前用户可退出登录，如图 9-5(d)所示。

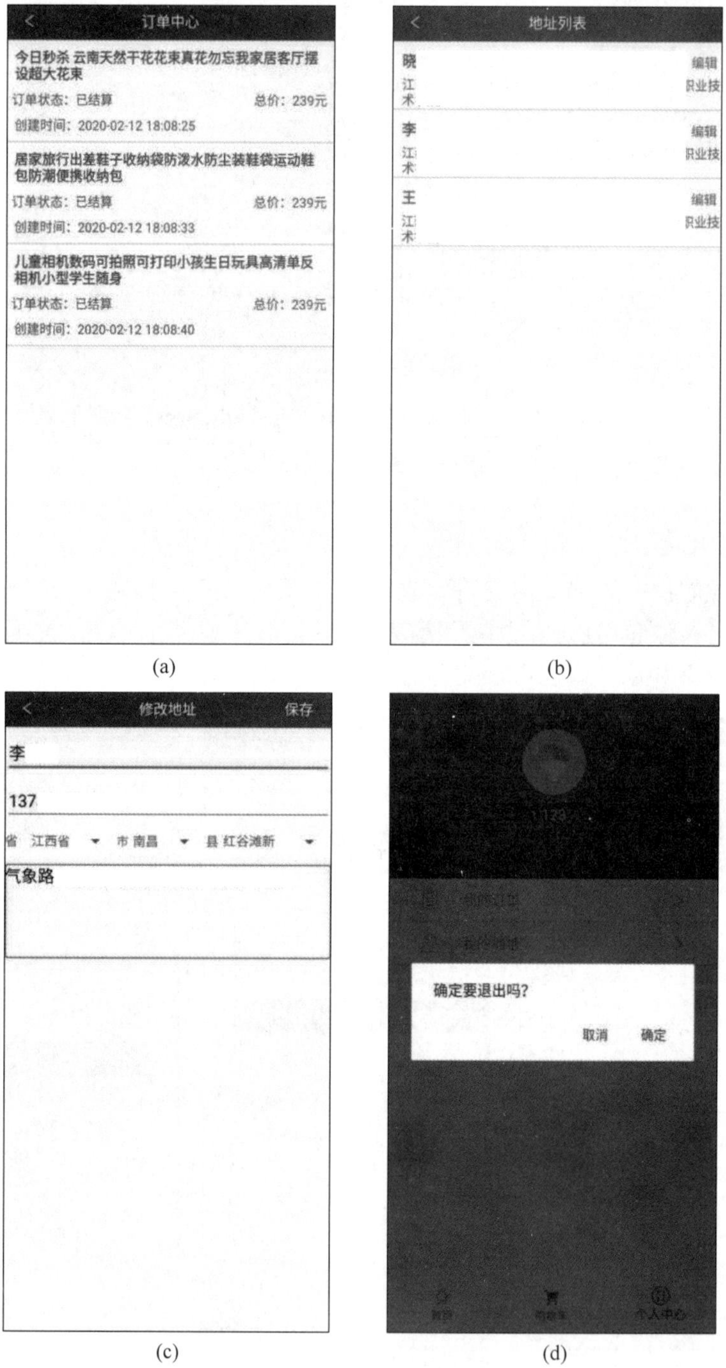

(a)　　(b)

(c)　　(d)

图 9-5　个人中心列表项展示

视频详解

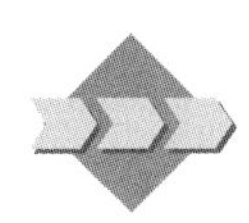

9.2 实现界面设计

整个项目以 MainActivity 为主页面。MainActivity 页面布局由两部分构成，上面一部分是 Fragement，底部是 BottomBar 导航栏。通过单击底部导航按钮跳转到不同的 Fragement。其中，首页对应的 Fragement 名字为 FragmentShouye，购物车对应的 Fragement 名字为 FragmentGouwuche，个人中心对应的 Fragement 名字为 FragmentMe。MainActivity 的布局如下。

程序清单 9-1：happyshop\app\src\main\res\layout\activity_main. xml

```
1    <?xml version = "1.0" encoding = "utf - 8"?>
2    < LinearLayout xmlns:android = "http://schemas.android.com/apk/res/android"
3    xmlns:app = "http://schemas.android.com/apk/res - auto"
4        xmlns:tools = "http://schemas.android.com/tools"
5        android:layout_width = "match_parent"
6        android:layout_height = "match_parent"
7        android:orientation = "vertical"
8        tools:context = ".MainActivity">
9
10        < FrameLayout
11            android:id = "@ + id/contentContainer"
12            android:layout_width = "match_parent"
13            android:layout_height = "0dp"
14            android:layout_weight = "1"
15            />
16
17        < com.roughike.bottombar.BottomBar
18            android:id = "@ + id/bottomBar"
19            android:layout_width = "match_parent"
20            android:layout_height = "56dp"
21            app:bb_tabXmlResource = "@xml/bottombar_menu"/>
22
23     </LinearLayout >
```

底部导航栏 bottombar_menu. xml 文件的代码如下。

程序清单 9-2：happyshop\app\src\main\res\xml\bottombar_menu. xml

```
1    <?xml version = "1.0" encoding = "utf - 8"?>
2    < tabs xmlns:android = "http://schemas.android.com/apk/res/android">
3        < tab
4            icon = "@mipmap/init"
5            id = "@ + id/tab_yi"
6            title = "首页" />
7        < tab
8            icon = "@mipmap/shop"
9            id = "@ + id/tab_er"
```

```
10          title = "购物车" />
11      < tab
12          icon = "@mipmap/self"
13          id = "@ + id/tab_san"
14          title = "个人中心" />
15  </tabs >
```

当程序启动之后，在 MainActivity 中首先加载 FragmentShouye. java，其中，FragmentShouye 的布局文件为 fragment_shoutye. xml，其代码如下。

程序清单 9-3：happyshop\app\src\main\res\layout\fragment_shoutye. xml

```
1   <?xml version = "1.0" encoding = "utf - 8"?>
2   < LinearLayout xmlns:android = "http://schemas.android.com/apk/res/android"
3       android:layout_width = "match_parent"
4       android:layout_height = "wrap_content"
5       android:orientation = "vertical">
6
7       < TextView
8           android:layout_width = "match_parent"
9           android:layout_height = "40dp"
10          android:background = "@color/colorPrimary"
11          android:gravity = "center"
12          android:text = "首页"
13          android:textColor = "#fff"
14          android:textSize = "20sp" />
15
16      < ScrollView
17          android:layout_width = "match_parent"
18          android:layout_height = "match_parent"
19          android:gravity = "center">
20
21          < LinearLayout
22              android:layout_width = "match_parent"
23              android:layout_height = "wrap_content"
24              android:orientation = "vertical">
25
26              < ImageView
27                  android:layout_width = "match_parent"
28                  android:layout_height = "150dp"
29                  android:scaleType = "fitXY"
32                  android:src = "@drawable/jhs" />
33
34              < LinearLayout
35                  android:layout_width = "match_parent"
36                  android:layout_height = "170dp">
37
38      < android.support.v7.widget.RecyclerView
```

```
39                android:id = "@ + id/recycle_View"
40                android:layout_width = "match_parent"
41                android:layout_height = "match_parent">
            </android.support.v7.widget.RecyclerView>
42          </LinearLayout>
43
44          <com.example.administrator.taobao.adapter.MyGridView
45                      android:id = "@ + id/gridView"
46                      android:layout_width = "match_parent"
47                      android:layout_height = "wrap_content"
48                      android:layout_marginTop = "5dp"
49                      android:gravity = "center"
50                      android:horizontalSpacing = "5dp"
51                      android:numColumns = "2"
52                      android:verticalSpacing = "5dp">
    </com.example.administrator.taobao.adapter.MyGridView>
53              </LinearLayout>
54          </ScrollView>
55
56      </LinearLayout>
```

在 MainActivity 中底部导航栏中单击“购物车”时，在 MainActivity 中加载 FragmentGouwuche.java，其对应的布局文件 fragment_buycar.xml 代码如下。

程序清单 9-4：happyshop\app\src\main\res\layout\fragment_buycar.xml

```
1       <?xml version = "1.0" encoding = "utf - 8"?>
2       <RelativeLayout xmlns:android = "http://schemas.android.com/apk/res/android"
3           android:layout_width = "match_parent"
4           android:layout_height = "match_parent">
5
6           <TextView
7               android:id = "@ + id/shoptitle"
8               android:layout_width = "match_parent"
9               android:layout_height = "40dp"
10              android:layout_alignParentTop = "true"
11              android:background = "@color/colorPrimary"
12              android:gravity = "center"
13              android:text = "购物车"
14              android:textColor = "#fff"
15              android:textSize = "20sp" />
16
17            <ListView
18              android:id = "@ + id/listview"
19              android:layout_width = "match_parent"
20              android:layout_height = "wrap_content"
21              android:layout_alignLeft = "@ + id/shoptitle"
22              android:layout_below = "@ + id/shoptitle"
23      android:layout_marginBottom = "50dp"></ListView>
```

```
24
25        <RelativeLayout
26            android:layout_width = "match_parent"
27            android:layout_height = "50dp"
28            android:layout_alignParentBottom = "true"
29            android:background = "@color/white">
30
31            <CheckBox
32                android:id = "@ + id/all_select_bx"
33                android:layout_width = "40dp"
34                android:layout_height = "40dp"
35                android:layout_centerVertical = "true"
36                android:layout_marginLeft = "10dp" />
37
38            <TextView
39                android:layout_width = "wrap_content"
40                android:layout_height = "wrap_content"
41                android:layout_centerVertical = "true"
42                android:layout_toRightOf = "@id/all_select_bx"
43                android:text = "全选"
44                android:textColor = "@color/cblack"
45                android:textSize = "18dp" />
46
47            <LinearLayout
48                android:layout_width = "wrap_content"
49                android:layout_height = "match_parent"
50                android:layout_alignParentRight = "true"
51                android:layout_marginRight = "10dp"
52                android:orientation = "horizontal">
53
54                <TextView
55                    android:layout_width = "wrap_content"
56                    android:layout_height = "wrap_content"
57                android:layout_gravity = "center_vertical"
58                    android:text = "合计:"
59                    android:textSize = "18dp" />
60
61                <TextView
62                    android:layout_width = "wrap_content"
63                    android:layout_height = "wrap_content"
64                android:layout_gravity = "center_vertical"
65                    android:text = " ¥ "
66                    android:textColor = "@color/red"
67                    android:textSize = "16dp" />
68
69                <TextView
70                    android:id = "@ + id/total_price"
71                    android:layout_width = "wrap_content"
72                    android:layout_height = "wrap_content"
73                android:layout_gravity = "center_vertical"
```

```
74                  android:layout_marginRight = "10dp"
75                  android:text = "0.0"
76                  android:textColor = "@color/red"
77                  android:textSize = "16dp" />
78
79              < Button
80                  android:id = "@ + id/pay"
81                  android:layout_width = "80dp"
82                  android:layout_height = "30dp"
83      android:layout_gravity = "center_vertical"
84                  android:background = "@drawable/button_red_border_style"
85                  android:text = "结算"
86                  android:textColor = "@color/white"
87                  android:textSize = "16dp" />
88          </LinearLayout >
89      </RelativeLayout >
90
91  </RelativeLayout >
```

9.3 获取数据

视频详解

项目中商品数据是存放在服务器上，在启动程序展示商品数据时，首先需要获取服务器的数据，访问服务器数据需要通过网络接口。在项目中采用 OkHttp 的方式访问服务器的数据。使用过程中首先在项目 utils 包下建立 http 包，并新建 RequestManger 类用于访问服务器数据，该类在项目中的位置如图 9-6 所示。

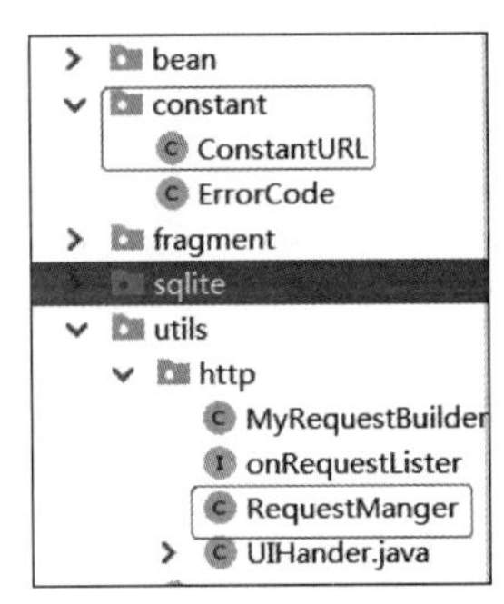

图 9-6 RequestManger 类在项目结构中的位置

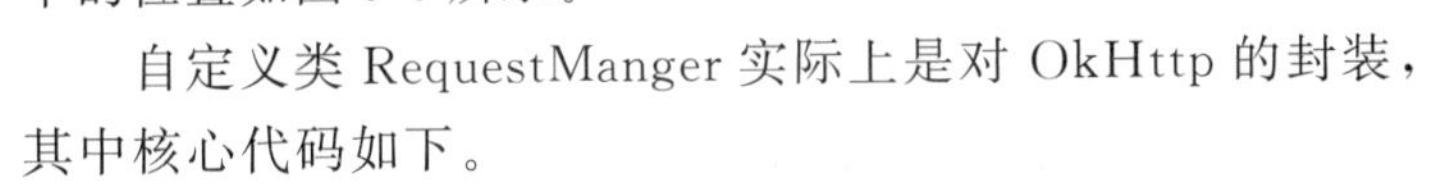

自定义类 RequestManger 实际上是对 OkHttp 的封装，其中核心代码如下。

程序清单 9-5：happyshop\app\src\main\java\com\example\administrator\happyshop\utils\http\RequestManger.java

```
1   public static void get(String url, final onRequestLister lister) throws IOException {
2       //在子线程中运行,结果反馈到子线程
3       Request.Builder builder = new Request.Builder();        →创建 Builder 对象
4       builder.url(url);
5       Request request = builder.build();
6       mOkHttpClient.newCall(request).enqueue(new Callback() {
7           @Override
8           public void onFailure(Request request, IOException e) {
9               LogUtils.setLog(mTag,e.getMessage());
10              lister.onFailed(0,e.getMessage());
11          }
12          @Override
```

```
13            public void onResponse(Response response) throws IOException {
14                String jsonStr = response.body().string();      →获取服务端返回数据
15
16    lister.onSucess(response.code(),jsonStr);
17            }
18        });
19    }
```

上述方法中，url 是服务器的接口地址，而接口的地址保存在 constant 包名下 ConstantURL 类中，如果访问成功则返回请求的字符串。ConstantURL 类的内容如下。

程序清单 9-6：happyshop\app\src\main\java\com\example\administrator\happyshop\constant\ConstantURL.java

```
1    public class ConstantURL {
2        public static final String youming = "http://IP 地址:8080";
3        public static final String severName = "/BookShop";
4        //登录接口
5        public static final String getShop = youming + severName + "/shopSer?keyword = ";
6    }
```

视频详解

9.4 操作数据

从服务器端获取的数据都是 JSON 格式，要操作这些数据之前首先需要对 JSON 数据进行解析并根据解析的结果建立实体类。因此，在项目 utils 包下添加新类 JsonUtil 用于解析 JSON 数据，详情见项目代码。在项目中添加 bean 包，在其目录下建立商品实体类，结构如图 9-7 所示。

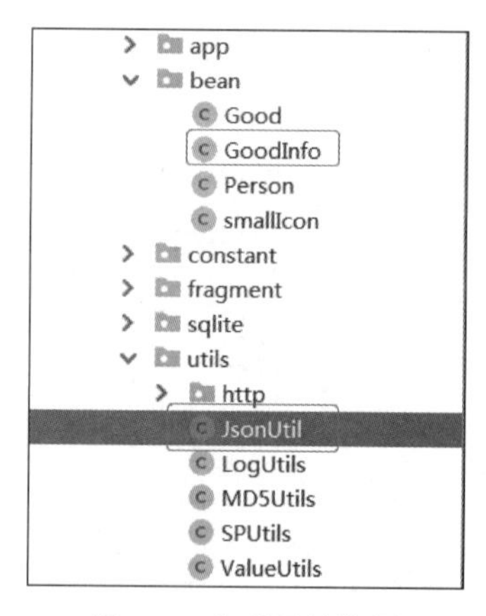

图 9-7　在项目结构图

实体类 GoodInfo 代码如下。

程序清单 9-7：happyshop\app\src\main\java\com\example\administrator\happyshop\bean\GoodInfo. java

```
1   public class GoodInfo implements Serializable {
2
3       private String image;
4       private String productLocation;
5       private String name;
6       private String price;
7       private String info;
8       private String volume;
9       private String shopName;
10      private String totalValue;
11      private String count;
12      private String dateTime;
13      private boolean beSelect;
14
15      public String getDateTime() {
16          return dateTime;
17      }
18
19      public void setDateTime(String dateTime) {
20          this.dateTime = dateTime;
21      }
22
23      public String getCount() {
24          return count;
25      }
26
27      public void setCount(String count) {
28          this.count = count;
29      }
30
31      public String getTotalValue() {
32          return totalValue;
33      }
34
35      public void setTotalValue(String totalValue) {
36          this.totalValue = totalValue;
37      }
38
39      public boolean isBeSelect() {
40          return beSelect;
41      }
42
43      public void setBeSelect(boolean beSelect) {
44          this.beSelect = beSelect;
45      }
46
47      public String getImage() {
48          return image;
```

```
49          }
50
51          public void setImage(String image) {
52              this.image = image;
53          }
54
55          public String getProductLocation() {
56              return productLocation;
57          }
58
59          public void setProductLocation(String productLocation) {
60              this.productLocation = productLocation;
61          }
62
63          public String getName() {
64              return name;
65          }
66
67          public void setName(String name) {
68              this.name = name;
69          }
70
71          public String getPrice() {
72              return price;
73          }
74
75          public void setPrice(String price) {
76              this.price = price;
77          }
78
79          public String getInfo() {
80              return info;
81          }
82
83          public void setInfo(String info) {
84              this.info = info;
85          }
86
87          public String getVolume() {
88              return volume;
89          }
90
91          public void setVolume(String volume) {
92              this.volume = volume;
93          }
94
95          public String getShopName() {
96              return shopName;
97          }
98
99          public void setShopName(String shopName) {
100             this.shopName = shopName;
101         }
102     }
```

当用户选择商品加入购物车后，需要把商品信息暂时保存在本地。项目中采用 SQLite 数据库来保存商品数据。因此在项目中新建包 sqlite，在包中新建两个类 SQLiteHelper 和 DBUtils。其中，SQLiteHelper 类用于生成数据库，而 DBUtils 类用于操作数据库中的表数据，包括对表数据的增、删、改、查，其结构代码如图 9-8 所示。

图 9-8　结构代码

SQLiteHelper 类代码如下。

程序清单 9-8：happyshop\app\src\main\java\com\example\administrator\happyshop\sqlite\SQLiteHelper. java

```
1    public class SQLiteHelper extends SQLiteOpenHelper {
2        private static final int DB_VERSION = 1;
3        private static final String DB_NAME = "goodInfo.db";
4
5        public SQLiteHelper(Context context) {
6            super(context, DB_NAME, null, DB_VERSION);
7            //TODO Auto - generated constructor stub
8        }
9
10        @Override
11        public void onCreate(SQLiteDatabase db) {
12            db.execSQL("create table if not exists" + " orderInfo" + "(" + "_id integer
13              primary key " + "autoincrement," + "goodsName varchar," + "goodsPrice
14              varchar," + "goodsLocation " + "varchar," + "goodsimage varchar," +
15              "goodsCount varchar," + "shopName varchar," + "goodsInfo varchar," +
                "totalValue varchar," + "dateTime varchar" + ")");
16        }
17        @Override
18        public void onUpgrade(SQLiteDatabase db, int oldVersion, int newVersion) {
19            onCreate(db);
20        }
21    }
```

在上述代码中，创建了一个 SQLiteHelper 类继承自 SQLiteOpenHelper，并重写该类的构造方法 SQLiteHelper()。在该方法中通过 super()调用父类 SQLiteOpenHelper 的构造方法，并传入 4 个参数，分别表示上下文对象、数据库名称、游标工厂(通常是 null)、数据库版本。然后重写了 onCreate() 和 onUpgrade()方法，在 onCreate()方法中创建表 orderInfo 用于保存订单信息，其中，onCreate()方法是在数据库第一次创建时调用的。

DBUtils 类的代码如下。

程序清单 9-9：happyshop\app\src\main\java\com\example\administrator\happyshop\sqlite\DBUtils.java

```
1    public class DBUtils {
2        private static SQLiteHelper helper;
3        private static SQLiteDatabase db;
4        private static DBUtils instance = null;
5        String iamgePath = "";
6        public DBUtils(Context context) {
7            helper = new SQLiteHelper(context);
8            db = helper.getWritableDatabase();
9            iamgePath = context.getFilesDir() + File.separator + "delault.jpg";
10       }
11       public static DBUtils getInstance(Context context) {
12           if (instance == null) {
13               instance = new DBUtils(context);
14           }
15           return instance;
16       }
17       //保存加入购物车的信息
18       public boolean saveGoodInfo(GoodInfo goodInfo) {
19           ContentValues values = new ContentValues();
20           values.put("goodsName", goodInfo.getName());
21           values.put("goodsPrice", goodInfo.getPrice());
22           if (goodInfo.getImage() == null) {
23               values.put("goodsimage", iamgePath);
24           } else {
25               values.put("goodsimage", goodInfo.getImage());
26           }
27           values.put("goodsLocation", goodInfo.getProductLocation());
28           values.put("goodsCount", "1");
29           values.put("shopName", goodInfo.getShopName());
30           values.put("totalValue", "239");
31           values.put("goodsInfo", goodInfo.getInfo());
32           SimpleDateFormat df = new SimpleDateFormat("yyyy - MM - dd HH:mm:ss");
33           values.put("dateTime", df.format(new Date()));
34
35           long i = db.insert("orderInfo", null, values);
36
37           if (i > 0) {
38               return true;
39           } else {
```

```
40                  return false;
41              }
42          }
43          //获取加入购物车的信息
44          public List<GoodInfo> getGoodsInfoList() {
45              List<GoodInfo> goodInfoList = new ArrayList<GoodInfo>();
46              GoodInfo goodInfo = null;
47              Cursor cursor = db.query("orderInfo", null, null, null, null, null, null);
48              if (cursor.getCount() > 0 && cursor != null) {
49                  while (cursor.moveToNext()) {
50                      goodInfo = new GoodInfo();
51                      goodInfo.setName(cursor.getString(cursor.getColumnIndex
    ("goodsName")));
52                      goodInfo.setPrice(cursor.getString(cursor.getColumnIndex
    ("goodsPrice")));
53                      goodInfo.setImage(cursor.getString(cursor.getColumnIndex
    ("goodsimage")));
54                      goodInfo.setProductLocation(cursor.getString(cursor.getColumnIndex
55                              ("goodsLocation")));
56                      goodInfo.setCount(cursor.getString(cursor.getColumnIndex
    ("goodsCount")));
57                      goodInfo.setShopName(cursor.getString(cursor.getColumnIndex
    ("shopName")));
58                      goodInfo.setTotalValue(cursor.getString(cursor.getColumnIndex
    ("totalValue")));
59                      goodInfo.setDateTime(cursor.getString(cursor.getColumnIndex
    ("dateTime")));
60                      goodInfoList.add(goodInfo);
61                  }
62                  cursor.close();
63                  return goodInfoList;
64              } else {
65                  return null;
66              }
67          }
68          //根据商品名称删除某条记录
69          public void deleteGoodsInfo(String name) {
70              db.delete("orderInfo", "goodsName = ?", new String[]{name});
71          }
72          //根据商品名称修改某条记录
73          public void updateGoodsInfo(String valueName, String value, String name) {
74              ContentValues values = new ContentValues();
75              values.put(valueName, value);
76              db.update("orderInfo", values, "goodsName = ?", new String[]{name});
77          }
78      }
```

在上述代码中定义了一个DBUtils类，实现对表orderInfo的增、删、改、查。其中，第11～16行代码定义getInstance()方法，实现单例模式。第18～42行代码定义

saveGoodInfo( )方法实现向数据库中添加一条记录。第44～67行代码定义getGoodsInfoList( )方法实现查询orderInfo表中的记录。第69～71行代码定义deleteGoodsInfo()方法实现根据id删除orderInfo表中的某条记录。第73～77行代码定义updateGoodsInfo()方法实现根据id修改orderInfo表中的某条记录。

9.5 各个模块实现

项目开发工具类在本章前面已经实现，本节主要任务是实现每个模块的功能，因为每个模块的功能大部分已经分解到前面章节中分别实现，再加上3个模块功能代码较多，因此在本节主要提供每个模块的重要代码，整个项目的代码可参考本书提供的项目案例。

视频详解

9.5.1 首页模块实现

当项目启动之后首先加载MainActivity页面，在MainActivity中默认加载FragmentShouye页面。MainActivity.java中的详细代码可参考项目案例。其中，FragmentShouye.java中的逻辑代码如下。

程序清单9-10：happyshop\app\src\main\java\com\example\administrator\happyshop\fragment\FragmentShouye.java

```
1    public class FragmentShouye extends Fragment {
2        private MyGridView gridView;
3        private RecyclerView recycleView;
4        private MyRecyclerViewAdapter mAdapter;
5        private View view;
6        private List<GoodInfo> goodInfos = new ArrayList<GoodInfo>();
7        private FragmentActivity activity;
8        private List<smallIcon> smallIconsList = new ArrayList<smallIcon>();
9        private int[] icno = {R.drawable.a, R.drawable.b, R.drawable.c, R.drawable.d, R.
 drawable.e, R
10                .drawable.f, R.drawable.g, R.drawable.h, R.drawable.i, R.drawable.k};
11       public FragmentShouye() {
12       }
13       @Override
14       public void onCreate(@Nullable Bundle savedInstanceState) {
15           super.onCreate(savedInstanceState);
16           activity = getActivity();
17       }
18       @Override
19       public View onCreateView(LayoutInflater inflater, ViewGroup container, Bundle
20               savedInstanceState) {
21           super.onCreateView(inflater, container, savedInstanceState);
22           view = inflater.inflate(R.layout.fragment_shoutye, container, false);
23           initSamllIcon();
24           getSmallIcon();
25           getDataFromSever();
26           return view;
```

```
27          }
28      private void getSmallIcon() {
29           recycleView = (RecyclerView) view.findViewById(R.id.recycle_View);
30            //设置 RecyclerView 管理器
31            StaggeredGridLayoutManager      staggeredGridLayoutManager      =      new
  StaggeredGridLayoutManager(5,
32              StaggeredGridLayoutManager.VERTICAL);
33            recycleView.setLayoutManager(staggeredGridLayoutManager);
34            //初始化适配器
35            mAdapter = new MyRecyclerViewAdapter(smallIconsList, activity);
36            //设置添加或删除 Item 时的动画,这里使用默认动画
37            recycleView.setItemAnimator(new DefaultItemAnimator());
38            //设置适配器
39            recycleView.setAdapter(mAdapter);
40        }
41        //请求网络
42        public void getDataFromSever() {
43             try {
44                   RequestManger.get(ConstantURL.getShop + "-1", new onRequestLister() {
45                    @Override
46                    public void onSucess(int code, String response) {
47                         goodInfos = JsonUtil.getInstance().jsonToArrayList(response,
  GoodInfo.class);
48                         //网络上获取到的数据
49                         if (goodInfos != null) {
50                                activity.runOnUiThread(new Runnable() {
51                                @Override
52                                public void run() {
53                                       gridView = (MyGridView) view.findViewById(R.id.
  gridView);
54                                       gridView.setAdapter(new GoodsAdapter(activity,
  goodInfos));
55                                       gridView.setOnItemClickListener(new AdapterView
56                                           .OnItemClickListener() {
57                   @Override
58                   public void onItemClick(AdapterView<?> parent, View view, int
59                              position, long id) {
60                              Intent intent = new Intent(activity, GoodDetailActivity.
  class);
61
62                             intent.putExtra("goodInfo", goodInfos.get(position));
63                              startActivity(intent);
64                                       }
65                                 });
66                            }
67                       });
68                  }
69             }
70
```

```
71                      @Override
72                      public void onFailed(int code, String message) {
73                          LogUtils.setLog("get data failed");
74                      }
75                  });
76              } catch (Exception e) {
77
78              }
79          }
80          //初始化小图标数据
81          private void initSamllIcon() {
82              smallIcon icon;
83              for (int i = 0; i < icno.length; i++) {
84                  icon = new smallIcon(icno[i]);
85                  smallIconsList.add(icon);
86              }
87          }
88      }
```

在上述代码中，第 9 行代码定义一个数组，保存小图标。第 28～40 行定义方法 getSmallIion()，在该方法中使用 RecyclerView 控件加载不同类别的小图标。

第 42～79 行代码定义方法 getDataFromServer()，在该方法中首先从服务器端获取数据，将获取的 JSON 数据解析成对象保存在集合中，之后为使用 GridView 控件通过适配器将集合中的数绑定在 GridView 控件上并展示，并且为 GridView 中的每项绑定单击事件，当单击每项时跳转到商品的详情页面，跳转的同时把商品的详细信息传递到商品详情页面。

第 81～88 行代码定义方法 initSamllIcon()，在该方法中将每个图标保存在对象中，并将所有的图标对象保存在集合中，完成小图标数据的初始化。

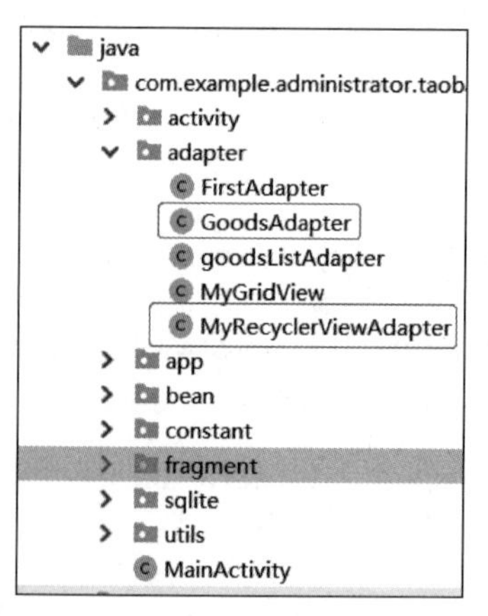

图 9-9　适配器在项目结构中的位置

其中，加载分类图标和商品列表时都用到了适配器，在加载小图标时用到 MyRecyclerViewAdapter 适配器，加载商品列表时用到 GoodsAdapter 适配器。这两个适配器在项目中的结构如图 9-9 所示，这两个类的详细代码可参考项目。

此外，在 GoodsAdapter 适配器中需要为列表项添加单击事件，单击每项的时候可跳转到商品详情页面的 GoodDetailActivity 的页面。其中，GoodDetailActivity 页面代码见项目。

在 MyRecyclerViewAdapter 适配器中也需要为每项注册单击事件，当单击每个分类小图标时，可根据类别跳转到相应类别下的 ProductListActivity 页面，其中，ProductListActivity 页面代码可参考项目。

在 MyRecyclerViewAdapter 注册单击事件的核心代码如下。

程序清单 9-11：happyshop\app\src\main\java\com\example\administrator\happyshop\adapter\MyRecyclerViewAdapter. java

```
1    public MyRecyclerViewAdapter.ViewHolder onCreateViewHolder(@NonNull ViewGroup
        viewGroup, int position) {
2        final ViewHolder viewHolder = new ViewHolder(View.inflate(mActivity,
            R.layout.smallicon_item, null));
3        viewHolder.functionView.setOnClickListener(new View.OnClickListener() {
4            @Override
5            public void onClick(View v) {
6                int adapterPosition = viewHolder.getAdapterPosition();
7                Log.d("My", "onClick: " + adapterPosition);
8                Intent intent = new Intent(mActivity, ProductListActivity.class);
9                intent.putExtra("id", adapterPosition);
10               mActivity.startActivity(intent);
11           }
12       });
13       return viewHolder;
14   }
```

9.5.2 购物车模块实现

视频详解

当单击底部导航栏中的“购物车”时，在 MainActivity 中展示 FragmentGouwuche 页面。在 FragmentGouwuche 页面首先读取数据库中已加入购物车的数据，并以列表的形式展示，具体代码如下。

程序清单 9-12：happyshop\app\src\main\java\com\example\administrator\happyshop\fragment\FragmentGouwuche. java

```
1    public class FragmentGouwuche extends Fragment {
2        private ListView listView;
3        private Context context;
4        private Activity activity;
5        private List<GoodInfo> goodArrayList = new ArrayList<GoodInfo>();
6        private FirstAdapter firstAdapter;
7        private CheckBox selectALL;
8        private TextView totalPrice;
9        private Button payMoney;
10       private View view;
11       @Override
12       public View onCreateView(LayoutInflater inflater, ViewGroup container, Bundle
   savedInstanceState) {
13           super.onCreateView(inflater, container, savedInstanceState);
14           view = inflater.inflate(R.layout.fragment_buycar, container, false);
15           activity = getActivity();
16           initView(view);
17           initData();
```

```
18            paydialog();
19            return view;
20        }
21        private void initView(View view) {
22            listView = view.findViewById(R.id.listview);
23            selectALL = view.findViewById(R.id.all_select_bx);
24            totalPrice = view.findViewById(R.id.total_price);
25            payMoney = view.findViewById(R.id.pay);
26        }
27        private void initData() {
28            goodArrayList = DBUtils.getInstance(activity).getGoodsInfoList();
29            if (goodArrayList != null) {
30                firstAdapter = new FirstAdapter(activity, goodArrayList, this);
31                listView.setAdapter(firstAdapter);
32                listView.setOnItemClickListener(new AdapterView.OnItemClickListener() {
33                    @Override
34                    public void onItemClick (AdapterView <? > parent, View view, int
position, long id) {
35                        Intent intent = new Intent();
36                        intent.setClass(activity, GoodDetailActivity.class);
37                        startActivity(intent);
38                    }
39                });
40                changeMoney();
41                selectALL. setOnCheckedChangeListener ( new CompoundButton.
OnCheckedChangeListener() {
42                    @Override
43                    public void onCheckedChanged ( CompoundButton buttonView, boolean
isChecked) {
44                        changeMoney();
45                        firstAdapter.selectALL(isChecked);
46                    }
47                });
48            } else {
49                Toast.makeText(activity, "还没有添加商品到购物车", Toast.LENGTH_
SHORT).show();
50            }
51        }
52        public void setContext(Context context) {
53            this.context = context;
54        }
55        public double getALLSelectMoney() {
56            double sum = 0;
57            for (GoodInfo good: goodArrayList) {
58                if (good.isBeSelect()) {
59                    sum = sum + Double.parseDouble(good.getPrice()) * Integer.
parseInt(good.getCount());
60                    sum = (double) Math.round(sum * 100) / 100;
61                }
62            }
```

```
63              return sum;
64          }
65          public void changeMoney() {
66              totalPrice.setText(getALLSelectMoney() + "");
67          }
68          public void paydialog(){
69              payMoney.setOnClickListener(new View.OnClickListener() {
70                  @Override
71                  public void onClick(View v) {
72                  final AlertDialog.Builder paybuilder = new AlertDialog.Builder
    (getContext());
73          View view = getLayoutInflater().inflate(R.layout.fragment_paydialog,null);
74                      TextView text = view.findViewById(R.id.sum);
75                      text.setText("需付款￥" + getALLSelectMoney() + "元");
76                      paybuilder.setView(view);
77                        paybuilder.setPositiveButton("确认支付", new DialogInterface.
    OnClickListener() {
78                          @Override
79                          public void onClick(DialogInterface dialog, int which) {
80                              AlertDialog.Builder sucbuilder = new AlertDialog.Builder
    (getContext());
81                              sucbuilder.setTitle("恭喜您支付成功!");
82                              sucbuilder.setPositiveButton("确定", new DialogInterface.
    OnClickListener() {
83                                  @Override
84                                  public void onClick(DialogInterface dialog, int
    which) {
85                                  }
86                              });
87                              sucbuilder.create().show();
88                          }
89                      });
90                      paybuilder.create().show();
91                  }
92              });
93          }
94          @Override
95          public void onResume() {
96              super.onResume();
97              initData();
98          }
99      }
```

在上述代码中，第27～51行代码定义方法initData()，在该方法中首先获取数据库中orderinfo表中订单数据，并通过FirstAdapter适配器将数据绑定在listView上展示。在该方法中调用changeMoney()方法获取当前所有选中商品的总价格。

第55～64行代码定义方法getALLSelectMoney()，在该方法中获取所有选中的列表项，并根据选中商品的价格和数量计算出所有商品的总价格。

第 68～93 行代码定义方法 paydialog()，在该方法中为“结算”按钮绑定单击事件，当单击“结算”按钮时，弹出对话框，在对话框中列举出付款方式；当用户单击“付款”按钮时，提示用户“恭喜您支付成功!”。

视频详解

9.5.3 个人中心模块实现

当单击底部导航栏中的“个人中心”，MainActivity 中展示 FragmentMe 页面，在 FragmentMe 页面会展示已登录用户的用户名，并显示“我的订单”“我的地址”和“退出登录”三个列表项。当用户未登录时，在“用户名”位置显示“登录”。当用户单击“登录”按钮时跳转到 LoginActivity 页面，当用户没有注册时，单击 LoginActivity 页面中的“立即注册”按钮可跳转到 RegisterActivity 页面进行注册。当用户登录成功之后，单击“我的订单”可跳转到 OrderListActivity 页面，展示用户的订单信息。当用户单击“我的地址”则跳转到 AddressActivity 页面，以列表的形式展示用户的收货地址，当用户单击某条收货地址时，可跳转到 EditAddressActivity 页面，可对收货地址进行修改。上述页面在项目中的结构位置如图 9-10 所示，每个页面的详细逻辑代码见项目。

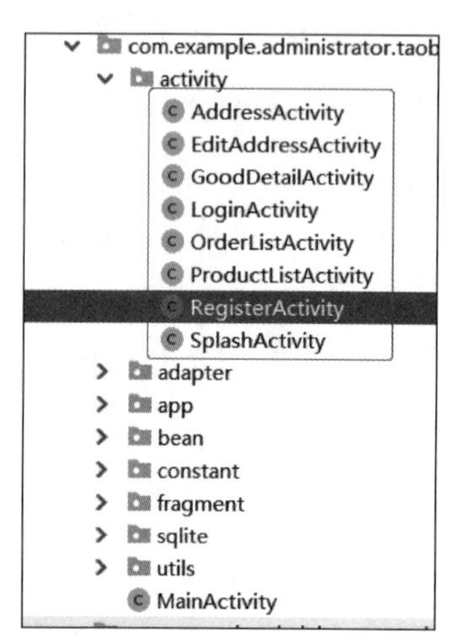

图 9-10　页面在项目结构中的位置

FragmentMe.java 中的逻辑代码如下。

程序清单 9-13：happyshop\app\src\main\java\com\example\administrator\happyshop\fragment\FragmentMe.java

```
1   public class FragmentMe extends Fragment {
2       private Activity mContext;
3       private View view;
4       public static final int REQUEST_HEAD_LOGIN = 1;
5       public static final int REQUEST_SETTING_LOGIN = 2;
6       private String name;
7       private Bitmap bm;
8       private TextView tvName;
9       private ImageButton ivHeadIcon;
10      private RelativeLayout lrMyorder;
11      private RelativeLayout lrMyaddress;
12      private RelativeLayout rlLoginout;
13      @Override
14      public void onCreate(Bundle savedInstanceState) {
15          super.onCreate(savedInstanceState);
16          mContext = getActivity();
17      }
18      @Override
19      public View onCreateView(LayoutInflater inflater, ViewGroup container,
20              Bundle savedInstanceState) {
```

```
21          view = inflater.inflate(R.layout.fragment_fragment_me, container, false);
22              initView(view);
23              initListener();
24              return view;
25          }
26          public void initView(View v) {
27              tvName = (TextView) v.findViewById(R.id.tv_user_name);
28              ivHeadIcon = (ImageButton) v.findViewById(R.id.iv_head_icons);
29              lrMyorder = (RelativeLayout) v.findViewById(R.id.rl_my_order);
30              lrMyaddress = (RelativeLayout) v.findViewById(R.id.rl_my_address);
31              rlLoginout = (RelativeLayout) v.findViewById(R.id.rl_loginout);
32          }
33          public void initListener() {
34              setLoginParams(SPUtils.readLoginStatus(mContext));
35              ivHeadIcon.setOnClickListener(new View.OnClickListener() {
36                  @Override
37                  public void onClick(View v) {
38              if (!SPUtils.readLoginStatus(mContext)) {
39                          Intent intent = new Intent(mContext, LoginActivity.class);
40                          mContext.startActivityForResult(intent, REQUEST_HEAD_LOGIN);
41                      }
42                  }
43              });
44              lrMyorder.setOnClickListener(new View.OnClickListener() {
45                  @Override
46                  public void onClick(View v) {
47               if (SPUtils.readLoginStatus(mContext)) {
48                          Intent intent = new Intent ( mContext, OrderListActivity.
class);
49                          startActivity(intent);
50                      } else {
51                          showMessageFromDialog();
52                      }
53                  }
54              });
55
56              lrMyaddress.setOnClickListener(new View.OnClickListener() {
57                  @Override
58                  public void onClick(View v) {
59                      if (SPUtils.readLoginStatus(mContext)) {
60                          Intent intent = new Intent(mContext, AddressActivity.class);
61                          startActivity(intent);
62                      } else {
63                          showMessageFromDialog();
64                      }
65                  }
66              });
67              rlLoginout.setOnClickListener(new View.OnClickListener() {
68                  @Override
69                  public void onClick(View v) {
```

```
70                      if (SPUtils.readLoginStatus(mContext)) {
71                          AlertDialog.Builder exitdialog = new AlertDialog.Builder
(mContext);
72                          exitdialog.setMessage("确定要退出吗?");
73                          exitdialog.setPositiveButton("确定", new DialogInterface.
OnClickListener() {
74
75                              @Override
76                              public void onClick(DialogInterface dialog, int which) {
77                                  SPUtils.SaveInfo(mContext, "", "", false);
78                                  setLoginParams(false);
79                              }
80                          });
81                  exitdialog.setNegativeButton("取消", null);
82                  AlertDialog alertDialog = exitdialog.create();
83                  alertDialog.show();
84                      } else {
85                          Toast.makeText(mContext, "还没有登录哦", Toast.LENGTH_SHORT).
show();
86                      }
87                  }
88              });
89          }
90          private void setLoginParams(boolean login) {
91              if (login) {
92                  name = SPUtils.getInfo(mContext).get("username");
93                  bm = BitmapFactory.decodeResource(mContext.getResources(), R.drawable.
head);
94                  tvName.setText(name);
95                  ivHeadIcon.setImageBitmap(bm);
96              } else {
97                  tvName.setText("单击登录");
98                  ivHeadIcon.setImageResource(R.drawable.head);
99              }
100         }
101         private void showMessageFromDialog() {
102             AlertDialog.Builder dialog = new AlertDialog.Builder(mContext);
103             dialog.setMessage("你还没有登录哎,请先登录");
104             dialog.setPositiveButton("去登录", new DialogInterface.OnClickListener() {
105
106                 public void onClick(DialogInterface dialog, int which) {
107                     Intent data = new Intent(mContext, LoginActivity.class);
108                     mContext.startActivityForResult(data, REQUEST_SETTING_LOGIN);
109                 }
110             });
111             dialog.setNegativeButton("不想登录", null);
112             AlertDialog alertDialog = dialog.create();
113             alertDialog.show();
114     alertDialog.getButton(AlertDialog.BUTTON_POSITIVE).setTextColor(getResources().
getColor(R.color.colorPrimary));
```

```
115     alertDialog.getButton(AlertDialog.BUTTON_NEGATIVE).setTextColor(getResources().
getColor(R.color.colorPrimary));
116         }
117     }
```

在上述代码中,第 26～32 行代码定义方法 initView(),该方法实现控件初始化。第 33～89 行代码定义方法 initListener(),在该方法中首先调用 setLoginParams()方法根据用户登录状态来判断是否加载用户头像和用户名。

第 35～43 行代码为头部绑定单击事件,如果用户没有登录,当用户单击时该区域就跳转到登录页面。第 44～54 行代码为列表中的“我的订单”绑定单击事件,当用户单击时跳转到订单详情页面。第 56～66 行代码为“我的地址”绑定单击事件,当用户单击时跳转到地址列表页面。第 67～89 行代码为“退出登录”项绑定单击事件,当用户单击时弹出对话框,提示用户是否要退出。

第 90～100 行代码定义方法 setLoginParams(),在该方法中根据用户登录状态加载用户信息。第 101～116 行代码定义方法 showMessageFromDialog(),封装弹出对话框的方法,在该方法中当用户没有登录时,单击个人中心页面中的列表项时调用此方法弹出对话框,提示用户登录。

本章小结

本章主要讲解开发“欢乐购商城”的流程,从总体的项目分析到具体数据获取再到数据存储操作以及各个功能模块的实现,是前面章节知识的综合运用。其中,商品数据是通过异步线程访问网络从 Tomcat 服务器上获取的,然后调用 Handler 将获取的信息发送到主线程并通过 JSON 解析将获取的数据显示到对应的界面上。在“欢乐购商城”项目的实现过程中用到了异步线程访问网络、Tomcat 服务器、Handler 消息通信、JSON 解析、界面控件的使用、Activity 之间的跳转和数据的传递、事件处理以及 SQLite 数据库的使用等知识点,这些知识点在以后开发项目中是必不可少的,因此希望读者认真分析每个模块的逻辑流程,并按照步骤完成项目,从而掌握 Android 应用程序开发的一般步骤、程序设计的原则。

自测习题

1. 简述“欢乐购商城”如何从服务器获取数据。
2. 简述 Android 中适配器的作用和意义。
3. 按照项目开发流程完成“欢乐购商城”项目。
4. 自主设计一个类似于“欢乐购商城”的项目并实现其功能。

第10章 鸿蒙操作系统开发入门

2019年8月9日，华为在东莞举行华为开发者大会，正式发布鸿蒙操作系统；2020年9月推出了鸿蒙2.0，全面使能全场景生态，具备跨设备、服务流转、极速直达、可视可说、隐私安全五大能力。在2021年6月2日的华为新品发布会中，鸿蒙手机操作系统正式登场，引起社会媒体的广泛关注，同时也掀起了国内更新鸿蒙操作系统和学习鸿蒙开发的浪潮。本章主要介绍鸿蒙操作系统开发入门，从鸿蒙操作系统特征、技术架构、技术特性等方面介绍什么是鸿蒙操作系统。开发方面介绍了开发环境搭建、创建鸿蒙应用、部署鸿蒙应用到模拟器。使用Java UI和JS UI两种方式完成实战演练页面跳转。

学习目标

本章要点

(1) 熟悉鸿蒙操作系统特征、技术架构和技术特性。

(2) 掌握DevEco Studio下载和安装方法。

(3) 掌握Node.js下载和安装方法。

(4) 掌握鸿蒙操作系统应用创建和部署。

(5) 掌握鸿蒙操作系统界面开发和页面跳转。

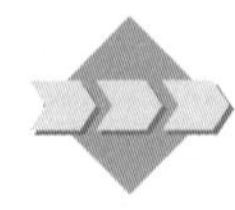

10.1 鸿蒙操作系统概述

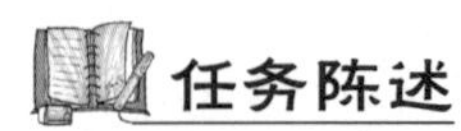

任务陈述

传说中国神话的远古时代，在盘古昆仑山开天辟地之前，世界是一团混沌的元气，这种自然的元气叫作鸿蒙，因此人们把那个时代称作鸿蒙时代，后来该词也常被用来泛指远古时代。2019年8月9日，华为在东莞举行华为开发者大会，正式发布鸿蒙操作系统(HarmonyOS或HongmengOS)。根据官方的定义，鸿蒙系统是一款面向未来、面向全场景(移动办公、运动健康、社交通信、媒体娱乐等)的分布式操作系统。

分析：鸿蒙操作系统与安卓操作系统比较相似，但也具有自身的优势。本节主要从系统特征、技术架构和技术特性等方面来分析鸿蒙操作系统。

相关知识

10.1.1 鸿蒙操作系统特征

华为鸿蒙操作系统的优点是流畅度较高，跨平台适用。这套系统主要是基于微内核的全场景分布式操作系统，可以按照需要进行扩展，由此来实现更为广泛的系统的安全性。它的主要特点是很低的时延，甚至可到毫秒级乃至亚毫秒级。在传统的单设备系统能力的基础上，鸿蒙操作系统提出了基于同一套系统能力、适配多种终端形态的分布式理念，能够支持多种终端设备。

鸿蒙操作系统具有以下三大特征。

(1) 搭载该操作系统的设备在系统层面融为一体，形成超级终端，让设备的硬件能力可以弹性扩展，实现设备之间硬件互助，资源共享。

对消费者而言，鸿蒙操作系统能够将生活场景中的各类终端进行能力整合，实现不同终端设备之间的快速连接、能力互助、资源共享，匹配合适的设备，提供流畅的全场景体验。

(2) 面向开发者，实现一次开发，多端部署。

对应用开发者而言，鸿蒙操作系统采用了多种分布式技术，使应用开发与不同终端设备的形态差异无关，从而让开发者能够聚焦上层业务逻辑，更加便捷、高效地开发应用。

(3) 一套操作系统可以满足不同能力的设备需求，实现统一操作系统，弹性部署。

对设备开发者而言，鸿蒙操作系统采用了组件化的设计方案，可根据设备的资源能力和业务特征灵活裁剪，满足不同形态终端设备对操作系统的要求。

鸿蒙操作系统提供了支持多种开发语言的 API，供开发者进行应用开发。支持的开发语言包括 Java、XML(Extensible Markup Language)、C/C++、JS(JavaScript)、CSS(Cascading Style Sheets)和 HML(HarmongOS Markup Language)。

10.1.2 鸿蒙操作系统技术架构

鸿蒙操作系统整体遵从分层设计，从下向上依次为：内核层、系统服务层、应用框架层和应用层。系统功能按照“系统→子系统→功能/模块”逐级展开，在多设备部署场景下，支持根据实际需求裁剪某些非必要的子系统或功能/模块。鸿蒙操作系统技术架构如图 10-1 所示。

1. 内核层

(1) 内核子系统。鸿蒙操作系统采用多内核设计，支持针对不同资源受限设备选用适合的操作系统内核。内核抽象层(Kernel Abstract Layer，KAL)通过屏蔽多内核差异，对上层提供基础的内核能力，包括进程/线程管理、内存管理、文件系统、网络管理和外设管理等。

(2) 驱动子系统。鸿蒙操作系统驱动框架(HDF)是鸿蒙操作系统硬件生态开放的基础，提供 统一外设访问能力和驱动开发、管理框架。

2. 系统服务层

系统服务层是鸿蒙操作系统的核心能力集合，通过框架层对应用程序提供服务。该层包含以下几个部分。

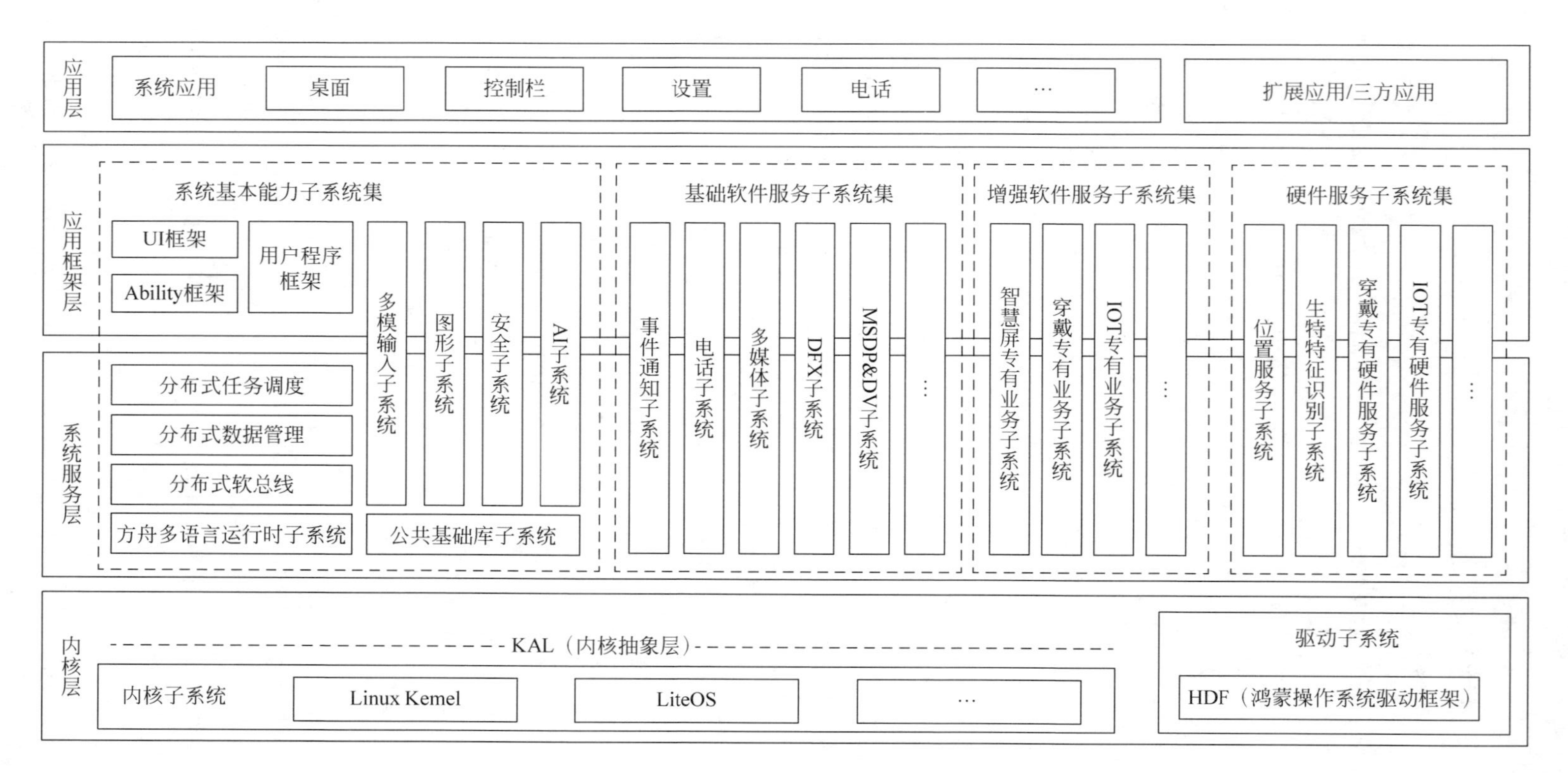

图 10-1　鸿蒙操作系统技术架构

(1) 系统基本能力子系统集：为分布式应用在鸿蒙操作系统多设备上的运行、调度、迁移等操作提供了基础能力，由分布式软总线、分布式数据管理、分布式任务调度、方舟多语言运行时、公共基础库、多模输入、图形、安全、AI 等子系统组成。其中，方舟运行时子系统提供了 C/C++/JS 多语言运行时和基础的系统类库，也为使用方舟编译器静态化的 Java 程序(即应用程序或框架层中使用 Java 语言开发的部分)提供运行时。

(2) 基础软件服务子系统集：为鸿蒙操作系统提供公共的、通用的软件服务，由事件通知、电话、多媒体、DFX、MSDP&DV 等子系统组成。

(3) 增强软件服务子系统集：为鸿蒙操作系统提供针对不同设备的、差异化的能力增强型软件服务，由智慧屏专有业务、穿戴专有业务、IoT 专有业务等子系统组成。

(4) 硬件服务子系统集：为鸿蒙操作系统提供硬件服务，由位置服务、生物特征识别、穿戴专 有硬件服务、IoT 专有硬件服务等子系统组成。

根据不同设备形态的部署环境，基础软件服务子系统集、增强软件服务子系统集、硬件服务子系统集内部可以按子系统粒度裁剪，每个子系统内部又可以按功能粒度裁剪。

3. 应用框架层

应用框架层为鸿蒙操作系统的应用程序提供了 Java/C/C++/JavaScript 等多语言的用户程序框架 Ability 框架，以及各种软硬件服务对外开放的多语言框架 API；同时为采用鸿蒙操作系统的设备提供了 C/C++/JavaScript 等多语言的框架 API，不同设备支持的 API 与系统的组件化裁剪程度相关。

4. 应用层

应用层包括系统应用和第三方非系统应用。鸿蒙操作系统的应用由一个或多个 FA (Feature Ability)或 PA(Particle Ability)组成。其中，FA 有 UI 界面，提供与用户交互的能力；而 PA 无 UI 界面，提供后台运行任务的能力以及统一的数据访问抽象。基于 FA/PA 开发的应用，能够实现特定的业务功能，支持跨设备调度与分发，为用户提供一致、高效的应用体验。

10.1.3　鸿蒙操作系统技术特性

1. 硬件互助，资源共享

1) 分布式软总线

分布式软总线是多种终端设备的统一基座，为设备之间的互连互通提供了统一的分布式通信能力，能够快速发现并连接设备，高效地分发任务和传输数据。分布式软总线示意图见图 10-2。

2) 分布式设备虚拟化

分布式设备虚拟化平台可以实现不同设备的资源融合、设备管理、数据处理，多种设备共同形成一个超级虚拟终端。针对不同类型的任务，为用户匹配并选择能力合适的执行硬件，让业务连续地在不同设备间流转，充分发挥不同设备的资源优势。分布式设备虚拟化示意图见图 10-3。

3) 分布式数据管理

分布式数据管理基于分布式软总线的能力，实现应用程序数据和用户数据的分布式管

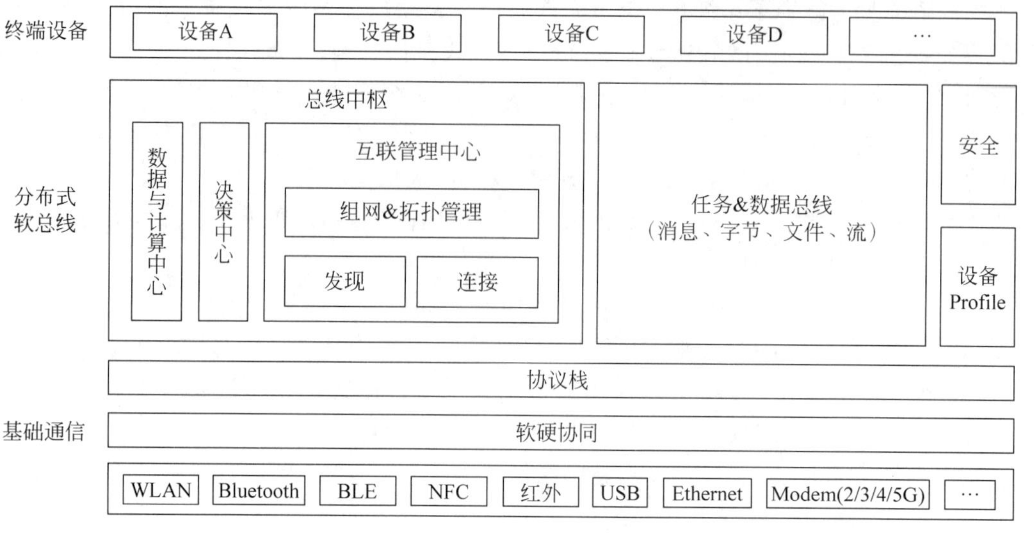

图 10-2　分布式软总线示意图

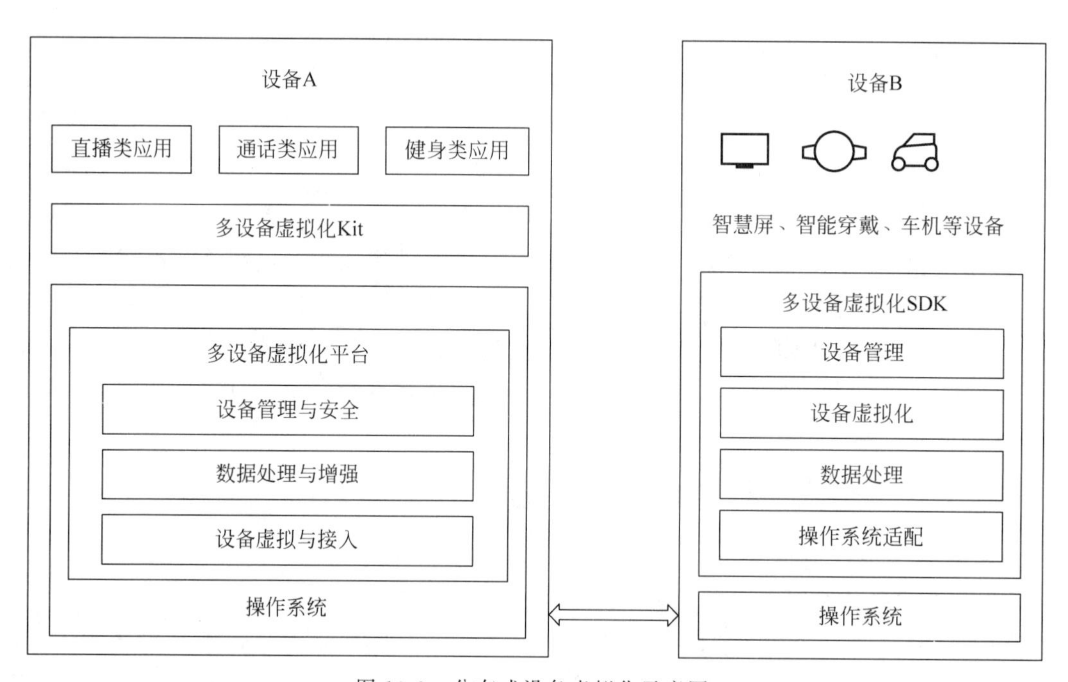

图 10-3　分布式设备虚拟化示意图

理。用户数据不再与单一物理设备绑定，业务逻辑与数据存储分离，应用跨设备运行时数据无缝衔接，为打造一致、流畅的用户体验创造了基础条件。分布式数据管理示意图见图 10-4。

4）分布式任务调度

分布式任务调度基于分布式软总线、分布式数据管理、分布式 Profile 等技术特性，构建统一的分布式服务管理（发现、同步、注册、调用）机制，支持对跨设备的应用进行远程启动、远程调用、远程连接以及迁移等操作，能够根据不同设备的能力、位置、业务运行状态、资源

使用情况，以及用户的习惯和意图，选择合适的设备运行分布式任务。图 10-5 以应用迁移为例，简要地展示了分布式任务调度能力。

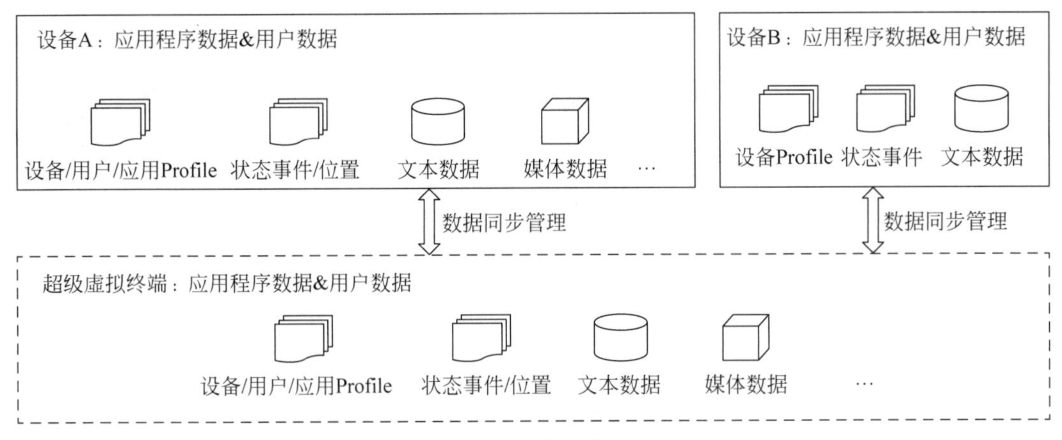

图 10-4 分布式数据管理示意图

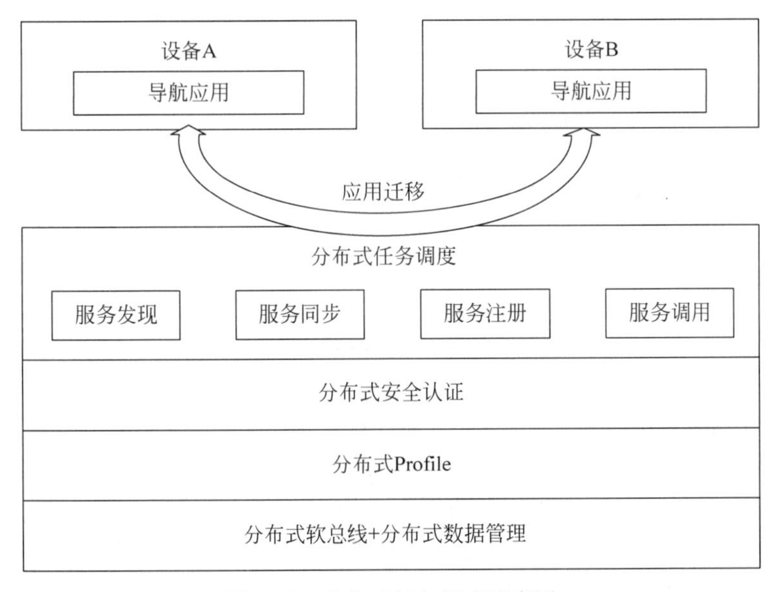

图 10-5 分布式任务调度示意图

2. 一次开发，多端部署

鸿蒙操作系统提供了用户程序框架、Ability 框架以及 UI 框架，支持应用开发过程中多终端的业务逻辑和界面逻辑进行复用，能够实现应用的一次开发、多端部署，提升了跨设备应用的开发效率。一次开发、多端部署示意图见图 10-6。

3. 统一操作系统，弹性部署

鸿蒙操作系统通过组件化和小型化等设计方法，支持多种终端设备按需弹性部署，能够适配不同类别的硬件资源和功能需求。支撑通过编译链关系去自动生成组件化的依赖关系，形成组件树依赖图，支撑产品系统的便捷开发，降低硬件设备的开发门槛。

(1) 支持各组件的选择(组件可有可无)。根据硬件的形态和需求，可以选择所需的

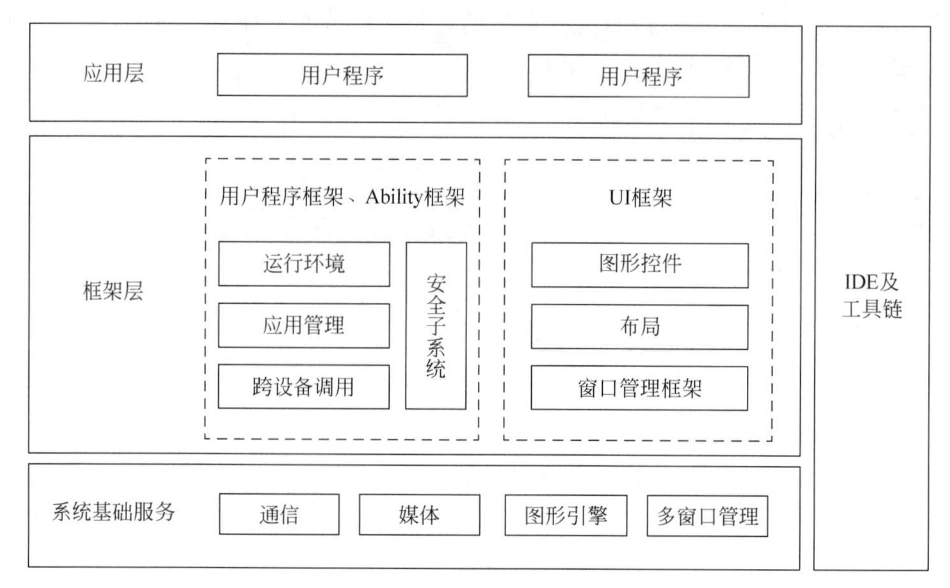

图 10-6　一次开发、多端部署示意图

组件。

（2）支持组件内功能集的配置（组件可大可小）。根据硬件的资源情况和功能需求，可以选择配置组件中的功能集。例如，选择配置图形框架组件中的部分控件。

（3）支持组件间依赖的关联（平台可大可小）。根据编译链关系，可以自动生成组件化的依赖关系。例如，选择图形框架组件，将会自动选择依赖的图形引擎组件等。

10.2　搭建鸿蒙操作系统环境

任务陈述

鸿蒙操作系统开发环境基于 Java 开发环境，为了便于开发，华为官网提供了集成工具 DevEco Studio。DevEco Studio 的编译构建依赖 JDK，DevEco Studio 预置了 Open JDK，版本为 1.8，安装过程中会自动安装 JDK。为保证 DevEco Studio 正常运行，建议计算机配置满足如下要求：操作系统为 Windows 10 64 位、内存 8GB 及以上、硬盘 100GB 及以上、分辨率 1280×800px 及以上。

分析：在开发过程中，UI 界面开发可以选择使用 JS UI 框架，所以鸿蒙操作系统环境搭建包括 DevEco Studio 和 Node.js 的下载和安装。

相关知识

10.2.1　Node.js 下载与安装

Node.js 应用于开发 JavaScript 应用和运行预览器功能，是开发鸿蒙操作系统应用过程中必备的软件。登录 Node.js 官方网站，下载 Node.js 软件包。选择 LTS 版本，Windows 64 位对应的软件包，见图 10-7。

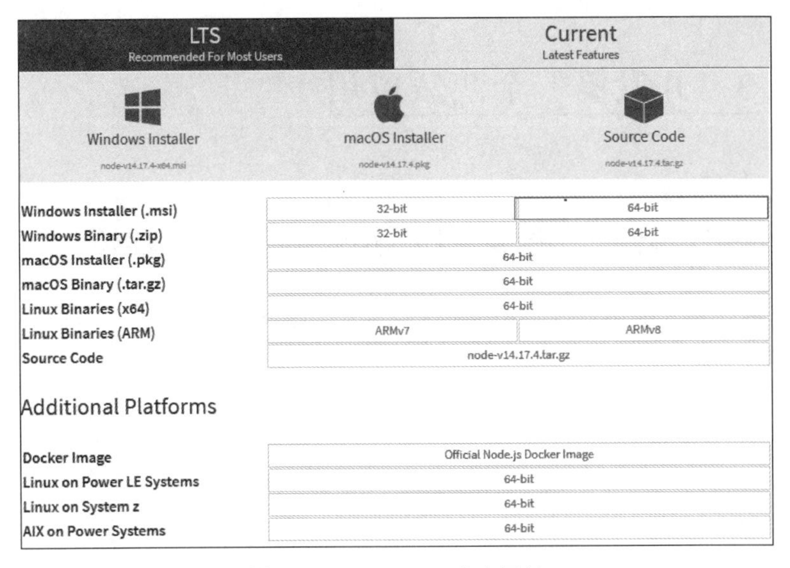

图 10-7　Node. js 官方网站

单击下载后的软件包进行安装，全部按照默认设置单击 Next 按钮，直至 Finish。安装过程中，Node. js 会自动在系统的 path 环境变量中配置 node. exe 的目录路径。

10.2.2 DevEco Studio 下载与安装

(1) 登录鸿蒙操作系统应用开发门户，单击右上角“注册”按钮，注册开发者账号，注册指导参考注册华为账号。如果已有华为开发者账号，直接单击“登录”按钮。

说明：使用 DevEco Studio 远程模拟器需要华为账号进行实名认证，建议在注册华为账号后，立即提交实名认证审核，认证方式包括“个人实名认证”和“企业实名认证”，详情请参考实名认证。

(2) 进入 HUAWEI DevEco Studio 产品页，下载 DevEco Studio 安装包。

(3) 双击下载的“deveco-studio-xxxx. exe”，进入 DevEco Studio 安装向导，在安装选项界面勾选 DevEco Studio launcher 复选框后，单击 Next 按钮，直至安装完成，如图 10-8 所示。

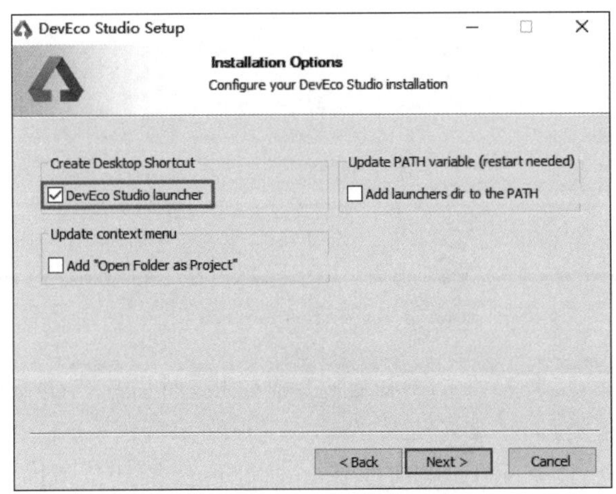

图 10-8　DevEco Studio 安装向导

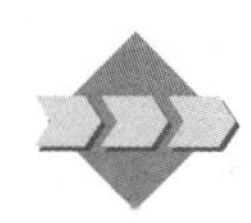

10.3 开发第一个鸿蒙应用

任务陈述

前面已经准备好了鸿蒙开发环境,接下来需要创建鸿蒙项目,掌握项目的创建过程以及配置。项目创建好后,需要把项目运行在模拟器上,鸿蒙的模拟和安卓模拟器有些不同,鸿蒙提供远程模拟器和本地模拟器,通过登录华为账号登录在线模拟器,使用 DevEco Studio 可将项目部署到远程模拟器中。

分析:使用 DevEco Studio 集成工具开发第一个鸿蒙应用,完成鸿蒙应用创建、鸿蒙模拟器创建、部署应用到模拟器。

相关知识

10.3.1 创建鸿蒙应用

打开 DevEco Studio,在欢迎页单击 Create HarmonyOS Project,创建一个新工程,如图 10-9 所示。

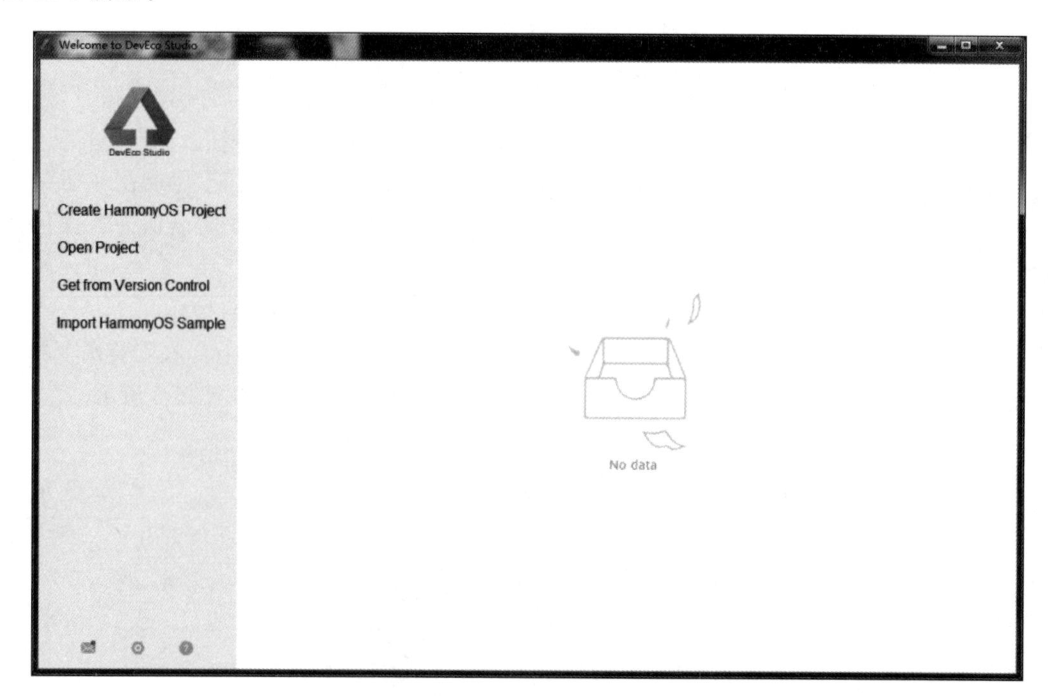

图 10-9 Create HarmonyOS Project

根据工程创建向导,选择需要的 Ability 工程模板,然后单击 Next 按钮。关于工程模板的介绍和支持的设备类型,请参考支持的设备模板和编程语言,如图 10-10 所示。

填写工程相关信息,Device Type 选择 Phone,其他保持默认值即可,单击 Finish 按钮,如图 10-11 所示。关于各个参数的详细介绍,请参考创建一个新的工程。

Create HarmonyOS Project
Choose your ability template
Empty Ability(JS)
Empty Ability(Java)
Native C++
[Lite]Empty Ability(JS)
Empty Ability(Java)
This Feature Ability template was developed in Java for phones, TVs, tablets, wearable devices and cars to display the basic Hello World functions.
Help
Cancel
Previous
Next

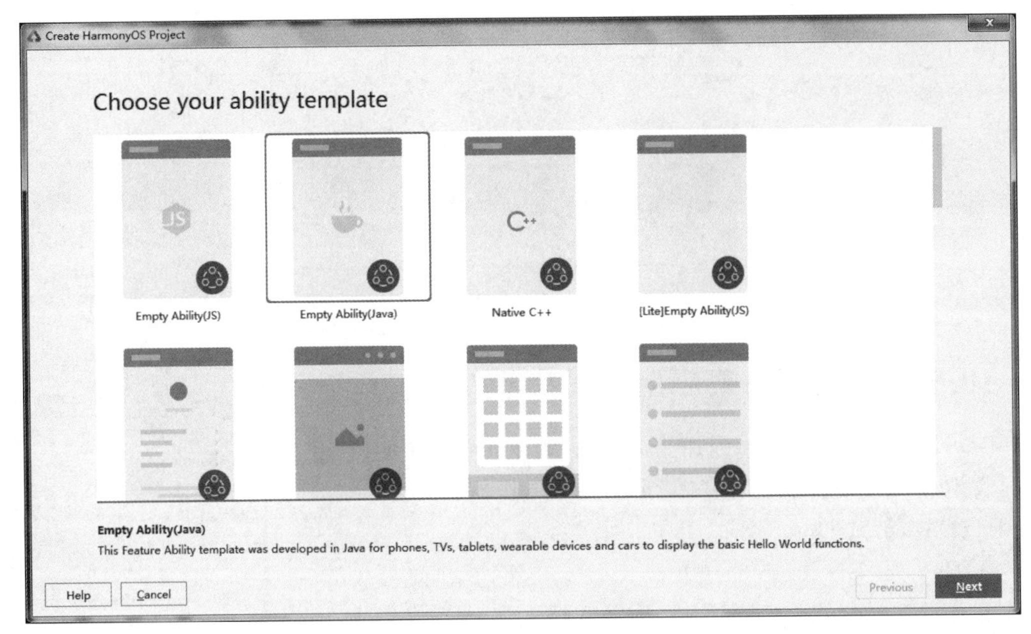

图 10-10 工程创建向导

Create HarmonyOS Project
Configure your project
Project Name
MyApplication
Project Type
Service
Application
Package Name
com.example.myapplication
Save Location
D:\project\DevEcoStudioProjects\MyApplication
Compatible API Version
SDK: API Version 4
Device Type
Phone
Tablet
TV
Wearable
Car
Help
Cancel
Previous
Finish

图 10-11 填写工程相关信息

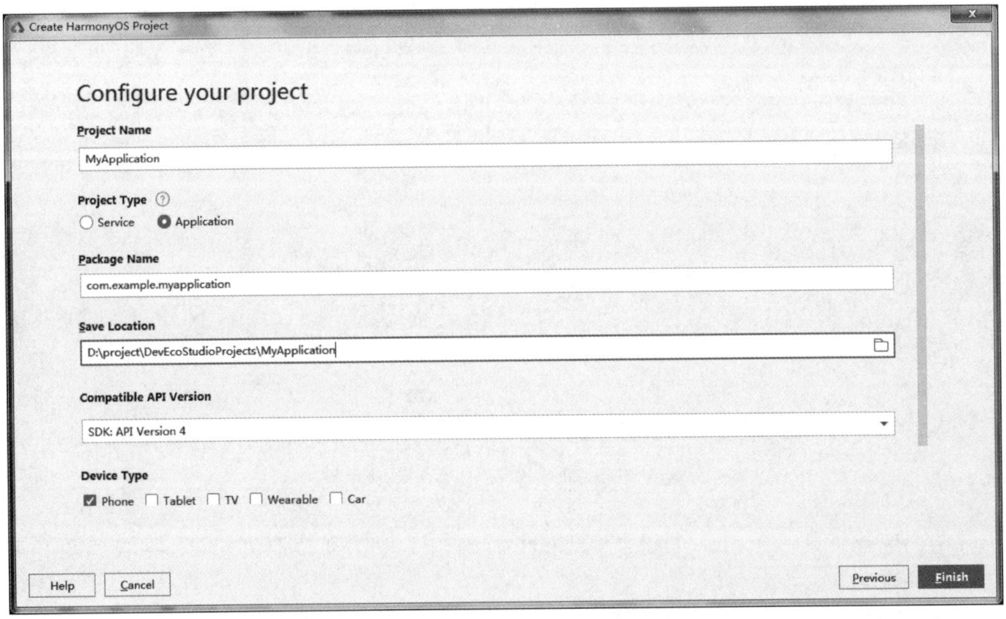

工程创建完成后，DevEco Studio会自动进行工程的同步，同步成功如图10-12所示。

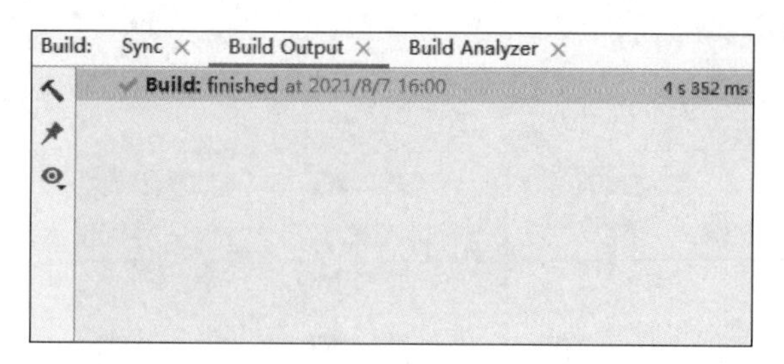

图10-12 工程自动同步

10.3.2 部署鸿蒙应用到模拟器

DevEco Studio提供远程模拟器和本地模拟器，本示例以远程模拟器为例进行说明。在DevEco Studio菜单栏，单击Tools→Device Manager，如图10-13所示。

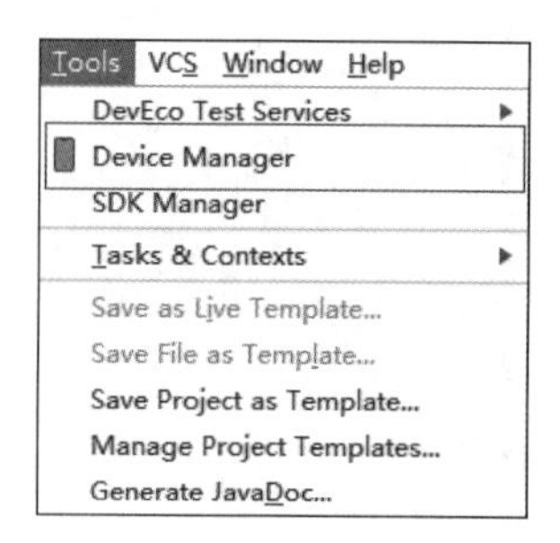

图10-13 打开设备管理器

在Remote Emulator页中单击Login按钮，在浏览器中弹出华为开发者联盟账号登录界面，请输入已实名认证的华为开发者联盟账号的用户名和密码进行登录(查看远程模拟器登录常见问题)，如图10-14所示。

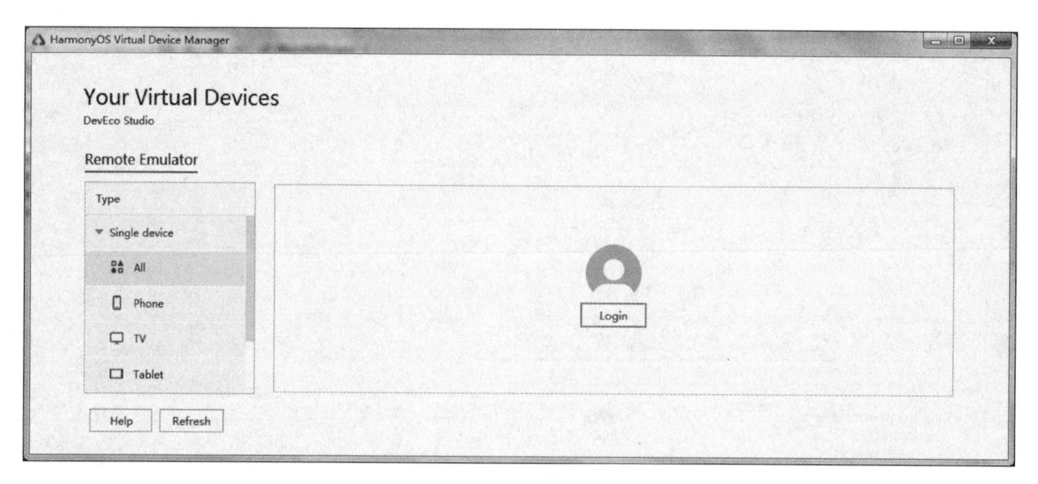

图10-14 登录华为开发者联盟账号

说明：推荐使用Chrome浏览器，如果使用Safari、360等其他浏览器，要取消阻止跨站跟踪和阻止所有Cookie功能。

登录后，请单击界面上的“允许”按钮进行授权，如图10-15所示。

在设备列表中，选择设备P40，并单击▶按钮，运行模拟器，如图10-16所示。

单击DevEco Studio工具栏中的▶按钮运行工程，或使用默认快捷键Shift+F10运行工程，如图10-17所示。

DevEco Studio会启动应用的编译构建，完成后应用即可运行在模拟器上，如图10-18所示。

图 10-15　华为账号授权

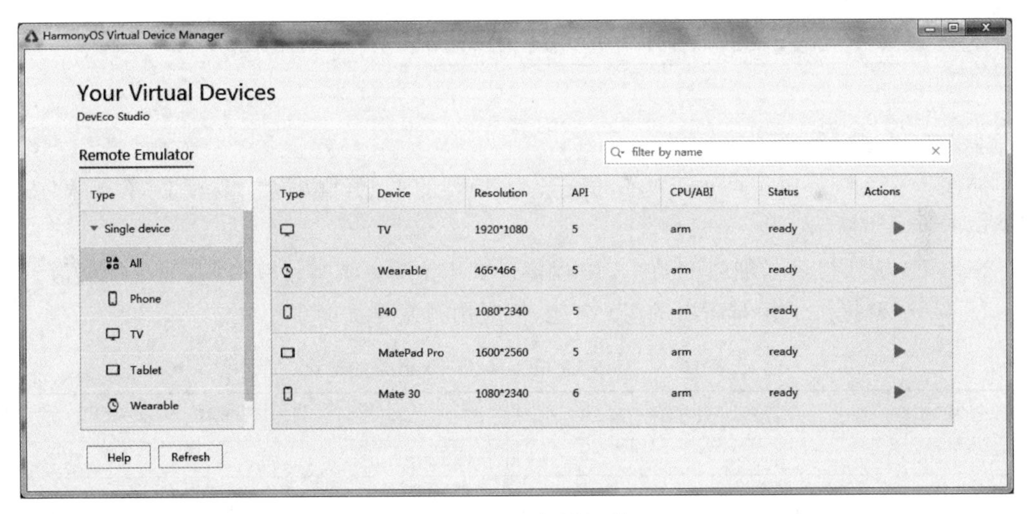

图 10-16　运行模拟器

图 10-17　运行工程到模拟器

图 10-18　工程运行成功页面

10.4　实战演练——页面跳转

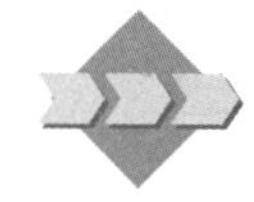

任务陈述

鸿蒙操作系统提供了两种 FA(Feature Ability)的 UI 开发框架：Java UI 框架和 JS UI 框架，如表 10-1 所示。

Java UI 框架提供了细粒度的 UI 编程接口，UI 元素更丰富，使应用开发更加灵活。

JS UI 框架提供了相对高层的 UI 描述，使应用开发更加简单。

表 10-1　Java UI 框架和 JS UI 框架的对比

比较项	Java UI 框架	JS UI 框架
语言生态	Java	JavaScript
接口方式	命令式	声明式
执行方式	开发者处理，基于 API 驱动的 UI 变更	框架层处理，基于数据驱动的 UI 自动变更
系统支持性	只有大型系统支持	覆盖平台更广，轻量系统、小型系统、标准系统、大型系统都支持
相对优势	UI 元素更丰富，开发更灵活	轻量化，开发更简便

分析：已成功运行了第一个鸿蒙操作系统应用，接下来通过 Java UI 方式和 JS UI 方式分别实现页面跳转，来了解鸿蒙操作系统应用的开发过程。

10.4.1　Java UI 方式实现页面跳转

1. 编写第一个页面

在 Java UI 框架中，提供了两种编写布局的方式：在 XML 中声明 UI 布局和在代码中创建布局。这两种方式创建出的布局没有本质差别，为了熟悉两种方式，将通过 XML 的方式编写第一个页面，通过代码的方式编写第二个页面。

在 Project 窗口中单击 entry→src→main→resources→base→layout，打开 ability_main.xml 文件，如图 10-19 所示。

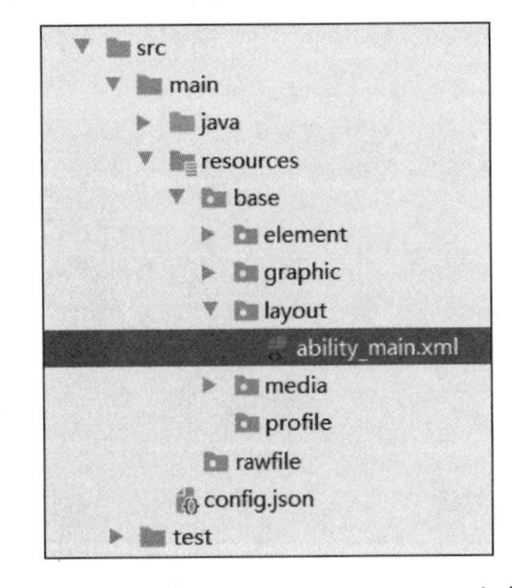

图 10-19　打开 ability_main.xml 文件

在 Project 窗口中单击 entry→src→main→resources→base→element，打开 string.json 文件，可在此文件内声明所需引用的资源内容。关于取值为 string 类型的文本属性，可以直接设置文本字符串，也可以引用 string 资源（推荐使用）。string.json 的示例代码如下。

程序清单 10-1：chart10\HMDemoJava\entry\src\main\resources\base\element\string.json

```
1     {
2         "string":[
3             {
4                 "name":"entry_MainAbility",
5                 "value":"entry_MainAbility"
6             },
7             {
8                 "name":"mainability_description",
9                 "value":"Java_Empty Ability"
10            },
11            {
12                "name":"mainability_HelloWorld",
13                "value":"Hello World"
14            },
15            {
16                "name":"button_Next",          →设置按钮标签名
17                "value":"Next"                 →设置按钮上显示内容
18            }
19        ]
20    }
```

第一个页面内有一个文本和一个按钮，使用 DependentLayout 布局，通过 Text 和 Button 组件来实现，其中，vp 和 fp 分别表示虚拟像素和字体像素。本次示例展示两个组件的显示文本分别采用直接设置文本字符串、引用 string 资源（推荐使用）的方式。ability_ main.xml 的示例代码如下。

程序清单 10-2：chart10\HMDemoJava\entry\src\main\resources\base\layout\ability_main.xml

```
1    <?xml version = "1.0" encoding = "utf - 8"?>
2    <DependentLayout
3        xmlns:ohos = "http://schemas.huawei.com/res/ohos"
4        ohos:width = "match_parent"
5        ohos:height = "match_parent">
6        <Text
7            ohos:id = "$ + id:text"
8            ohos:width = "match_content"
9            ohos:height = "match_content"
10           ohos:text = "Hello World"
11           ohos:text_color = "#000000"
12           ohos:text_size = "32fp"
13           ohos:center_in_parent = "true"/>
14        <Button
15           ohos:id = "$ + id:button"
16           ohos:width = "match_content"
17           ohos:height = "match_content"
18           ohos:text = "$ string:button_Next"
19           ohos:text_size = "19fp"
20           ohos:text_color = "#FFFFFF"
21           ohos:top_padding = "8vp"
22           ohos:bottom_padding = "8vp"
23           ohos:right_padding = "70vp"
24           ohos:left_padding = "70vp"
25           ohos:center_in_parent = "true"
26           ohos:below = "$ id:text"
27           ohos:margin = "10vp"/>
28   </DependentLayout>
```

按钮的背景是蓝色胶囊样式，可以通过 graphic 目录下的 XML 文件来设置。

右击 graphic 文件夹，选择 New→File，命名为“background_button.xml”，按 Enter 键，生成文件如图 10-20 所示。

程序清单 10-3：chart10\HMDemoJava\entry\src\main\resources\base\graphic\background_button.xml

```
1    <?xml version = "1.0" encoding = "utf - 8"?>
2    <shape
3        xmlns:ohos = "http://schemas.huawei.com/res/ohos"
4        ohos:shape = "rectangle">
5        <corners
6            ohos:radius = "100"/>
7        <solid
8            ohos:color = "#007DFF"/>
9    </shape>
```

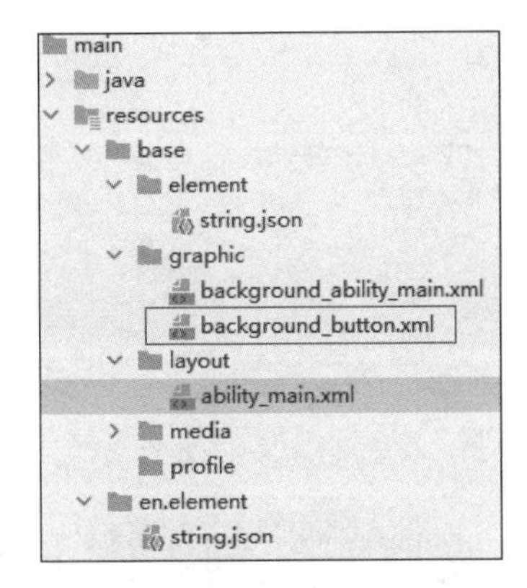

图 10-20 创建 background_button. xml 文件

在 layout 目录下的 ability_main. xml 文件中，使用 background_element＝" $ graphic：background_button"的方式引用 background_button. xml 文件，如图 10-21 所示。

```
<Button
    ohos:id="$+id:button"
    ohos:width="match_content"
    ohos:height="match_content"
    ohos:text="Next"
    ohos:text_size="19fp"
    ohos:text_color="#FFFFFF"
    ohos:top_padding="8vp"
    ohos:bottom_padding="8vp"
    ohos:right_padding="70vp"
    ohos:left_padding="70vp"
    ohos:center_in_parent="true"
    ohos:below="$id:text"
    ohos:margin="10vp"
    ohos:background_element="$graphic:background_button"/>
</DirectionalLayout>
```

图 10-21 引用 background_button. xml 文件

在 XML 文件中添加组件后，需要在 Java 代码中加载 XML 布局。

在 Project 窗口中选择 entry→src→main→java→com. example. myapplication→slice，打开 MainAbilitySlice. java 文件，使用 setUIContent()方法加载 ability_main. xml 布局。此外，运行代码前需采用 import(可使用 Alt＋Enter 快捷键)引入对应类，否则会产生报错提示。

说明：鸿蒙操作系统提供了 Ability 和 AbilitySlice 两个基础类，一个有界面的 Ability 可以由一个或多个 AbilitySlice 构成，AbilitySlice 主要用于承载单个页面的具体逻辑实现和界面 UI，是应用显示、运行和跳转的最小单元。

程序清单 10-4：chart10\HMDemoJava\entry\src\main\resources\base\graphic\background_button. xml

```
1    package com.example.myapplication.slice;
2    import com.example.myapplication.ResourceTable;
3    import ohos.aafwk.ability.AbilitySlice;
4    import ohos.aafwk.content.Intent;
```

```
5    public class MainAbilitySlice extends AbilitySlice {
6        @Override
7        public void onStart(Intent intent) {
8            super.onStart(intent);
9            super.setUIContent(ResourceTable.Layout_ability_main);      →加载 XML 布局
         }
     }
```

使用预览器或模拟器运行项目，效果如图 10-22 所示。

图 10-22　运行项目效果(一)

2. 创建另一个页面

前面用 XML 的方式编写了一个包含文本和按钮的页面。为了帮助开发者熟悉在代码中创建布局的方式，接下来使用代码的方式编写第二个页面。

在 Project 窗口中打开 entry→src→main→java→com. example. myapplication，右击 slice 文件夹，选择 New→Java Class，命名为 SecondAbilitySlice，按 Enter 键。

第二个页面上有一个文本。打开 SecondAbilitySlice. java 文件，添加一个 Text，示例代码如下。

程序清单 10-5：chart10\HMDemoJava\entry\src\main\java\com\example\myapplication\slice\SecondAbilitySlice. java

```
1    public class SecondAbilitySlice extends AbilitySlice {
2        @Override
3        public void onStart(Intent intent) {
4            super.onStart(intent);
```

```
5          DependentLayout myLayout = new DependentLayout(this);      →声明布局
6          myLayout.setWidth(LayoutConfig.MATCH_PARENT);             →设置布局宽高
7          myLayout.setHeight(LayoutConfig.MATCH_PARENT);
8          ShapeElement background = new ShapeElement();             →设置布局背景为白色
9          background.setRgbColor(new RgbColor(255, 255, 255));
10         myLayout.setBackground(background);
11         //创建一个文本
12         Text text = new Text(this);
13         text.setText("Hi there");
14         text.setWidth(LayoutConfig.MATCH_PARENT);
15         text.setTextSize(100);
16         text.setTextColor(Color.BLACK);
17         DependentLayout.LayoutConfig textConfig = new
           DependentLayout.LayoutConfig
     (LayoutConfig.MATCH_CONTENT, LayoutConfig.MATCH_CONTENT);        →设置文本的布局
18         textConfig.addRule(LayoutConfig.CENTER_IN_PARENT);
19         text.setLayoutConfig(textConfig);
20         myLayout.addComponent(text);
21         super.setUIContent(myLayout);
22     }
23 }
```

3. 实现页面跳转

打开第一个页面的 MainAbilitySlice.java 文件，添加按钮的响应逻辑，实现单击按钮跳转到下一页，示例代码如下。

程序清单 10-6：chart10\HMDemoJava\entry\src\main\java\com\example\myapplication\slice\MainAbilitySlice.java

```
1  public class MainAbilitySlice extends AbilitySlice {
2      @Override
3      public void onStart(Intent intent) {
4          super.onStart(intent);
5          super.setUIContent(ResourceTable.Layout_ability_main);
6          Button button = (Button)
          findComponentById(ResourceTable.Id_button);
7          button.setClickedListener(listener -> present
              (new SecondAbilitySlice(), new Intent()));  →单击按钮跳转至第二个页面
8      }
9  }
```

再次运行项目，效果如图 10-23 所示。

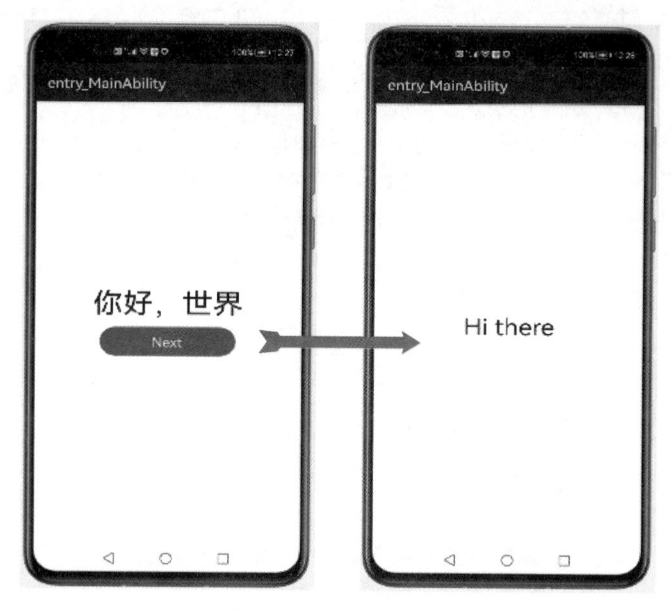

图 10-23　运行项目效果(二)

10.4.2　JS UI 方式实现页面跳转

1. 编写第一个页面

第一个页面内有一个文本和一个按钮，通过 Text 和 Button 组件来实现。

在 Project 窗口中选择 entry→src→main→js→default→pages.index，打开 index.hml 文件，添加一个文本和一个按钮，示例代码如下。

程序清单 10-7：chart10\HMDemoJS\entry\src\main\js\default\pages\index\index.hml

```
1    <div class = "container">
2        <!-- 添加一个文本 -->
3        <text class = "text">
4            Hello World
5        </text>
6        <!-- 添加一个按钮，按钮样式设置为胶囊型，文本显示为 Next，绑定 launch 事件 -->
7        <button class = "button" type = "capsule" value = "Next" onclick = "launch"></button>
8    </div>
```

打开 index.css 文件，设置文本和按钮的样式，示例代码如下。

程序清单 10-8：chart10\HMDemoJS\entry\src\main\js\default\pages\index\index.css

```
1    /* index.css */
2    .container {
3        flex-direction: column;      /* 设置容器内的项目纵向排列 */
4        justify-content: center;     /* 设置项目位于容器主轴的中心 */
5        align-items: center;         /* 项目在交叉轴居中 */
6    }
```

```
7      /* 对 class = "text"的组件设置样式 */
8      .text{
9          font-size: 42px;
10     }
11     /* 对 class = "button"的组件设置样式 */
12     .button {
13         width: 240px;
14         height: 60px;
15         background-color: #007dff;
16         font-size: 30px;
17         text-color: white;
18         margin-top: 20px;
19     }
```

使用预览器或模拟器运行项目，效果如图 10-24 所示。

2. 创建另一个页面

在 Project 窗口中打开 entry→src→main→js→default，右击 pages. index 文件夹，选择 New→JS Page，命名为 details，按 Enter 键。

创建完成后，可以看到 pages. index 文件夹下的文件目录结构如图 10-25 所示。

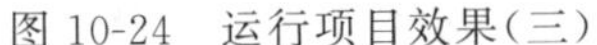
图 10-24　运行项目效果（三）

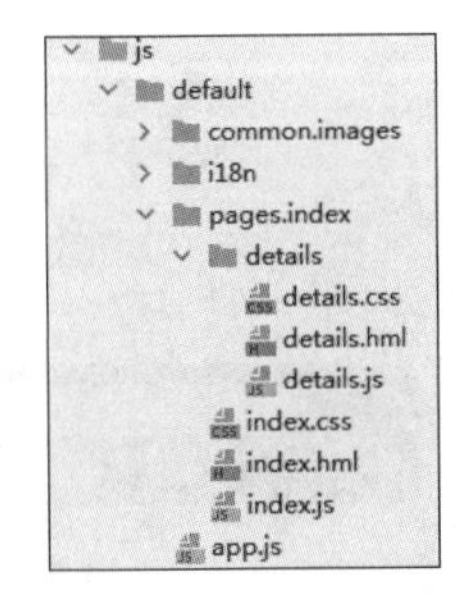

图 10-25　pages. index 目录结构

打开 details. hml 文件，添加一个文本，示例代码如下。

程序清单 10-9：chart10\HMDemoJS\entry\src\main\js\default\pages\index\details\details. hml

```
1    <!-- details.hml -->
2    <div class = "container">
3      <text class = "text">
4        Hi there
5      </text>
6    </div>
```

打开 details. css 文件，设置文本的样式，示例代码如下。

程序清单 10-10：chart10\HMDemoJS\entry\src\main\js\default\pages\index\details\details. hml

```
1    /* details.css */
2    .container {
3        flex-direction: column;
4        justify-content: center;
5        align-items: center;
6    }
7    .text {
8        font-size: 42px;
9        text-align: center;
10    }
```

3. 实现页面跳转

打开第一个页面的 index. js 文件，导入 router 模块，页面路由 router 根据页面的 uri 来找到目标页面，从而实现跳转。示例代码如下。

程序清单 10-11：chart10HMDemoJS\entry\src\main\js\default\pages\index\index. js

```
1    //index.js
2    import router from '@system.router';
3    export default {
4      launch() {
5        router.push ({
6          uri:'pages/index/details/details', //指定要跳转的页面
7        })
8      }
9    }
```

再次运行项目，效果如图 10-26 所示。

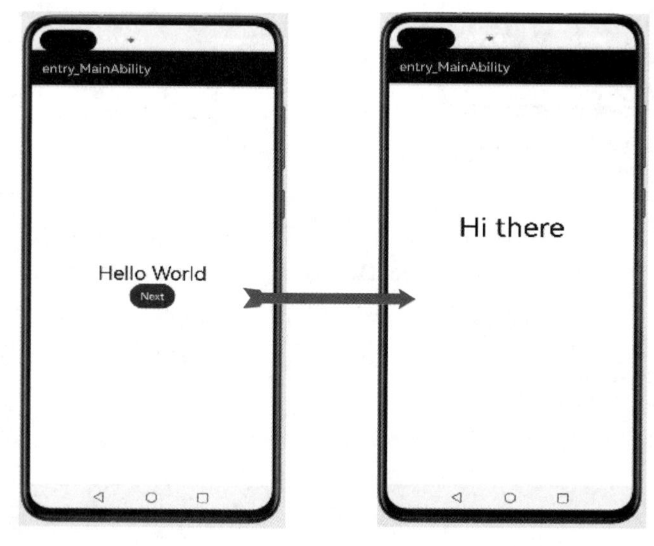

图 10-26　运行项目效果(四)

本章小结

本章主要带领读者进入鸿蒙操作系统的开发世界，通过系统特征、技术架构、技术特性三个方面了解鸿蒙操作系统。在鸿蒙操作系统环境搭建完成的前提下，完成鸿蒙应用的搭建、鸿蒙模拟器创建和应用部署。鸿蒙操作系统提供Java UI和JS UI两种界面开发方式，本章使用两种方式实现页面跳转，带领读者开发入门。

自测习题

1. 鸿蒙操作系统技术架构分为哪几个层次？与安卓操作系统技术架构的区别是什么？
2. 鸿蒙操作系统模拟器分为哪几种类型？分别如何使用？
3. 鸿蒙实现UI有哪几种方式？各自的优缺点是什么？

图书资源支持

感谢您一直以来对清华版图书的支持和爱护。为了配合本书的使用，本书提供配套的资源，有需求的读者请扫描下方的"书圈"微信公众号二维码，在图书专区下载，也可以拨打电话或发送电子邮件咨询。

如果您在使用本书的过程中遇到了什么问题，或者有相关图书出版计划，也请您发邮件告诉我们，以便我们更好地为您服务。

我们的联系方式：

地　　址：北京市海淀区双清路学研大厦 A 座 714

邮　　编：100084

电　　话：010-83470236　010-83470237

客服邮箱：2301891038@qq.com

QQ：2301891038（请写明您的单位和姓名）

资源下载：关注公众号"书圈"下载配套资源。

资源下载、样书申请

书圈

图书案例

清华计算机学堂

观看课程直播